Data Engineering with AWS Cookbook

A recipe-based approach to help you tackle data engineering problems with AWS services

Trâm Ngọc Phạm

Gonzalo Herreros González

Viquar Khan

Huda Nofal

Data Engineering with AWS Cookbook

Group Product Manager: Apeksha Shetty

Publishing Product Manager: Nilesh Kowadkar

Book Project Manager: Urvi Sharma

Senior Editor: Rohit Singh

Technical Editor: Kavyashree K S

Copy Editor: Safis Editing

Proofreader: Rohit Singh

Indexer: Manju Arasan

Production Designer: Shankar Kalbhor

Senior DevRel Marketing Executive: Nivedita Singh

First published: November 2024

Production reference: 1301024

Published by Packt Publishing Ltd.

Grosvenor House

11 St Paul's Square

Birmingham

B3 1RB, UK.

ISBN 978-1-80512-728-4

www.packtpub.com

To my mother, Ngoc Truong, for her love and sacrifices, and for exemplifying the power of determination. To my family members and friends, who always offer support and kindness throughout my life journey.

– Trâm Ngọc Phạm

Contributors

About the authors

Trâm Ngọc Phạm is a senior data architect with over a decade of hands-on experience working in the big data and AI field, from playing a lead role in tailoring cloud data platforms to BI and analytics use cases for enterprises in Vietnam. While working as a Senior Data and Analytics consultant for the AWS Professional Services team, she specialized in guiding finance and telco companies across Southeast Asian countries to build enterprise-scale data platforms and drive analytics use cases that utilized AWS services and big data tools.

Gonzalo Herreros González is a principal data architect. He holds a bachelor's degree in computer science and a master's degree in data analytics. He has experience of over a decade in big data and two decades of software development, both in AWS and on-premises.

Previously, he worked at MasterCard where he achieved the first PCI-DSS Hadoop cluster in the world. More recently, he worked at AWS for over 6 years, building data pipelines for the internal network data, and later, as an architect in the AWS Glue service team, building transforms for AWS Glue Studio and helping large customers succeed with AWS data services.

Viquar Khan is a senior data architect at AWS Professional Services and brings over 20 years of expertise in finance and data analytics, empowering global financial institutions to harness the full potential of AWS technologies. He designs cutting-edge, customized data solutions tailored to complex industry needs. A polyglot developer skilled in Java, Scala, Python, and other languages, Viquar has excelled in various technical roles. As an expert group member of JSR368 (JavaTM Message Service 2.1), he has shaped industry standards and actively contributes to open source projects such as Apache Spark and Terraform. His technical insights have reached and benefited over 6.7 million users on Stack Overflow.

Huda Nofal is a seasoned data engineer with over 7 years of experience at Amazon, where she has played a key role in helping internal business teams achieve their data goals. With deep expertise in AWS services, she has successfully designed and implemented data pipelines that power critical decision-making processes across various organizations. Huda's work primarily focuses on leveraging Redshift, Glue, data lakes, and Lambda to create scalable, efficient data solutions.

About the reviewers

Saransh Arora is a seasoned data engineer with more than 6 years of experience in the field. He has developed proficiency in Python, Java, Spark, SQL, and various data engineering tools, enabling him to address a wide range of data challenges. He has expertise in data orchestration, management, and analysis, with a strong emphasis on leveraging big data technologies to generate actionable insights. Saransh also possesses significant experience in machine learning and predictive analytics. Currently serving as a data engineer at AWS, he is dedicated to driving innovation and delivering business value. As an expert in data engineering, Saransh has also been working on the integration of generative AI into data engineering practices.

Haymang Ahuja specializes in ETL development, cloud computing, big data technologies, and cutting-edge AI. He is adept at creating robust data pipelines and delivering high-performance data solutions, backed by strong software development skills and proficiency in programming languages such as Python and SQL. His expertise includes big data technologies such as Spark, Apache Hudi, Airflow, Kylin, HDFS, and HBase. With a combination of technical knowledge, problem-solving skills, and a commitment to leveraging emerging technologies, he helps organizations achieve their strategic objectives and stay competitive in the dynamic digital landscape.

Table of Contents

2

Sharing Your Data Across Environments and Accounts   31

3

Ingesting and Transforming Your Data with AWS Glue   69

4

A Deep Dive into AWS Orchestration Frameworks   107

5

Running Big Data Workloads with Amazon EMR   161

6

Governing Your Platform   205

7

Data Quality Management   237

8

DevOps – Defining IaC and Building CI/CD Pipelines   265

9

Monitoring Data Lake Cloud Infrastructure   299

10

Building a Serving Layer with AWS Analytics Services 335

11

Migrating to AWS – Steps, Strategies, and Best Practices for Modernizing Your Analytics and Big Data Workloads 371

12

Harnessing the Power of AWS for Seamless Data Warehouse Migration   405

13

Strategizing Hadoop Migrations – Cost, Data, and Workflow Modernization with AWS   439

Preface

Hello and welcome! In today's rapidly evolving data landscape, managing, migrating, and governing large-scale data systems are among the top priorities for data engineers. This book serves as a comprehensive guide to help you navigate these essential tasks, with a focus on three key pillars of modern data engineering:

- **Hadoop and data warehouse migration**: Organizations are increasingly moving from traditional Hadoop clusters and on-premises data warehouses to more scalable, cloud-based data platforms. This book walks you through the best practices, methodologies, and how to use the tools for migrating large-scale data systems, ensuring data consistency, minimal downtime, and scalable performance.

- **Data lake operations**: Building and maintaining a data lake in today's multi-cloud, big data environment is complex and demands a strong operational strategy. This book covers how to ingest, transform, and manage data at scale using AWS services such as S3, Glue, and Athena. You will learn how to structure and maintain a robust data lake architecture that supports the varied needs of data analysts, data scientists, and business users alike.

- **Data lake governance**: Managing and governing your data lake involves more than just operational efficiency; it requires stringent security protocols, data quality controls, and compliance measures. With the explosion of data, it's more important than ever to have clear governance frameworks in place. This book delves into the best practices for implementing governance strategies using services such as AWS Lake Formation, Glue, and other AWS security frameworks. You'll also learn about setting up policies that ensure your data lake is compliant with industry regulations while maintaining scalability and flexibility.

This cookbook is tailored to data engineers who are looking to implement best practices and take their cloud data platforms to the next level. Throughout this book, you'll find practical examples, detailed recipes, and real-world scenarios from the authors' experience of working with complex data environments across different industries.

By the end of this journey, you will have a thorough understanding of how to migrate, operate, and govern your data platforms at scale, all while aligning with industry best practices and modern technological advancements.

So, let's dive in and build the future of data engineering together!

Who this book is for

This book is designed for data engineers, data platform engineers, and cloud practitioners who are actively involved in building and managing data infrastructure in the cloud. If you're involved in designing, building, or overseeing data solutions on AWS, this book will be ideal as it provides proven strategies for addressing challenges in large-scale data environments. Data engineers and big data professionals aiming to enhance their understanding of AWS features for optimizing their workflow, even if they're new to the platform, will find value. Basic familiarity with AWS security (users and roles) and command shell is recommended. This book will provide you with practical guidance, hands-on recipes, and advanced techniques for tackling real-world challenges.

What this book covers

Chapter 1, *Managing Data Lake Storage*, covers the fundamentals of managing S3 buckets. We'll focus on implementing robust security measures through data encryption and access control, managing costs by optimizing storage tiers and applying retention policies, and utilizing monitoring techniques to ensure timely issue resolution. Additionally, we'll cover other essential aspects of S3 bucket management.

Chapter 2, *Sharing Your Data Across Environments and Accounts*, presents methods for securely and efficiently sharing data across different environments and accounts. We will explore strategies for load distribution and collaborative analysis using Redshift data sharing and RDS replicas. We will implement fine-grained access control with Lake Formation and manage Glue data sharing through both Lake Formation and Resource Access Manager (RAM). Additionally, we will discuss real-time sharing via event-driven services, temporary data sharing with S3, and sharing operational data from CloudWatch.

Chapter 3, *Ingesting and Transforming Your Data with AWS Glue*, explores different features of AWS Glue when building data pipelines and data lakes. It covers the multiple tools and engines provided for the different kinds of users, from visual jobs with little or no code to managed notebooks and jobs using the different data handling APIs provided.

Chapter 4, *A Deep Dive into AWS Orchestration Frameworks*, explores the essential services and techniques for managing data workflows and pipelines on AWS. You'll learn how to define a simple workflow using AWS Glue Workflows, set up event-driven orchestration with Amazon EventBridge, and create data workflows with AWS Step Functions. We also cover managing data pipelines using Amazon MWAA, monitoring their health, and setting up a data ingestion pipeline with AWS Glue to bring data from a JDBC database into a catalog table.

Chapter 5, *Running Big Data Workloads with Amazon EMR*, teaches how to make the most of your AWS EMR clusters and explore the service features that enable them to be customizable, efficient, scalable, and robust.

Chapter 6, *Governing Your Platform*, presents the key aspects of data governance within AWS. This includes data protection techniques such as data masking in Redshift and classifying sensitive information using Maice. We will also cover ensuring data quality with Glue quality checks. Additionally, we will discuss resource governance to enforce best practices and maintain a secure, compliant infrastructure using AWS Config and resource tagging.

Chapter 7, *Data Quality Management*, covers how to use AWS Glue Deequ and AWS DataBrew to automate data quality checks and maintain high standards across your datasets. You will learn how to define and enforce data quality rules and monitor data quality metrics. This chapter also provides practical examples and recipes for integrating these tools into your data workflows, ensuring that your data is accurate, complete, and reliable for analysis.

Chapter 8, *DevOps – Defining IaC and Building CI/CD Pipelines*, explores multiple ways to automate AWS services and CI/CD deployment pipelines, the pros and cons of each tool, and examples of common data product deployments to illustrate DevOps best practices.

Chapter 9, *Monitoring Data Lake Cloud Infrastructure*, provides a comprehensive guide to the day-to-day operations of a cloud-based data platform. It covers key topics such as monitoring, logging, and alerting using AWS services such as CloudWatch, CloudTrail, and X-Ray. You will learn how to set up dashboards to monitor the health and performance of your data platform, troubleshoot issues, and ensure high availability and reliability. This chapter also discusses best practices for cost management and scaling operations to meet changing demands, making it an essential resource for anyone responsible for the ongoing maintenance and optimization of a data platform.

Chapter 10, *Building a Serving Layer with AWS Analytics Services*, guides you through the process of building an efficient serving layer using AWS Redshift, Athena, and QuickSight. The serving layer is where your data becomes accessible to end-users for analysis and reporting. In this chapter, you will learn how to load data from your data lake into Redshift, query it using Redshift Spectrum and Athena, and visualize it using QuickSight. This chapter also covers best practices for managing different QuickSight environments and migrating assets between them. By the end of this chapter, you will have the knowledge to create a powerful and user-friendly analytics layer that meets the needs of your organization.

Chapter 11, *Migrating to AWS – Steps, Strategies, and Best Practices for Modernizing Your Analytics and Big Data Workloads*, presents a theoretical framework for migrating data and workloads to AWS. It explores key concepts, strategies, and best practices for planning and executing a successful migration. You'll learn about various migration approaches—rehosting, replatforming, and refactoring—and how to choose the best option for your organization's needs. The chapter also addresses critical challenges and considerations, such as data security, compliance, and minimizing downtime, preparing you to navigate the complexities of cloud migration with confidence.

Chapter 12, Harnessing the Power of AWS for Seamless Data Warehouse Migration, explores the key strategies for efficiently migrating data warehouses to AWS. You'll learn how to generate a migration assessment report using the AWS Schema Conversion Tool (SCT), extract and transfer data with AWS Database Migration Service (DMS), and handle large-scale migrations with the AWS Snow Family. You'll also learn how to streamline your data migration, ensuring minimal disruption and maximum efficiency while transitioning to the cloud.

Chapter 13, Strategizing Hadoop Migrations – Cost, Data, and Workflow Modernization with AWS, guides you through essential recipes for migrating your on-premises Hadoop ecosystem to AWS, covering a range of critical tasks. You'll learn about cost analysis using the AWS Total Cost of Ownership (TCO) calculators and the Hadoop Migration Assessment tool. You'll also learn how to choose the right storage solution, migrate HDFS data using AWS DataSync, and transition key components such as the Hive Metastore and Apache Oozie workflows to AWS EMR. We also cover setting up a secure network connection to your EMR cluster, seamless HBase migration to AWS, and transitioning HBase to DynamoDB.

To get the most out of this book

To follow the recipes in this book, you will need the following:

Software/hardware covered in the book	OS requirements
AWS CLI	Windows, macOS X, and Linux (any)
Access to AWS services such as EMR, Glue, Redshift, QuickSight, and Lambda	
Python (for scripting and SDK usage)	

In addition to these requirements, you will also need a basic knowledge of data engineer terminology.

If you are using the digital version of this book, we advise you to type the code yourself or access the code via the GitHub repository (link available in the next section). Doing so will help you avoid any potential errors related to the copying and pasting of code.

Download the example code files

You can download the example code files for this book from GitHub at https://github.com/PacktPublishing/Data-Engineering-with-AWS-Cookbook. In case there's an update to the code, it will be updated on the existing GitHub repository.

We also have other code bundles from our rich catalog of books and videos available at https://github.com/PacktPublishing/. Check them out!

Conventions used

There are a number of text conventions used throughout this book.

`Code in text`: Indicates code words in text, database table names, folder names, filenames, file extensions, pathnames, dummy URLs, user input, and Twitter handles. Here is an example: "Make sure you replace `<your_bucket_name>` with the actual name of your S3 bucket."

A block of code is set as follows:

```
{
    Sid: DenyListBucketFolder,
    Action: [s3:*],
    Effect: Deny,
    Resource: [arn:aws:s3:::<bucket-name>/<folder-name>/*]
}
```

Any command-line input or output is written as follows:

```
CREATE DATASHARE datashare_name;
```

Bold: Indicates a new term, an important word, or words that you see onscreen. For example, words in menus or dialog boxes appear in the text like this. Here is an example: "Choose **Policies** from the navigation pane on the left and choose **Create policy**."

> **Tips or important notes**
> Appear like this.

Sections

In this book, you will find several headings that appear frequently (*Getting ready*, *How to do it...*, *How it works...*, *There's more...*, and *See also*).

To give clear instructions on how to complete a recipe, use these sections as follows:

Getting ready

This section tells you what to expect in the recipe and describes how to set up any software or any preliminary settings required for the recipe.

How to do it...

This section contains the steps required to follow the recipe.

How it works...

This section usually consists of a detailed explanation of what happened in the previous section.

There's more...

This section consists of additional information about the recipe in order to make you more knowledgeable about the recipe.

See also

This section provides helpful links to other useful information for the recipe.

Get in touch

Feedback from our readers is always welcome.

General feedback: If you have questions about any aspect of this book, mention the book title in the subject of your message and email us at `customercare@packtpub.com`.

Errata: Although we have taken every care to ensure the accuracy of our content, mistakes do happen. If you have found a mistake in this book, we would be grateful if you would report this to us. Please visit `www.packtpub.com/support/errata`, select your book, click on the Errata Submission Form link, and enter the details.

Piracy: If you come across any illegal copies of our works in any form on the Internet, we would be grateful if you would provide us with the location address or website name. Please contact us at `copyright@packt.com` with a link to the material.

If you are interested in becoming an author: If there is a topic that you have expertise in and you are interested in either writing or contributing to a book, please visit `authors.packtpub.com`.

Share Your Thoughts

Once you've read *Data Engineering with AWS Cookbook*, we'd love to hear your thoughts! Scan the QR code below to go straight to the Amazon review page for this book and share your feedback.

https://packt.link/r/1-805-12728-4

Your review is important to us and the tech community and will help us make sure we're delivering excellent quality content.

Download a free PDF copy of this book

Thanks for purchasing this book!

Do you like to read on the go but are unable to carry your print books everywhere?

Is your eBook purchase not compatible with the device of your choice?

Don't worry, now with every Packt book you get a DRM-free PDF version of that book at no cost.

Read anywhere, any place, on any device. Search, copy, and paste code from your favorite technical books directly into your application.

The perks don't stop there, you can get exclusive access to discounts, newsletters, and great free content in your inbox daily

Follow these simple steps to get the benefits:

1. Scan the QR code or visit the link below

https://packt.link/free-ebook/9781805127284

2. Submit your proof of purchase
3. That's it! We'll send your free PDF and other benefits to your email directly

1

Managing Data Lake Storage

Amazon Simple Storage Service (**Amazon S3**) is a highly scalable and secure cloud storage service. It allows you to store and retrieve any amount of data at any time from anywhere in the world. S3 buckets aim to help enterprises and individuals achieve their data backup and delivery needs and serve a variety of use cases, including but not limited to web and mobile applications, big data analytics, data lakes, and data backup and archiving.

In this chapter, we will learn how to keep data secure in S3 buckets and configure your buckets in a way that best serves your use case from performance and cost perspectives.

The following recipes will be covered in this chapter:

- Controlling access to S3 buckets
- Storage types in S3 for optimized storage costs
- Enforcing encryption of S3 buckets
- Setting up retention policies for your objects
- Versioning your data
- Replicating your data
- Monitoring your S3 buckets

Technical requirements

The recipes in this chapter assume you have an S3 bucket with admin permission. If you don't have admin permission to the bucket, you will need to configure the permission for each recipe as needed.

You can find the code files for this chapter in this book's GitHub repository: `https://github.com/PacktPublishing/Data-Engineering-with-AWS-Cookbook/tree/main/Chapter01`.

Controlling access to S3 buckets

Controlling access to S3 buckets through policies and IAM roles is crucial for maintaining the security and integrity of your objects and data stored in Amazon S3. By defining granular permissions and access controls, you can ensure that only authorized users or services have the necessary privileges to interact with your S3 resources. You can restrict permissions according to your requirements by precisely defining who can access your data, what actions they can take, and under what conditions. This fine-grained access control helps protect sensitive data, prevent unauthorized modifications, and mitigate the risk of accidental or malicious actions.

AWS **Identity and Access Management (IAM)** allows you to create an entity referred to as an IAM identity, which is granted specific actions on your AWS account. This entity can be a person or an application. You can create this identity as an IAM role, which is designed to be attached to any entity that needs it. Alternatively, you can create IAM users, which represent individual people and are usually used for granting long-term access to specific users. IAM users can be grouped into an IAM group, allowing permissions to be assigned at the group level and inherited by all member users. IAM policies are sets of permissions that can be attached to the IAM identity to grant specific access rights.

In this recipe, we will learn how to create a policy so that we can view all the buckets in the account, give read access to one specific bucket content, and then give write access to one of its folders.

Getting ready

For this recipe, you need to have an IAM user, role, or group to which you want to grant access. You also need to have an S3 bucket with a folder to grant access to.

To learn how to create IAM identities, go to https://docs.aws.amazon.com/IAM/latest/UserGuide/id.html.

How to do it...

1. Sign in to the AWS Management Console (https://console.aws.amazon.com/console/home?nc2=h_ct&src=header-signin) and navigate to the IAM console.

2. Choose **Policies** from the navigation pane on the left and choose **Create policy**.

3. Choose the **JSON** tab to provide the policy in JSON format and replace the existing JSON with this policy:

```
{
  "Version": "2012-10-17",
  "Statement": [
    {
        "Sid": "AllowListBuckets",
        "Effect": "Allow",
        "Action": [
```

```
                "s3:ListAllMyBuckets"
            ],
            "Resource": "*"
        },
        {
            "Sid": "AllowBucketListing",
            "Effect": "Allow",
            "Action": [
                "s3:ListBucket",
                "s3:GetBucketLocation"
            ],
            "Resource": [
                "arn:aws:s3:::<bucket-name>"
            ]
        },
        {
            "Sid": "AllowFolderAccess",
            "Effect": "Allow",
            "Action": [
                "s3:GetObject",
                "s3:PutObject",
                "s3:DeleteObject"
            ],
            "Resource": [
                "arn:aws:s3:::<bucket-name>/<folder-name>/*"
            ]
        }
    ]
}
```

4. Provide a policy name and, optionally, a description of the policy in the respective fields.

5. Click on **Create Policy**.

Now, you can attach this policy to an IAM role, user, or group. However, exercise caution and ensure access is granted only as necessary; avoid providing admin access policies to regular users.

How it works...

An IAM policy comprises three key elements:

- Effect: This specifies whether the policy allows or denies access

- Action: This details the specific actions being allowed or denied

- Resource: This identifies the resources to which the actions apply

A single statement can apply multiple actions to multiple resources. In this recipe, we've defined three statements:

- The `AllowListBuckets` statement gives access to list all buckets in the AWS account
- The `AllowBucketListing` statement gives access to list the content of a specific S3 bucket
- The `AllowFolderAccess` gives access to upload, download, and delete objects from a specific folder

There's more...

If you want to make sure that no access is given to a specific bucket or object in your bucket, you can use a deny statement, as shown here:

```
{
    "Sid":"DenyListBucketFolder",
        "Action":[
            "s3:*"
        ],
        "Effect":"Deny",
        "Resource":[
            "arn:aws:s3:::<bucket-name>/<folder-name>/*"
}
```

Instead of using an IAM policy to set up permissions to your bucket, you can use S3 bucket policies. These can be located in the **Permission** tab of the bucket. Bucket policies can be used when you're trying to set up access at the bucket level, regardless of the IAM role or user.

See also

- AWS provides a set of policies that are managed and administered by AWS, all of which can be used for many common use cases. You can learn more about these policies at https://docs.aws.amazon.com/AmazonS3/latest/userguide/security-iam-awsmanpol.html.
- To learn how to set up cross-account access to S3 buckets, go to https://docs.aws.amazon.com/AmazonS3/latest/userguide/example-walkthroughs-managing-access-example2.html.

Storage types in S3 for optimized storage costs

Amazon S3 offers different tiers or classes of storage that allow you to optimize for cost and performance based on your access pattern and data requirements. The default storage class for S3 buckets is S3 Standard, which offers high availability and low latency. For less frequently accessed data, S3 Standard-IA and S3 One Zone-IA can be used. For rare access, Amazon S3 offers archiving classes called Glacier, which are the lowest-cost classes. If you're not sure how frequently your data will be accessed, S3 Intelligent-Tiering would be optimal for you as it will automatically move objects between the classes based on the access patterns. However, be aware that additional costs may be incurred when you're moving objects to a higher-cost storage class.

These storage classes provide users with the flexibility to choose the right trade-off between storage costs and access performance based on their specific data storage and retrieval requirements. You can choose the storage class based on your access patterns, durability requirements, and budget considerations. Configuring storage classes at the object level allows for a mix of storage classes within the same bucket. Objects from diverse storage classes, including S3 Standard, S3 Intelligent-Tiering, S3 Standard-IA, and S3 One Zone-IA, can coexist in a single bucket.

In this recipe, we will learn how to enforce the S3 Intelligent-Tiering storage class for an S3 bucket through a bucket policy.

Getting ready

For this recipe, you only need to have an S3 bucket for which you will enforce the storage class.

How to do it...

1. Open the AWS Management Console (https://console.aws.amazon.com/console/home?nc2=h_ct&src=header-signin) and navigate to the S3 service.

2. Locate and select the S3 bucket on which you want to enable S3 Intelligent-Tiering and navigate to the **Permissions** tab.

3. Under the **Bucket Policy** section, click on **Edit**.

4. In the bucket policy editor, add the following statement. Make sure you replace <your_bucket_name> with the actual name of your S3 bucket:

```
{
  "Version": "2012-10-17",
  "Statement": [
    {
      "Sid": "EnableIntelligentTiering",
      "Effect": "Deny",
      "Principal": {
        "AWS": "*"
      },
      "Action": "s3:PutObject",
      "Resource": "arn:aws:s3:::<your-bucket-name>/*",
      "Condition": {
        "StringNotEquals": {
```

```
                    "s3:x-amz-storage-class": "INTELLIGENT_TIERING"
                }
            }
        }
    ]
}
```

5. Save the bucket policy by clicking on **Save changes**.

How it works...

The policy will ensure that objects are stored via the Intelligent-Tiering class by allowing the PUT operation to be used on the bucket for all users (`Principal: *`), but only if the storage class is set to INTELLIGENT_TIERING. You can do this by choosing it from the storage class list in the **Object properties** section. If you're using the console or the S3 API, add the `x-amz-storage-class: INTELLIGENT_TIERING` header. Use the `-storage-class INTELLIGENT_TIERING` parameter when using the AWS CLI.

There's more...

Intelligent-Tiering will place newly uploaded objects in the S3 Standard class (Frequent Access class). If the object hasn't been accessed in 30 consecutive days, it will be moved to the Infrequent Access tier; if it hasn't been accessed in 90 consecutive days, it will be moved to the Archive Instant Access tier. For further cost savings, you can enable INTELLIGENT_TIERING to move your object to the Archive Access tier and Deep Archive Access tier if they have not been accessed for a longer period. To do this, follow these steps:

1. Navigate to the **Properties** tab for the bucket.

2. Scroll down to Intelligent-Tiering Archive configurations and click on **Create configuration**.

3. Name the configuration and specify whether you want to enable it for all objects in the bucket or on a subset based on a filter and/or tags.

4. Under **Status**, click on **Enable** to enable the configuration directly after you create it.

5. Under **Archive rule actions**, enable the **Archive Access** tier and specify the number of days in which the objects should be moved to this class if they're not being accessed. The value must be between 90 and 730 days. Similarly, enable the **Deep Archive Access** tier and set the number of days to a minimum of 180 days. It's also possible to enable only one of these classes:

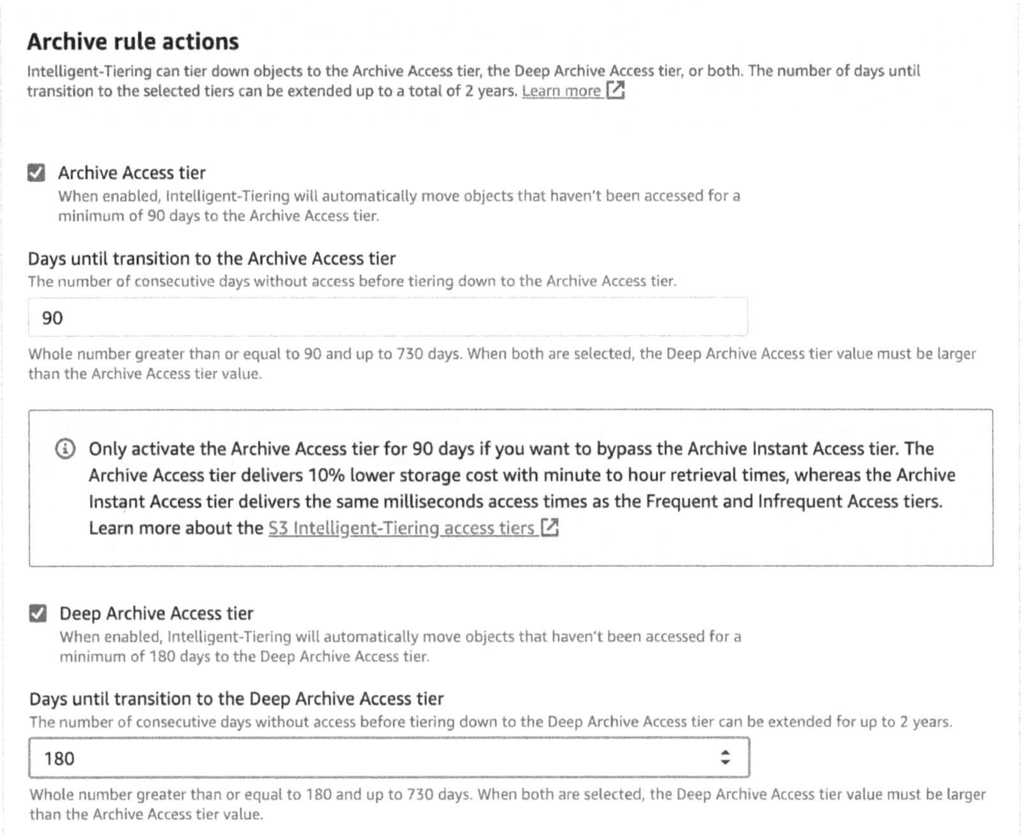

Archive rule actions

Intelligent-Tiering can tier down objects to the Archive Access tier, the Deep Archive Access tier, or both. The number of days until transition to the selected tiers can be extended up to a total of 2 years. Learn more ⬚

☑ Archive Access tier

   When enabled, Intelligent-Tiering will automatically move objects that haven't been accessed for a
   minimum of 90 days to the Archive Access tier.

Days until transition to the Archive Access tier
The number of consecutive days without access before tiering down to the Archive Access tier.

90

Whole number greater than or equal to 90 and up to 730 days. When both are selected, the Deep Archive Access tier value must be larger than the Archive Access tier value.

> ⓘ Only activate the Archive Access tier for 90 days if you want to bypass the Archive Instant Access tier. The
> Archive Access tier delivers 10% lower storage cost with minute to hour retrieval times, whereas the Archive
> Instant Access tier delivers the same milliseconds access times as the Frequent and Infrequent Access tiers.
> Learn more about the S3 Intelligent-Tiering access tiers ⬚

☑ Deep Archive Access tier

   When enabled, Intelligent-Tiering will automatically move objects that haven't been accessed for a
   minimum of 180 days to the Deep Archive Access tier.

Days until transition to the Deep Archive Access tier
The number of consecutive days without access before tiering down to the Deep Archive Access tier can be extended for up to 2 years.

180	⬍

Whole number greater than or equal to 180 and up to 730 days. When both are selected, the Deep Archive Access tier value must be larger than the Archive Access tier value.

Figure 1.1 – Intelligent-Tiering Archive rule action

6. Click on **Create** to create the configuration.

See also

- A detailed comparison of storage classes: `https://docs.aws.amazon.com/AmazonS3/latest/userguide/storage-class-intro.html`

- S3 storage classes pricing: `https://aws.amazon.com/s3/pricing/`

Enforcing encryption on S3 buckets

Amazon S3 encryption increases the level of security and privacy of your data; it helps ensure that only authorized parties can read it. Even if an unauthorized person gains logical or physical access to that data, the data is unreadable if they don't get a hold of the key to unencrypt it.

S3 supports encrypting data both at transit (as it travels to and from S3) and at rest (while it's stored on disks in S3 data centers).

For protecting data at rest, you have two options. The first is **server-side encryption** (**SSE**), in which Amazon S3 will be handling the heavy encryption operation on the server side in AWS. By default, Amazon S3 encrypts your data using **SSE-S3**. However, you can change this to **SSE-KMS**, which uses KMS keys for encryption, or to **SSE-C**, where you can provide and manage your own encryption key. Alternatively, you can encrypt your data using **client-side encryption**, where Amazon S3 doesn't play any role in the encryption process rather; you are responsible for all the encryption operations.

In this recipe, we'll learn how to enforce SSE-KMS server-side encryption using customer-managed keys.

Getting ready

For this recipe, you need to have a KMS key in the same region as your bucket to use for encryption. KMS provides a managed key for S3 (aws/s3) that can be utilized for encryption. However, if you desire greater control over the key properties, such as modifying its policies or performing key rotation, you can create a customer-managed key. To do so, follow these steps:

1. Sign in to the AWS Management Console (`https://console.aws.amazon.com/console/home?nc2=h_ct&src=header-signin`) and navigate to the AWS **Key Management Service** (**AWS KMS**) service.

2. In the navigation pane, choose **Customer managed keys** and click on **Create key**.

3. For **Key type**, choose **Symmetric**, while for **Key usage**, choose **Encrypt and decrypt**. Click on **Next**:

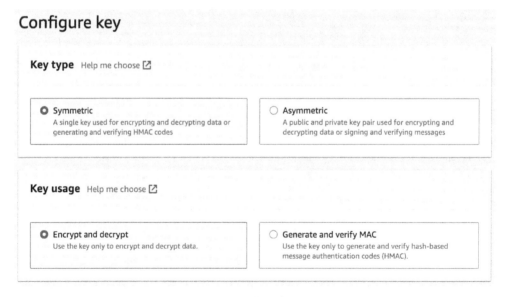

Figure 1.2 – KMS configuration

4. Click on **Next**.

5. Type an **Alias** value for the KMS key. This will be the display name. Optionally, you can provide **Description** and **Tags** key-value pairs for the key.

6. Click on **Next**. Optionally, you can provide **Key administrators** to administer the key. Click on **Finish** to create the key.

How to do it...

1. Sign in to the AWS Management Console (`https://console.aws.amazon.com/console/home?nc2=h_ct&src=header-signin`) and navigate to the **S3** service.

2. In the **Buckets** list, choose the name of the bucket that you want to change the encryption for and navigate to the **Properties** tab.

3. Click on **Edit** in the **Default encryption** section.

4. For **Encryption type**, choose **Server-side encryption with AWS Key Management Service keys (SSE-KMS)**.

5. For **AWS KMS key**, you can select **Enter AWS KMS key ARN** to enter the key you have created or browse it using **Choose from your AWS KMS keys**.

6. Keep **Bucket Key** enabled and save your changes:

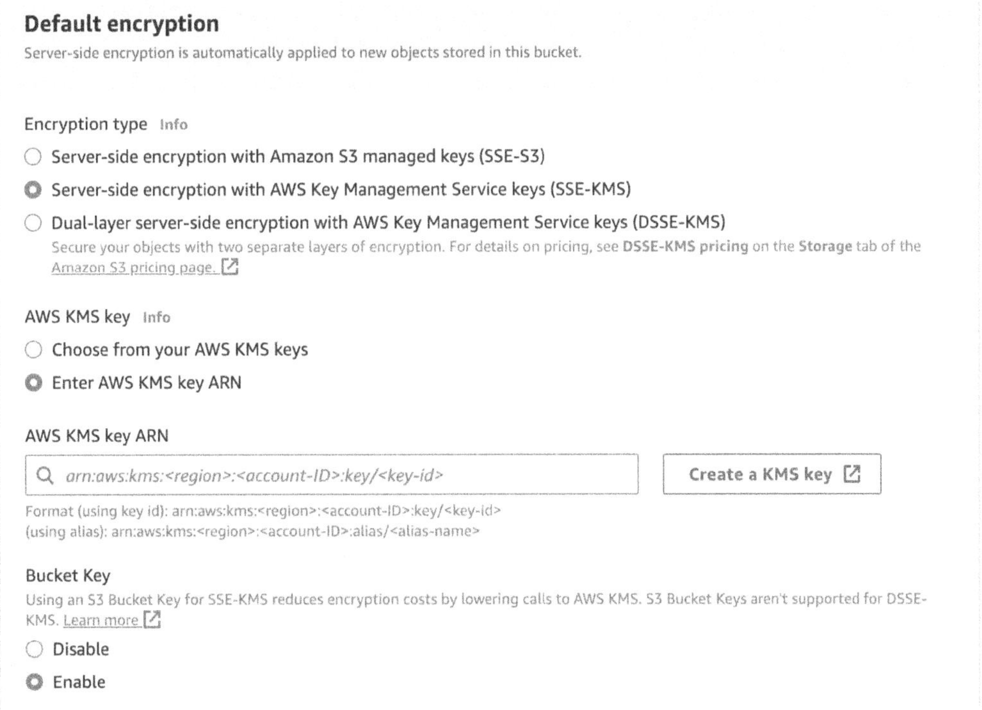

Default encryption

Server-side encryption is automatically applied to new objects stored in this bucket.

Encryption type Info

○ Server-side encryption with Amazon S3 managed keys (SSE-S3)

◉ Server-side encryption with AWS Key Management Service keys (SSE-KMS)

○ Dual-layer server-side encryption with AWS Key Management Service keys (DSSE-KMS)

   Secure your objects with two separate layers of encryption. For details on pricing, see **DSSE-KMS pricing** on the **Storage** tab of the Amazon S3 pricing page. 🗗

AWS KMS key Info

○ Choose from your AWS KMS keys

◉ Enter AWS KMS key ARN

AWS KMS key ARN

🔍 *arn:aws:kms:<region>:<account-ID>:key/<key-id>* | Create a KMS key 🗗 |

Format (using key id): arn:aws:kms:<region>:<account-ID>:key/<key-id>
(using alias): arn:aws:kms:<region>:<account-ID>:alias/<alias-name>

Bucket Key

Using an S3 Bucket Key for SSE-KMS reduces encryption costs by lowering calls to AWS KMS. S3 Bucket Keys aren't supported for DSSE-KMS. Learn more 🗗

○ Disable

◉ Enable

Figure 1.3 – Changing the default encryption

How it works...

By changing the default encryption for your bucket, all newly uploaded objects to your bucket, which don't have an encryption setting, will be encrypted using the KMS you have provided. Already existing objects in your bucket will not be affected. Enabling the bucket key leads to cost savings in KMS service calls associated with the encryption or decryption of individual objects. This is achieved by KMS generating a key at the bucket level rather than generating a separate KMS key for each encrypted object. S3 uses this bucket-level key to generate distinct data keys for objects within the bucket, thereby eliminating the need for additional KMS requests to complete encryption operations.

There's more...

By following this recipe, you can encrypt your objects with SSE-KMS but only if they don't have encryption configured. You can enforce your objects to have an SSE-KMS encryption setting in the PUT operation using a bucket policy, as shown here:

1. Navigate to the bucket's **Permissions** tab.

2. Go to the **Bucket Policy** section and click on **Edit**.

3. Paste the following policy. Make sure you replace `<your-bucket-name>` with the actual name of your S3 bucket and `<your-kms-key-arn>` with the **Amazon Resource Name (ARN)** of your KMS key:

```
{
  "Version": "2012-10-17",
  "Id": "EnforceSSE-KMS",
  "Statement": [
      {
          "Sid": "DenyNonKmsEncrypted",
          "Effect": "Deny",
          "Principal": "*",
          "Action": "s3:PutObject",
          "Resource": "arn:aws:s3:::<your-bucket-name>/*",
          "Condition": {
              "StringNotEquals": {
                  "s3:x-amz-server-side-encryption": "aws:kms"
              }
          }
      },
      {
          "Sid": "AllowKmsEncrypted",
          "Effect": "Allow",
          "Principal": "*",
          "Action": "s3:PutObject",
```

```
            "Resource": "arn:aws:s3:::<your-bucket-name>/*",
            "Condition": {
                "StringEquals": {
                    "s3:x-amz-server-side-encryption": "aws:kms",
                    "s3:x-amz-server-side-encryption-aws-kms-
    key-id": "<your-kms-key-arn>"
                }
            }
        }
    ]
}
```

4.   Save your changes.

This policy contains two statements. The first statement (DenyNonKmsEncrypted) denies the s3:PutObject action for any request that does not include SSE-KMS encryption. The second statement (AllowKmsEncrypted) only allows the s3:PutObject action when the request includes SSE-KMS encryption and the specified KMS key.

See also

- SSE-C: https://docs.aws.amazon.com/AmazonS3/latest/userguide/ ServerSideEncryptionCustomerKeys.html

- Client-side encryption: https://docs.aws.amazon.com/AmazonS3/latest/ userguide/UsingClientSideEncryption.html

- Enforcing encryption in transit with TLS1.2 or higher with Amazon S3: https://aws. amazon.com/blogs/storage/enforcing-encryption-in-transit-with- tls1-2-or-higher-with-amazon-s3/

- Encrypting existing S3 objects: https://aws.amazon.com/blogs/storage/ encrypting-existing-amazon-s3-objects-with-the-aws-cli/

Setting up retention policies for your objects

Amazon S3's storage lifecycle allows you to manage the lifecycle of objects in an S3 bucket based on predefined rules. The lifecycle management feature consists of two main actions: transitions and expiration. Transitions involve automatically moving objects between different storage classes based on a defined duration. This helps in optimizing costs by storing less frequently accessed data in a cheaper storage class. Expiration, on the other hand, allows users to set rules to automatically delete objects from an S3 bucket. These rules can be based on a specified duration. Additionally, you can apply a combination of transitions and expiration actions to objects. Amazon S3's storage lifecycle provides flexibility and ease of management for users and it helps organizations optimize storage costs while ensuring that data is stored according to its relevance and access patterns.

In this recipe, we will learn how to set up a lifecycle policy to archive objects in S3 Glacier after a certain period and then expire them.

Getting ready

To complete this recipe, you need to have a Glacier vault, which is a separate storage container that can be used to store archives, independent from S3. You can create one by following these steps:

1. Open the AWS Management Console (`https://console.aws.amazon.com/console/home?nc2=h_ct&src=header-signin`) and navigate to the Glacier service.

2. Click on **Create vault** to start creating a new Glacier vault.

3. Provide a unique and descriptive name for your vault in the **Vault name** field.

4. Optionally, you can choose to receive notifications for events by clicking **Turn on notifications** under the **Event notifications** section.

5. Click on **Create** to create the vault.

How to do it...

1. Open the AWS Management Console (`https://console.aws.amazon.com/console/home?nc2=h_ct&src=header-signin`) and navigate to the S3 service.

2. Select the desired bucket for which you want to configure the lifecycle policy and navigate to the **Management** tab.

3. In the left panel, select **Lifecycle** and click on **Create lifecycle rule**.

4. Under **Rule name**, name the lifecycle rule to identify it.

5. Under **Choose a rule scope**, you can choose **Apply to all objects in the bucket** or **Limit the scope of this rule using one or more filters** to specify the objects for which the rule will be applied. You can use one of the following filters or a combination of them:

    - Filter objects based on prefixes (for example, logs)

    - Filter objects based on tags; you can add multiple key-value pair tags to filter on

    - Filter objects based on object size by setting **Specify minimum object size** and/or **Specify maximum object size** and specifying the size value and unit

The following screenshot shows a rule that's been restricted to a set of objects based on a prefix:

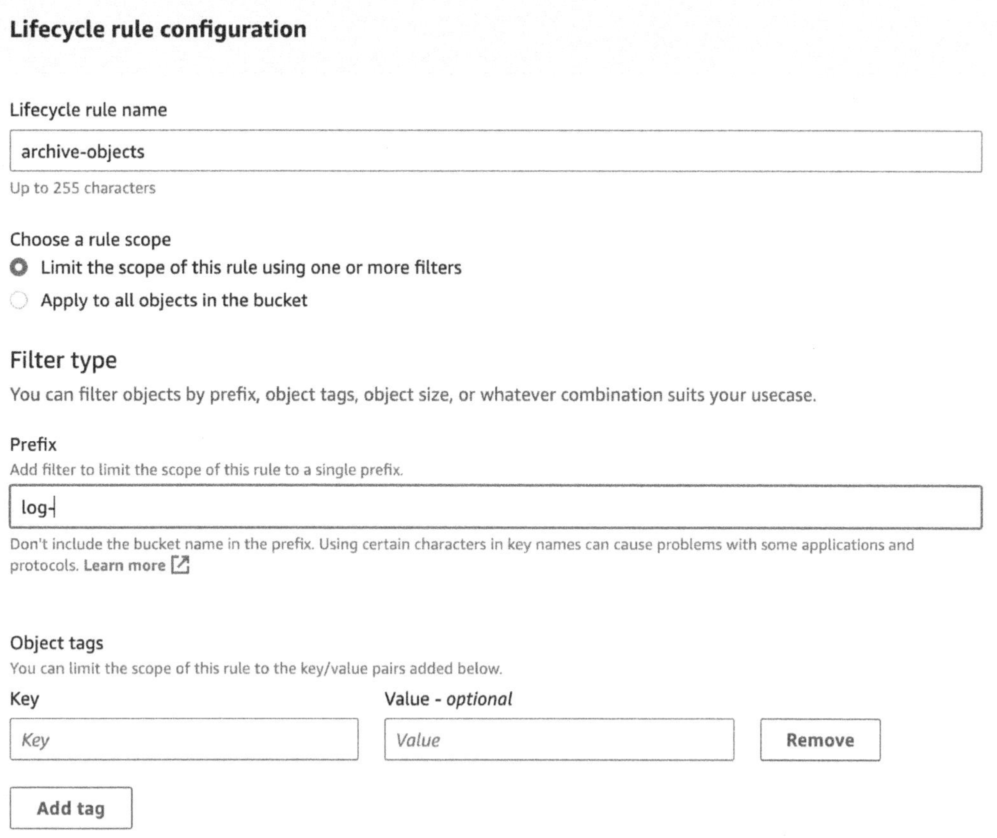

Lifecycle rule configuration

Lifecycle rule name

archive-objects

Up to 255 characters

Choose a rule scope
- ◉ Limit the scope of this rule using one or more filters
- ○ Apply to all objects in the bucket

Filter type

You can filter objects by prefix, object tags, object size, or whatever combination suits your usecase.

Prefix

Add filter to limit the scope of this rule to a single prefix.

log-|

Don't include the bucket name in the prefix. Using certain characters in key names can cause problems with some applications and protocols. **Learn more** ☑

Object tags

You can limit the scope of this rule to the key/value pairs added below.

Key	Value - *optional*	
Key	Value	Remove

Add tag

Figure 1.4 – Lifecycle rule configuration

6.   Under **Lifecycle rule actions**, select the following options:

- **Move current versions of objects between storage classes**. Then, choose one of the Glacier classes and set **Days after object creation** in which the object will be transitioned (for example, 60 days).

- **Expire current versions of objects**. Then, set **Days after object creation** in which the object will expire. Choose a value higher than the one you set for transitioning the object to Glacier (for example, 100).

Review the transition and expiration actions you have set and click on **Create rule** to apply the lifecycle policy to the bucket:

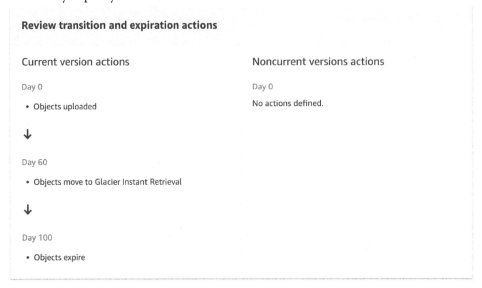

Figure 1.5 – Reviewing the lifecycle rule

> **Note**
>
> It may take some time for the lifecycle rule to be applied to all the selected objects, depending on the size of the bucket and the number of objects. The rule will affect existing files, not just new ones, so ensure that no applications are accessing files that will be archived or deleted as they will no longer be accessible via direct S3 retrieval.

How it works...

After you save the lifecycle rule, Amazon S3 will periodically evaluate it to find objects that meet the criteria specified in the lifecycle rule. In this recipe, the object will remain in its default storage type for the specified period (for example, 60 days) after which it will automatically be moved to the Glacier storage class. This transition is handled transparently, and the object's metadata and properties remain unchanged. Once the objects are transitioned to Glacier, they are stored in a Glacier vault and become part of the Glacier storage infrastructure. Objects will then remain in Glacier for the remaining period of expiry (for example, 40 days), after which they will expire and be permanently deleted from your S3 bucket.

Please note that once the objects have expired, they will be queued for deletion, so it might take a few days after the object reaches the end of its lifetime for it to be deleted.

There's more...

Lifecycle configuration can be specified as an XML when using the S3 API or AWS console, which can be helpful if you are planning on using the same lifecycle rules on multiple buckets. You can read more on setting this up at `https://docs.aws.amazon.com/AmazonS3/latest/userguide/intro-lifecycle-rules.html`.

See also

- Setting up notifications for events related to your lifecycle rule: `https://docs.aws.amazon.com/AmazonS3/latest/userguide/lifecycle-configure-notification.html`

- Supported lifecycle transitions and related constraints: `https://docs.aws.amazon.com/AmazonS3/latest/userguide/lifecycle-transition-general-considerations.html`

Versioning your data

Amazon S3 versioning refers to maintaining multiple variants of an object at the same time in the same bucket. Versioning provides you with an additional layer of protection by giving you a way to recover from unintended overwrites and accidental deletions as well as application failures.

S3 Object Versioning is not enabled by default and has to be explicitly enabled for each bucket. Once enabled, versioning cannot be disabled and can only be suspended. When versioning is enabled, you will be able to preserve, retrieve, and restore any version of an object stored in the bucket using the version ID. Every version of an object is the whole object, not the delta from the previous version, and you can set permissions at the version level. So, you can set different permissions for different versions of the same object.

In this recipe, we'll learn how to delete the current version of an object to make the previous one the current version.

Getting ready

For this recipe, you need to have a version-enabled bucket with an object that has at least two versions.

You can enable versioning for your bucket by going to the bucket's **Properties** tab, editing the **Bucket Versioning** area, and setting it to **Enable**:

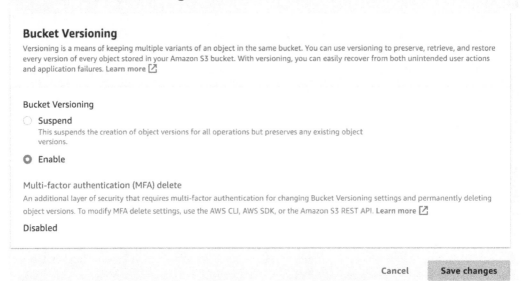

Figure 1.6 – Enabling bucket versioning

You can create a new version of an object by simply uploading a file with the same name to the versioning-enabled bucket.

It's important to note that enabling versioning for a bucket is irreversible. Once versioning is enabled, it will be applied to all existing and future objects in that bucket. So, before enabling versioning, make sure that your application or workflow is compatible with object versioning.

Enabling versioning for the first time will take time to take effect, so we recommend waiting 15 minutes before performing any write operation on objects in the bucket.

How to do it...

1. Sign in to the AWS Management Console (https://console.aws.amazon.com/console/home?nc2=h_ct&src=header-signin) and navigate to the S3 service.

2. In the **Buckets** list, select the S3 bucket that contains the object for which you want to set the previous version as the current one.

3. In the **Objects** tab, click on **Show versions**. Here, you can view all your object versions:

Figure 1.7 – Object versions

4. Select the current version of the object that you want to delete. It's the top-most version with the latest modified date.

5. Click on the **Delete** button and write `permanently delete` as prompted on the next screen.

   After deleting the current version, the previous version will automatically become the latest version:

	Name	▲	Type		Version ID	Last modified	
☐	🗋 D73H62		-		ONDxepWD 9sCLJWXph sAlZFk8v84i xWXX	September 4, 2023, 00:07:42 (UTC+03:00)	
☐	└🗋 D73H62		-		z42zAHR8Si mCvifm.OR 2pvdJJ0KAK Zbi	September 2, 2023, 23:28:48 (UTC+03:00)	

Figure 1.8 – Object versions after version deletion

6. Verify that the previous version is now the latest version by checking the **Last modified** timestamps or verifying this through object listing, metadata, or download.

How it works...

Once you enable bucket versioning, each object in the bucket will have a version ID that uniquely identifies the object in the bucket, and the non-version-enabled buckets will have their version IDs set to null for their objects. The older versions of an object become non-current but continue to exist and remain accessible. When you delete the current version of the object, it will be permanently removed and the S3 versioning mechanism will automatically promote the previous version as the current one after deletion. If you delete an object without specifying the version ID, Amazon S3 doesn't delete it permanently; instead, it inserts a delete marker into it and it becomes the current object version. However, you can still restore its previous versions:

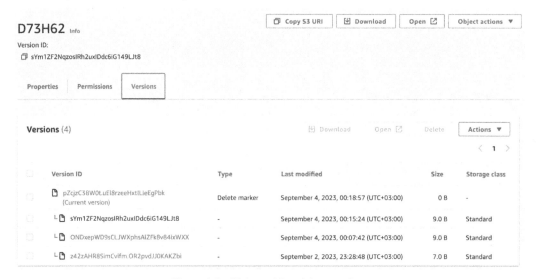

Figure 1.9 – Object with a delete marker

There's more...

S3 rates apply to every version of an object that's stored and requested, so keeping non-current versions of objects can increase your storage cost. You can use lifecycle rules to archive the non-current versions or permanently delete them after a certain period and keep the bucket clean from unnecessary object versions.

Follow these steps to add a lifecycle rule to delete non-current versions after a certain period:

1.   Go to the bucket's **Management** tab and click on the **Lifecycle** configuration.

2.   Click on the **Add lifecycle rule** button to create a new rule.

3.   Provide a unique name for the rule.

4. Under **Apply rule to**, select the appropriate resources (for example, the entire bucket or specific prefixes).

5. Set the action to **Permanently delete non-current versions**.

6. Specify **Days after objects become noncurrent** in which the delete will be executed. Optionally, you can specify **Number of newer versions to retain**, which means it will keep the said number of versions for the object and all others will be deleted when they are eligible for deletion based on the specified period.

7. Click on **Save** to save the lifecycle rule.

See also

- You can prevent accidental or malicious deletion of objects in a versioned bucket by enabling **Multi-Factor Authentication (MFA) Delete**. You can learn how to implement this at `https://docs.aws.amazon.com/AmazonS3/latest/userguide/MultiFactorAuthenticationDelete.html`.

- Setting permissions at the version level: `https://docs.aws.amazon.com/AmazonS3/latest/userguide/VersionedObjectPermissionsandACLs.html`.

Replicating your data

AWS S3 replication is an automatic asynchronous process that involves copying objects to one or multiple destination buckets. Replication can be configured across buckets in the same AWS region with **Same-Region Replication**, which can be useful for scenarios such as isolating different workloads, segregating data for different teams, or achieving compliance requirements. Replication can also be configured for buckets across different AWS regions with **Cross-Region Replication** (**CRR**), which helps in reducing latency for accessing data, especially for enterprises with a large number of locations, by maintaining multiple copies of the objects in different geographies or different regions. It provides compliance and data redundancy for improved performance, availability, and disaster recovery capabilities.

In this recipe, we'll learn how to set up replication between two buckets in different AWS regions and the same AWS account.

Getting ready

You need to have an S3 bucket in the destination AWS region to act as a target for the replication. Also, S3 versioning must be enabled for both the source and destination buckets.

How to do it...

1. Sign in to the AWS Management Console (`https://console.aws.amazon.com/console/home?nc2=h_ct&src=header-signin`) and navigate to the S3 service.

2. In the **Buckets** list, choose the source bucket you want to replicate.

3. Go to the **Management** tab and select **Create replication rule** under **Replication rules**.

4. Under **Replication rule name** in the **Replication rule configuration** section, give your rule a unique name.

5. Under **Status**, either keep it **Enabled** for the rule to take effect once you save it or change it to **Disabled** to enable it later as required:

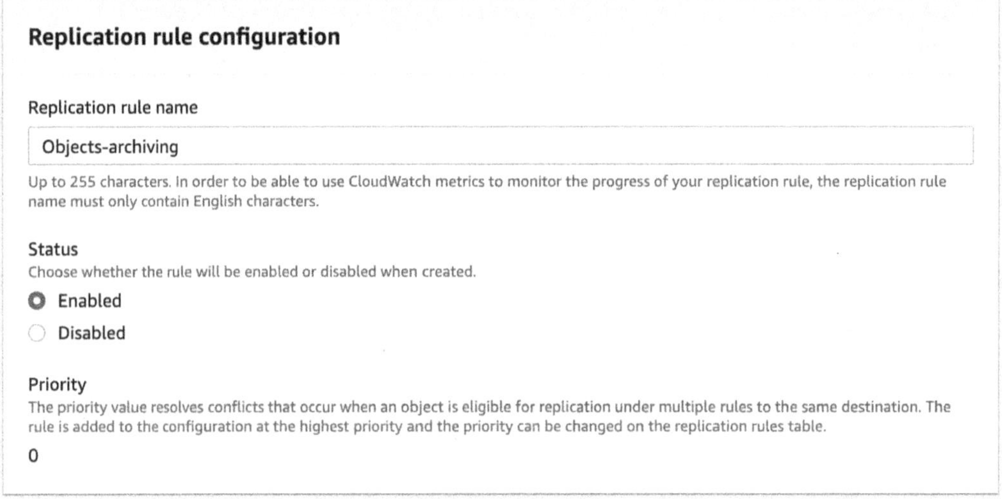

Replication rule configuration

Replication rule name

> Objects-archiving

Up to 255 characters. In order to be able to use CloudWatch metrics to monitor the progress of your replication rule, the replication rule name must only contain English characters.

Status
Choose whether the rule will be enabled or disabled when created.

- ● Enabled
- ○ Disabled

Priority
The priority value resolves conflicts that occur when an object is eligible for replication under multiple rules to the same destination. The rule is added to the configuration at the highest priority and the priority can be changed on the replication rules table.

0

Figure 1.10 – Replication rule configuration

6. If this is the first replication rule for the bucket, **Priority** will be set to **0**. Subsequent rules that are added will be assigned higher priorities. When multiple rules share the same destination, the rule with the highest priority takes precedence during execution, typically the one created last. If you wish to control the priority for each rule, you can achieve this by setting the rule using XML. For guidance on how to configure this, refer to the *See also* section.

7. In the **Source bucket** section, you have the option to replicate all objects in the bucket by selecting **Apply to all objects in the bucket** or you can narrow it down to specific objects by selecting **Limit the scope of this rule using one or more filters** and specifying a **Prefix** value (for example, `logs_` or `logs/`) to filter objects. Additionally, you have the option to replicate objects based on their tags. Simply choose **Add tag** and input key-value pairs. This process can be repeated so that you can include multiple tags:

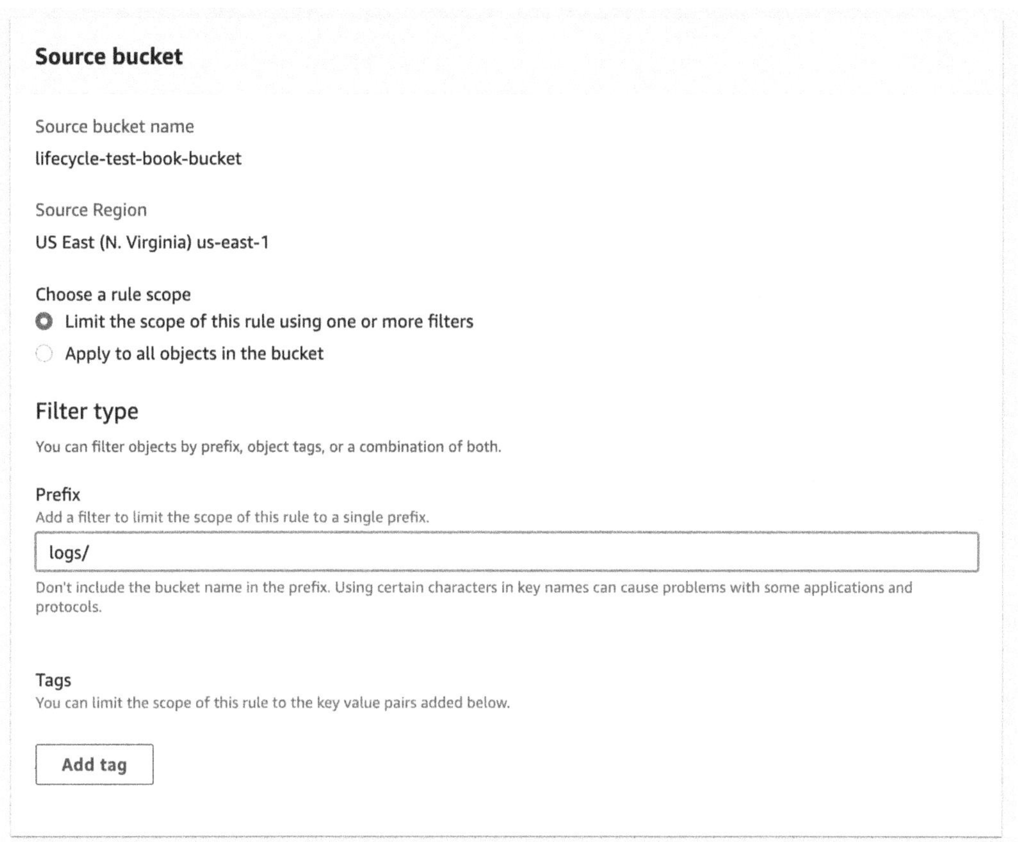

Source bucket

Source bucket name
lifecycle-test-book-bucket

Source Region
US East (N. Virginia) us-east-1

Choose a rule scope
⦿ Limit the scope of this rule using one or more filters
◯ Apply to all objects in the bucket

Filter type

You can filter objects by prefix, object tags, or a combination of both.

Prefix
Add a filter to limit the scope of this rule to a single prefix.

logs/

Don't include the bucket name in the prefix. Using certain characters in key names can cause problems with some applications and protocols.

Tags
You can limit the scope of this rule to the key value pairs added below.

Add tag

Figure 1.11 – Source bucket configuration

8. Under **Destination**, select **Choose a bucket in this account** and enter or browse for the destination bucket name.

9. Under **IAM role**, select **Choose from existing IAM roles,** then choose **Create new role** from the drop-down list.

10. Under **Destination storage class**, you can select **Change the storage class for the replicated objects** and choose one of the storage classes to be set for the replicated objects in the destination bucket.

11. Click on **Save** to save your changes.

How it works...

By adding this replication rule, you grant the source bucket permission to replicate objects to the destination bucket in the said region. Once the replication process is complete, the destination bucket will contain a copy of the objects from the source bucket. The objects in the destination bucket will have the same ownership, permissions, and metadata as the source objects. When you enable replication to your bucket, several background processes occur to facilitate this process. S3 continuously monitors changes to objects in your source bucket. Once a change is detected, S3 generates a replication request for the corresponding objects and initiates the process of transferring the data from the source to the destination bucket.

There's more...

There are additional options that you can enable while setting the replication rule under **Additional replication options**. The **Replication metrics** option enables you to monitor the replication progress with S3 Replication metrics. It does this by tracking bytes pending, operations pending, and replication latency. **The Replication Time Control (RTC)** option can be beneficial if you have a strict **service-level agreement (SLA)** for data replication as it will ensure that approximately 99% of your objects will be replicated within a 15-minute timeframe. It also enables replication metrics to notify you of any instances of delayed object replication. The **Delete marker replication** option will replicate object versions with a delete marker. Finally, the **Replica modification sync** option will replicate the metadata changes of objects.

See also

- Replicating buckets in different AWS accounts: `https://docs.aws.amazon.com/AmazonS3/latest/userguide/replication-walkthrough-2.html`

- Replication configuration: `https://docs.aws.amazon.com/AmazonS3/latest/userguide/replication-add-config.html`

Monitoring your S3 bucket

Enabling and monitoring S3 metrics allows you to proactively manage your S3 resources, optimize performance, ensure appropriate security and compliance measures are in place, identify cost-saving opportunities, and ensure the operational readiness of your S3 infrastructure. S3 offers various methods for monitoring your buckets, including S3 server access logs, CloudTrail, CloudWatch metrics, and S3 event notifications. S3 server access logs can be enabled to log each request made to the bucket. CloudTrail captures actions taken on S3 or API calls on the bucket, allowing you to monitor and audit actions, including object-level operations such as uploads, downloads, and deletions. CloudWatch metrics track specific metrics for your buckets and allow you to set up alarms so that you receive notifications when certain thresholds are met. S3 event notifications enable you to set up notifications for specific S3 events and configure actions in response to those events. In this recipe, we will cover enabling CloudTrail for your S3 buckets and configuring CloudWatch metrics to monitor high-volume data transfer based on these logs.

Getting ready

To proceed with this recipe, you need to enable CloudTrail so that it can log S3 data events and insights. Follow these steps:

1. Open the AWS Management Console (`https://console.aws.amazon.com/console/home?nc2=h_ct&src=header-signin`) and navigate to the CloudTrail service.

2. Click on **Trails** in the left navigation pane and click on **Create trail** to create a new trail.

3. Provide a name for the trail in the **Trail name** field.

4. For **Storage location**, you need to provide an S3 bucket for storing CloudTrail logs. You can select **Use existing S3 bucket** or **Create new S3 bucket.**

5. Optionally, you can enable **Log file SSE-KMS encryption** and choose the KMS key.

6. Under **CloudWatch Logs**, choose **Yes** for **Send CloudTrail events to CloudWatch Logs**.

7. Configure the **CloudWatch Logs** settings as per your requirements. For example, you can select an existing CloudWatch Logs group or create a new one:

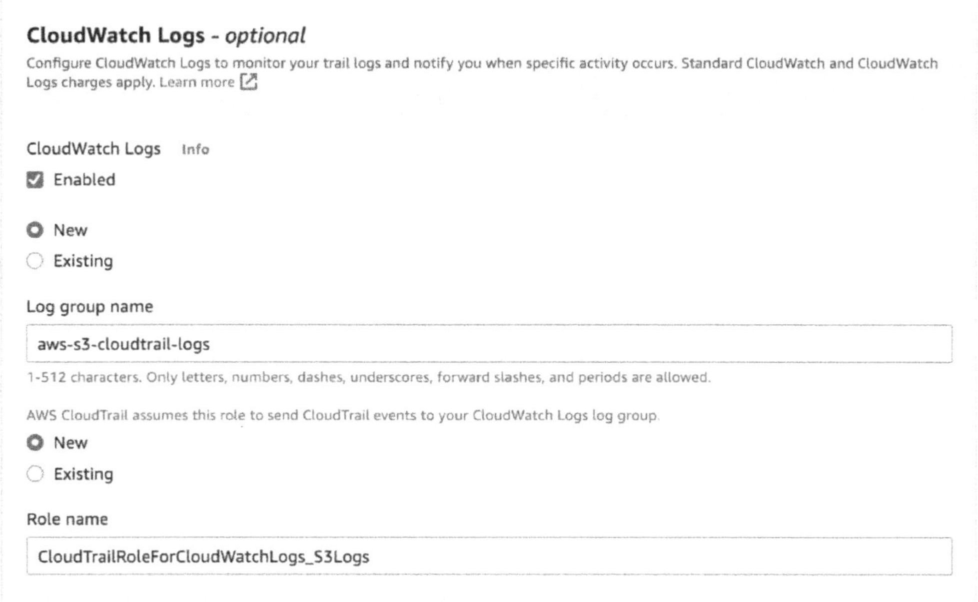

CloudWatch Logs - *optional*

Configure CloudWatch Logs to monitor your trail logs and notify you when specific activity occurs. Standard CloudWatch and CloudWatch Logs charges apply. Learn more ☑

CloudWatch Logs   Info
☑ Enabled

● New
○ Existing

Log group name

```
aws-s3-cloudtrail-logs
```

1-512 characters. Only letters, numbers, dashes, underscores, forward slashes, and periods are allowed.

AWS CloudTrail assumes this role to send CloudTrail events to your CloudWatch Logs log group.

● New
○ Existing

Role name

```
CloudTrailRoleForCloudWatchLogs_S3Logs
```

Figure 1.12 – Enabling CloudWatch Logs

8. For **Role name**, choose to create a new one and give it a name.

9. Review the other trail settings, such as log file validation and tags, make adjustments if needed, and click on **Next**.

10. Under the **Events** section, enable **Data events** and **Insight events** in addition to **Management events**, which is already enabled.

11. Under **Management events**, select **Read** and **Write**:

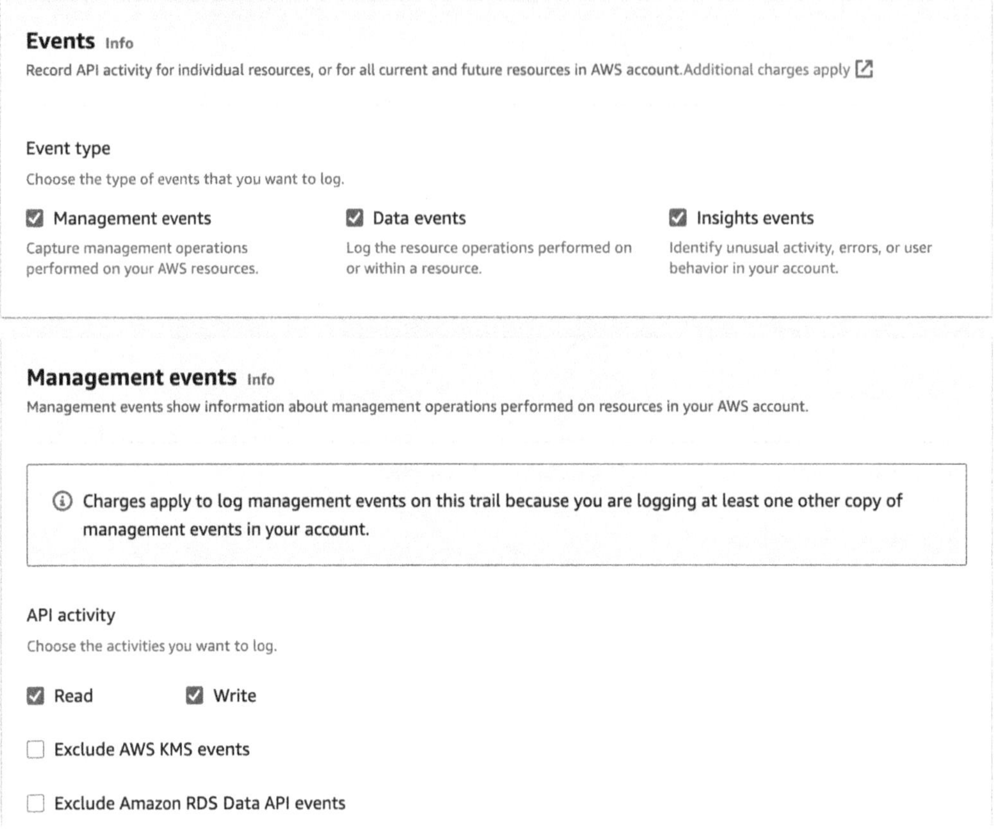

Events Info

Record API activity for individual resources, or for all current and future resources in AWS account. Additional charges apply

Event type

Choose the type of events that you want to log.

☑ **Management events**

Capture management operations performed on your AWS resources.

☑ **Data events**

Log the resource operations performed on or within a resource.

☑ **Insights events**

Identify unusual activity, errors, or user behavior in your account.

Management events Info

Management events show information about management operations performed on resources in your AWS account.

> ⓘ Charges apply to log management events on this trail because you are logging at least one other copy of management events in your account.

API activity

Choose the activities you want to log.

☑ Read   ☑ Write

☐ Exclude AWS KMS events

☐ Exclude Amazon RDS Data API events

Figure 1.13 – Configuring Events

12. Under **Data events**, choose S3 for **Data event type** and **Log all events** for the **Log selector** template.

13. Under **Insights events**, select **API call rate** and **API error rate**.

14. Click on **Next** and then click on **Create trail** to create the trail.

Once the trail has been created, CloudTrail will start capturing S3 data events and storing the logs in the specified S3 bucket. Simultaneously, the logs will be sent to the CloudWatch Logs group specified during trail creation.

How to do it...

1. Open the AWS Management Console (https://console.aws.amazon.com/console/home?nc2=h_ct&src=header-signin) and navigate to the CloudWatch console.

2. Go to **Log groups** from the navigation pane on the left and select the CloudTrail log group you just created.

3. Click on **Create Metric Filter** from the **Action** drop-down list.

4. Provide { ($.eventName = CopyObject) || ($.eventName = PutObject) || ($.eventName = CompleteMultipartUpload) && $.request.bytes_ transferred > 500000000} as the filter pattern. This filter pattern will capture events related to copying or uploading objects to S3 that are larger than 500 MB. The threshold value should be set based on your bucket access patterns.

    You can test your pattern by specifying one of the log files or providing a custom log in the **Test pattern** section. Then, you can click on **Test pattern** and validate the result:

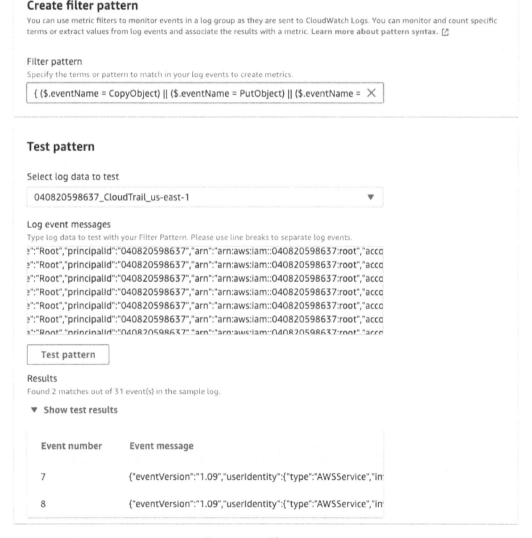

Figure 1.14 – Filter pattern

5.   Click on **Next**.

6.   Under the **Create filter name** field, specify a filter name.

7.   Under the **Metric Details** section, specify a **Metric namespace** value (for example, S3Metrics) and provide a name for the metric itself under **Metric name** (for example, HighVolumeTransfers).

8.   Set **Unit** to **Count** for your metric and set **Metric value** to 1 to indicate that a transfer event has occurred. Finally, set **Default value** to 0:

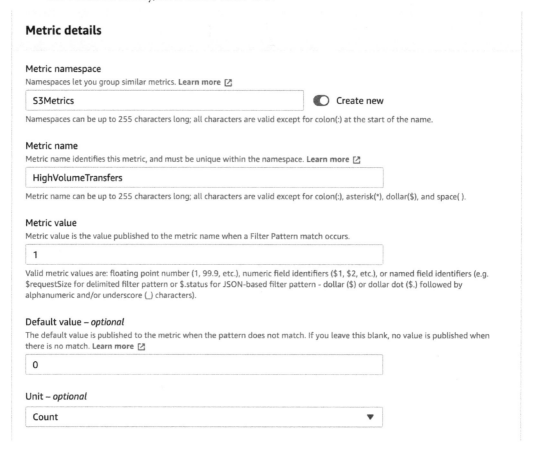

Figure 1.15 – Metric details

9.   Click on **Create metric filter**.

How it works...

By enabling CloudTrail in your AWS account and ensuring that the logs are delivered to CloudWatch, the S3 API activities can be accessed and analyzed within your AWS environment. By creating a metric filter with a customized filter pattern that matches S3 transfer events, relevant information from the CloudTrail logs can be extracted. Once the metric filter is created, CloudWatch generates a custom metric based on the filter's configuration. This metric represents the occurrence of high-volume S3 transfers. You can then view this metric in the CloudWatch console, where you can gain insights into your S3 transfer activity and take the necessary actions.

There's more...

Once your metric has been created, you can create alarms based on the metric's value to notify you when a high volume of S3 transfers has been detected.

To create an alarm for the metric you have created based on high-volume S3 transfers from CloudTrail logs on CloudWatch, follow these steps:

1. Go to the CloudWatch console and select the **Alarms** tab.

2. Click on **Create Alarm**. In the **Create Alarm** wizard, select the metric you created. You can find it by navigating the namespace and finding the metric name you configured earlier.

3. Under the **Metric** section, set **Statistic** to **Sum** and **Period** to **15 minutes**. This can be changed as per your needs:

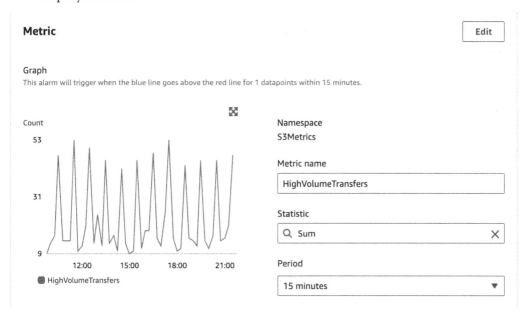

Figure 1.16 – Metric statistics

4. Under the **Conditions** section, Set **Threshold type** to **Static** and choose **Greater than** for the alarm condition. This indicates a high volume transfer on your bucket, as per your observations. Optionally, you can choose how many data points within the evaluation period must be breached to cause the alarm to go to the alarm state by expanding **Additional configuration**. This will help you avoid false positives caused by transient spikes in the metric values:

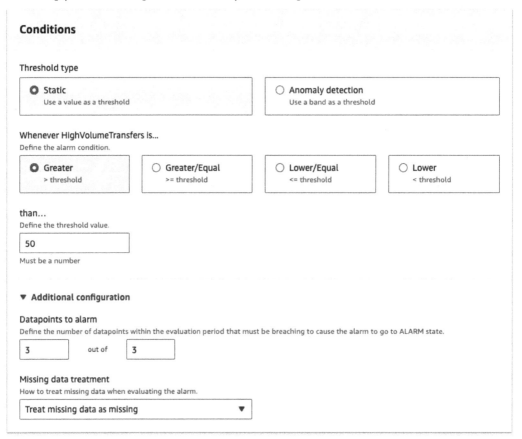

Figure 1.17 – Metric conditions

5. Click on **Next**.

6. Under the **Notification** section, choose **In alarm** to send a notification when the metric is in the alarm state. Choose **Create a new topic**, provide a name for it and the email endpoints that will receive the SNS notification, and click on **Create topic** or choose an existing SNS topic if you have one that's been configured. You can configure other actions to be executed if the alarm state is triggered, such as executing a Lambda function or performing automated scaling actions:

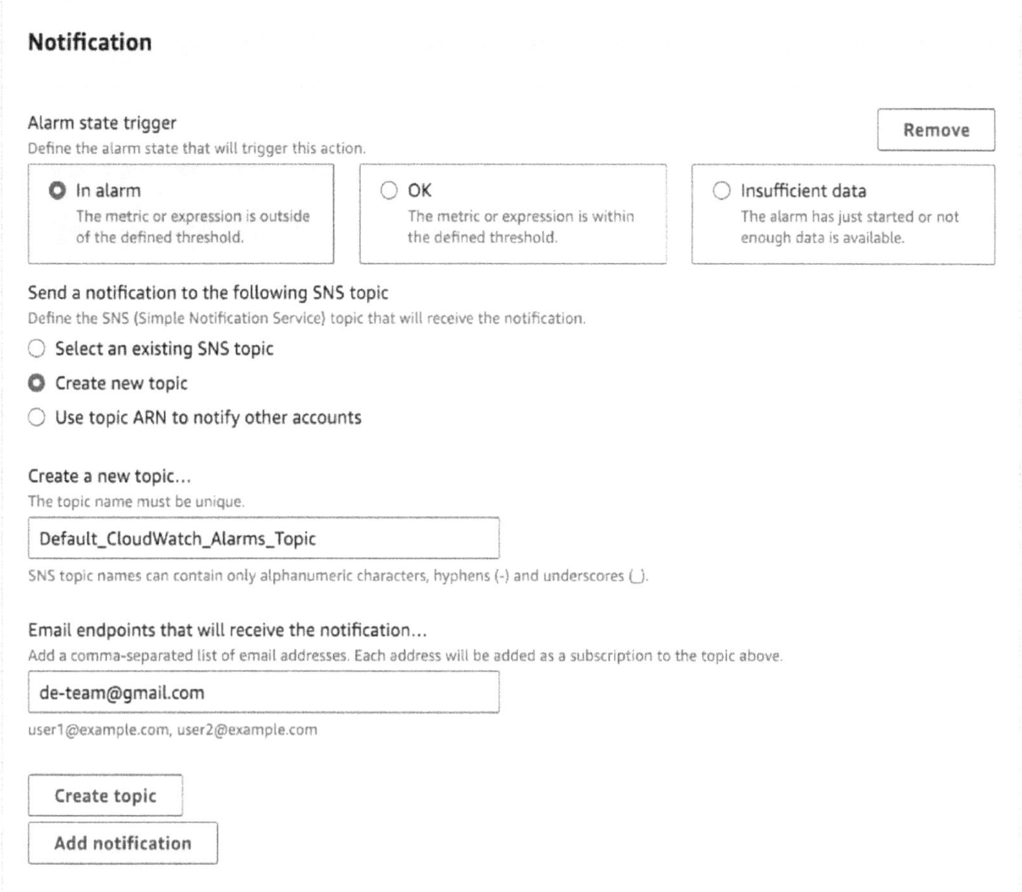

Notification

Alarm state trigger
Define the alarm state that will trigger this action.

Remove

⦿ In alarm
The metric or expression is outside of the defined threshold.

○ OK
The metric or expression is within the defined threshold.

○ Insufficient data
The alarm has just started or not enough data is available.

Send a notification to the following SNS topic
Define the SNS (Simple Notification Service) topic that will receive the notification.

○ Select an existing SNS topic

⦿ Create new topic

○ Use topic ARN to notify other accounts

Create a new topic...
The topic name must be unique.

Default_CloudWatch_Alarms_Topic

SNS topic names can contain only alphanumeric characters, hyphens (-) and underscores (_).

Email endpoints that will receive the notification...
Add a comma-separated list of email addresses. Each address will be added as a subscription to the topic above.

de-team@gmail.com

user1@example.com, user2@example.com

Create topic

Add notification

Figure 1.18 – Metric notification settings

7. Provide a name for the alarm so that it can be identified with ease.

8. Review the alarm settings and click on **Create Alarm** to create the alarm.

Once the alarm has been created, it will start monitoring the metric for high-volume S3 transfers based on the defined conditions. If the threshold is breached for the specified duration (there are more than 150 data transfer requests of more than 500 MB within 45 minutes), the alarm state will be triggered, and an SNS notification will be sent. This allows you to receive timely notifications and take appropriate remedial actions in case of high-volume S3 transfers, ensuring that any potential issues are addressed proactively.

See also

- S3 monitoring tools: `https://docs.aws.amazon.com/AmazonS3/latest/userguide/monitoring-automated-manual.html`
- Logging options for S3: `https://docs.aws.amazon.com/AmazonS3/latest/userguide/logging-with-S3.html`

2

Sharing Your Data Across Environments and Accounts

Data sharing plays a pivotal role in today's data-driven world, enabling organizations to unlock the full potential of their data resources. It encourages collaboration and drives innovation. Sharing data securely and efficiently is a fundamental requirement for businesses across various industries. This chapter explores various AWS solutions designed to facilitate the secure sharing of data with both internal and external stakeholders. These solutions recognize the significance of data sharing and provide robust options to meet diverse sharing needs.

We will start by discussing methods for sharing database resources by creating read-only replicas for RDS, which ensures reliable relational data sharing and reduces the operational load on the primary RDS instance. Similarly, we will explore Redshift live data sharing that allows multiple clusters to access the same data, enhancing collaborative analysis and distributing the load across different clusters.

Next, we will leverage Lake Formation to synchronize the Glue Data Catalog across different AWS accounts, providing a unified view of metadata for data discovery. We'll also examine enforcing fine-grained permissions on **Simple Storage Service** (**S3**) data using Lake Formation to enhance data lake security.

For dynamic data sharing, we will cover real-time S3 data sharing using SNS and Lambda for immediate access, as well as temporary sharing using presigned URLs. Finally, we'll discuss sharing operational data by providing read-only access to CloudWatch data with other AWS accounts.

This chapter contains the following recipes:

- Creating read-only replicas for RDS
- Redshift live data sharing among your clusters
- Synchronizing Glue Data Catalog to a different account
- Enforcing fine-grained permissions on S3 data sharing using Lake Formation

- Sharing your S3 data temporarily using a presigned URL

- Real-time sharing of S3 data

- Sharing read-only access to your CloudWatch data with another AWS account

Technical requirements

Many recipes in this chapter presume the availability of an S3 bucket, Glue databases and tables, and an IAM role/user, in addition to the one you are utilizing to implement the recipes, to serve as the grantee for data sharing. Furthermore, certain recipes assume the presence of two AWS accounts to streamline the process of sharing data with other AWS accounts.

You can find the code files for this chapter on GitHub at `https://github.com/PacktPublishing/Data-Engineering-with-AWS-Cookbook/tree/main/Chapter02`.

Creating read-only replicas for RDS

Amazon **Relational Database Service (RDS)** provides a functionality to establish read-only replicas, replicating your database for multiple users' access and efficient data sharing. These replicas can reside in the same region as, or a different one from, the primary one. This proves beneficial when encountering heavy read loads that could potentially impact RDS performance; replicas can accommodate the heavy read loads without impacting the primary instance's performance. Additionally, you can create a replica in a region that is closer to your users, which can also help in the event of primary region disruption where the replica can be promoted to a standalone instance, thus allowing seamless operation continuation and disaster recovery.

RDS offers two primary replication modes for read replicas:

- In **single-primary mode**, a single primary RDS instance manages both read and write operations, while one or more read replicas can be added to offload read traffic, making it ideal for applications with high read and low write loads.

- In contrast, **multi-primary mode** features multiple primary instances that handle both read and write operations, with synchronous replication among them. This setup ensures that if one primary instance fails, the others can seamlessly take over, making it suitable for applications with high read and write demands that require high availability.

In this recipe, we will learn how to create a read-only replica for RDS.

Getting ready

For this recipe, you need to have an RDS instance that can be of the MySQL, PostgreSQL, or MariaDB type, with automatic backup enabled. You can enable this by modifying your RDS instance and setting the backup retention period to a value greater than zero.

How to do it...

1. Log in to the AWS Management Console (`https://console.aws.amazon.com/console/home?nc2=h_ct&src=header-signin`) and navigate to the **RDS** service.

2. Select **Databases** from the left navigation pane, locate your RDS, and select **Create read replica** from the **Actions** menu.

3. Enter a unique name for your replica under the **DB instance identifier** field.

4. In the **Instance configuration** section, choose a class for your replica that matches the primary RDS class or a bigger one that aligns with your requirements.

5. Choose the destination region for your replica. You can choose the same region as your primary replica or a different one.

6. When configuring storage, it's advisable to allocate a type and size equivalent to or larger than what is allocated to the primary RDS. Under **Storage autoscaling**, you can check the box next to **Enable storage autoscaling** and choose the maximum value that the replica can auto-scale to.

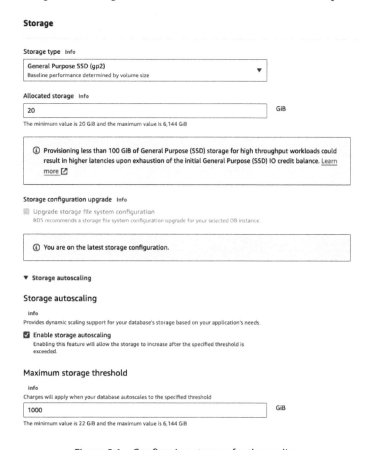

Figure 2.1 – Configuring storage for the replica

7. In the **Availability** section, select from three options according to your requirements:

    - **Multi-AZ DB Cluster**, which deploys one primary DB and two readable standby instances across different availability zones

    - **Multi-AZ DB Instance**, which launches one primary instance and one standby DB in separate availability zones

    - **Single DB instance**, which creates a single DB instance

8. For the **Connectivity** section, choose the protocol used to communicate with your database, **IPv4** or **Dual-stack mode**, which will allow communications from both IPv6 and IPv4 protocols. Ensure that the replica is assigned to the same DB subnet group as the primary RDS. Determine whether the replica will allow public access (this is usually disabled in production settings) and specify the security group governing access. Lastly, set the port number for replica connections.

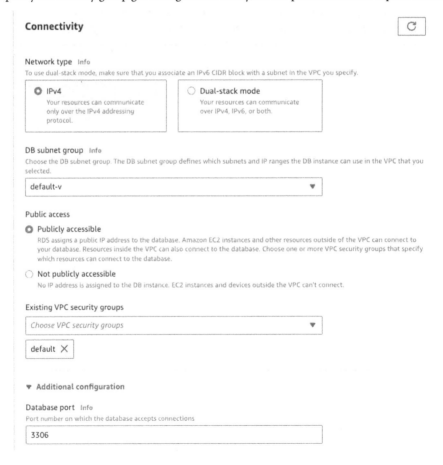

Figure 2.2 – Configuring connectivity for the replica

9. Select the preferred authentication method for the replica.

10. Configure additional settings as required in the **Additional configuration** section. Note that encryption settings must match those of the primary RDS.

11. Click on **Create read replica**.

How it works...

While configuring the read replica, you need to make several key decisions to align with your performance, availability, and security requirements. First, choose the DB instance class and storage to match or exceed the specifications of the primary RDS, especially if you anticipate a high load on the replica. Next, decide on the region for your replica. Placing the replica in the same region as the primary RDS results in lower latency for data replication, while placing it in a different region enhances disaster recovery capabilities but introduces higher latency and potentially higher data transfer costs. For availability, you must strike a balance between higher availability options and associated costs. For connectivity, you can choose between IPv4, which supports communication over the traditional IP version 4, or dual-stack mode, which allows communication over both IPv4 and IPv6, offering greater flexibility. Once the replica creation is initiated, AWS generates a snapshot of the primary RDS that it uses to create the replica. The duration of replica creation depends on the data size of the primary RDS. Upon completion, the replica becomes available and the snapshot is deleted. Asynchronous replication governs data transfer from the primary RDS to the read replica. Data changes in the primary RDS are logged in the binary log, periodically retrieved by the replication process, and transmitted to the read replica. Replica lag, a potential delay between the primary and replica RDS instances, may occur due to various factors including network latency, workload on the primary instance, and the magnitude of data changes.

There's more...

For enhanced scalability, it's possible to create up to five replicas per database and implement load balancing among them via AWS Route 53 to evenly distribute the workload between them. Furthermore, you can establish read replicas from these replicas; however, it's important to note that this may result in increased lag due to dual-level synchronization. In the event of disruption in the primary RDS, replicas can be manually promoted to standalone instances, ensuring seamless operational continuity and facilitating disaster recovery.

See also

- How can I distribute read requests across multiple Amazon RDS read replicas?: `https://repost.aws/knowledge-center/requests-rds-read-replicas`

Redshift live data sharing among your clusters

Redshift data sharing is a feature provided by Amazon Redshift. It enables the sharing of data within and across AWS accounts, allowing different teams or users to access the same data sets for their analytics or reporting needs. Datashare reflects updates in the producer cluster in real time with the consumers ensuring that all users have access to the most up-to-date information without the need to create and maintain duplicate copies of the same data.

In this recipe, we will see how to share Redshift data between two Redshift clusters in different AWS accounts.

Getting ready

This recipe assumes that you have two encrypted Redshift clusters using Ra3 node types. Also, the tables to be shared from the producer cluster should not have interleaved sort keys. If you want to follow the recipe with clusters in the same account, you can skip the authorization and association steps. Instead, directly grant usage of the datashare to the cluster namespace rather than to the AWS account.

How to do it...

The following actions must be performed in the Redshift producer cluster:

1. **Create the datashare**:

    I.   Connect to your Redshift cluster using a super user.

    II.   Create a datashare using the following command:

    ```
    CREATE DATASHARE datashare_name;
    ```

    Add PUBLICACCESSIBLE = TRUE to the command if your consumer cluster has public access enabled.

    III.   Add a schema to the datashare:

    ```
    ALTER DATASHARE datashare_name ADD SCHEMA schema_name;
    ```

    IV.   Add tables under the schema that you have added to the datashare:

    ```
    ALTER DATASHARE datashare_name ADD TABLE schema_name.table_
    name;
    ```

You can add all the tables within the schema using the command that follows:

```
ALTER DATASHARE datashare_name ADD ALL TABLES in schema
schema_name;
```

To add all new tables in the schema, add `SET INCLUDENEW = TRUE;` to the previous command.

V.   Repeat the previous two steps for all the schemas you want to share.

VI.   Grant permission for the consumer cluster AWS account:

```
GRANT USAGE ON DATASHARE datashare_name TO ACCOUNT 'account_
ID';
```

2. **Authorize the datashare**:

I.   Log in to AWS Management Console (`https://console.aws.amazon.com/console/home?nc2=h_ct&src=header-signin`) and navigate to the **Redshift** service.

II.   From the navigation menu on the left, go to **Datashares**.

III. Go to the **In my account** tab and open the datashare you have created.

IV.   In the **Data consumers** section, choose the consumer you gave access to and then click on **Authorize.**

Figure 2.3 – Authorizing a consumer

Actions in the consumer Redshift cluster include the following:

1. **Associate the datashare with consumer clusters:**

I.   Log in to AWS Management Console and navigate to the **Redshift** service.

II.   From the navigation menu on the left, go to **Datashares**.

III.   Choose the **From other accounts** tab. Select the datashare you have created in the producer cluster and click on **Associate**.

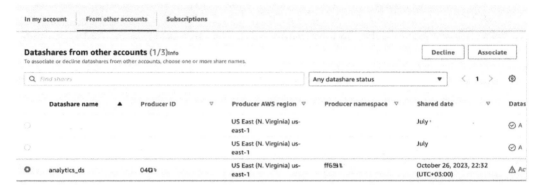

Figure 2.4 – Associating the datashare

IV.   Under **Association type**, select **Specific AWS Regions and namespaces**, then choose **Add Region**.

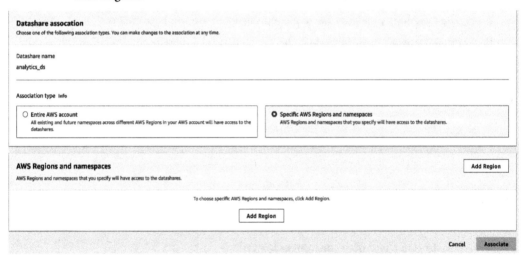

Figure 2.5 – Choosing the association type

V.   Select the region your cluster resides in, then select **Add specific cluster namespaces**, select one or more cluster namespaces, and click on **Add AWS Region**. Alternatively, you can choose to associate the datashare with all the clusters in your account by choosing the entire AWS account or all clusters in a specific region by choosing **Entire AWS account** or **Add all cluster namespaces**.

Figure 2.6 – Defining the AWS region to associate

VI.   Click on **Associate**.

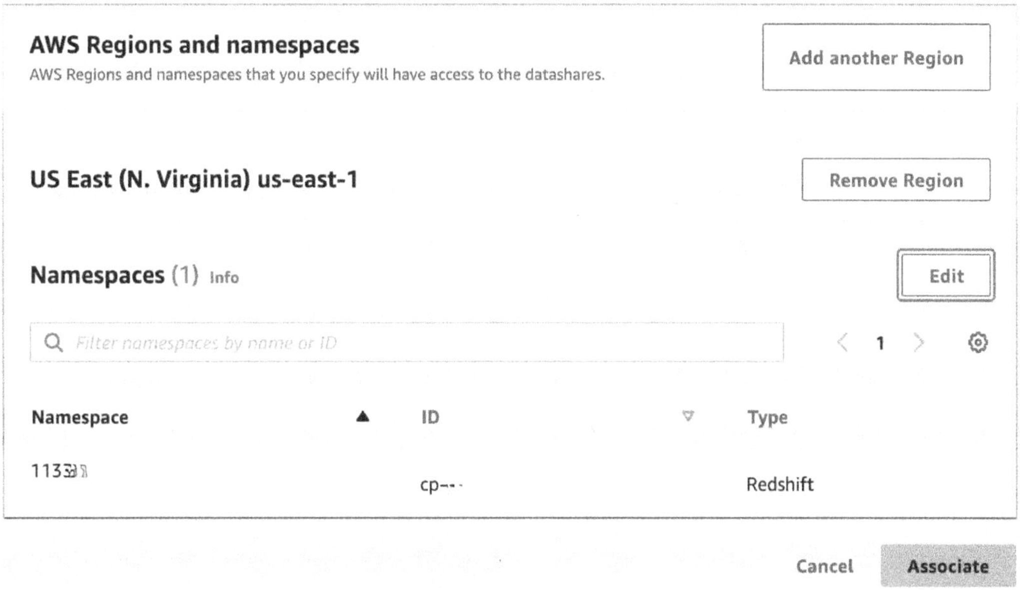

Figure 2.7 – Association with specific AWS region

2. **Create a database for the datashare**:

    I.   Connect to your Redshift cluster.

    II.   Create a local database that references the datashare, specifying the namespace and account ID:

    ```
    CREATE DATABASE consumer_database_name
    FROM DATASHARE datashare_name
        OF ACCOUNT 'producer_account_ID'
        NAMESPACE 'producer_cluster_namespace';
    ```

    III. You can now query the tables as follows:

    ```
    SELECT * FROM consumer_database_name.schema_name.table_name;
    ```

How it works...

First, you initiate the recipe by creating the datashare within the producer cluster and including the relevant databases and tables for the consumer's access. Afterward, you grant the consumer cluster's AWS account access to the datashare, initially resulting in a **pending authorization** status. Upon authorization, the status shifts to **Authorized**. At this point, the consumer perceives the data share's status as **Available** (with action required within the Amazon Redshift console). To complete the process and activate the datashare in both the producer and consumer clusters, you associate the datashare with the consumer cluster.

This enables the consumer to create a database associated with the datashare. This enables direct querying of the tables in the data share using `consumer_database_name.schema_name.table_name`.

There's more...

You can extend the same datashare to multiple consumers by providing access to each consumer cluster's namespace if it's within the same AWS account, or to its AWS account if it belongs to a different account. You can achieve this by executing separate Redshift `GRANT` commands on the datashare.

To address security concerns, it's possible to share your views without sharing the underlying tables. Sharing a view follows the same procedure as sharing a table.

See also

- Managing existing datashares: `https://docs.aws.amazon.com/redshift/latest/dg/manage-datashare-existing-console.html`

- Consumer cluster administrator actions: `https://docs.aws.amazon.com/redshift/latest/dg/consumer-cluster-admin.html`

- Amazon Redshift database encryption: `https://docs.aws.amazon.com/redshift/latest/mgmt/working-with-db-encryption.html`

Synchronizing Glue Data Catalog to a different account

As a centralized repository, a data lake allows you to store all your structured and unstructured data at any scale without needing upfront data schema definitions. AWS Lake Formation simplifies data lake management on AWS by acting as a layer on top of existing data storage services such as S3. It automatically catalogs data, manages access, and integrates with various AWS services such as Glue and Athena. One of its key features is its integration with AWS Glue. Lake formation in conjunction with AWS **Resource Access Manager** (**RAM**) offers a secure and simple method for the sharing of the AWS Glue Data Catalog across multiple AWS accounts. This promotes collaboration and data accessibility across different teams or organizations, allowing them to make the most of their data lakes and extract valuable insights from their shared resources. However, you need to carefully plan and test your data-sharing process. Before sharing the catalog, conduct a comprehensive analysis of existing applications and workflows to identify any dependencies on specific catalog resources or permissions that may require adjustments. Existing applications may slow down due to the increased read load from the shared account. Ensure that permissions are set up correctly so that your applications can access the intended resources, whether they are original or shared, and avoid unauthorized exposure.

To ensure that your resources can handle the additional load, consider conducting load testing and monitoring performance metrics, adjusting resource allocation as necessary. Be aware that delays in synchronization may lead to data consistency issues. To mitigate these risks, consider sharing the data within the same region to reduce latency or select regions that have low latency between them.

In this recipe, we will learn how to share the Glue Data Catalog with another AWS account using Lake Formation.

Getting ready

This recipe assumes the existence of two AWS accounts and a Glue database within one of those accounts, which you intend to share with the other. Ensure that you perform this recipe with a user or role with administrative privileges in AWS Lake Formation by attaching the **AWSLakeFormationDataAdmin** managed policy to its permissions. To be able to use Lake Formation to control access to the Glue Data Catalog, you need to switch to the Lake Formation permission model. If you already have Glue tables permission managed by IAM identities, list down these permissions, and once you finish the outlined steps, grant the same permissions using the Lake Formation:

1. Revoke the super access from the **IAMAllowedPrincipals** principle, an automatically generated entity containing IAM users and roles authorized to access your Data Catalog resources in accordance with your IAM policies.

    I.   Navigate to Lake Formation and select **Data Lake permissions** under **Permissions** in the navigation pane on the left.

    II.   Search for the permission for the **IAMAllowedPrincipals** principal on your table. Select the permission and click on **Revoke**.

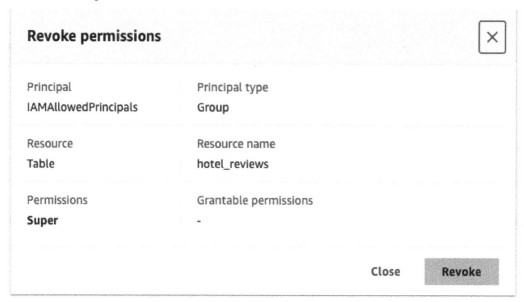

Figure 2.8 – Revoking the IAMAllowedPrincipals permission

2. Disable the default setting of controlling access to the new tables added to the databases only through IAM access control for the databases you share:

    I.   Go to **Databases** under **Data Catalog** in the left navigation pane on the Lake Formation service.

    II.   Select the database you will be sharing and click on **Edit**. Disable **Use only IAM access control for new tables in this database** and **Use only IAM access control for new databases** under **Default permissions**.

    III.   Repeat these steps for all the databases you will be sharing.

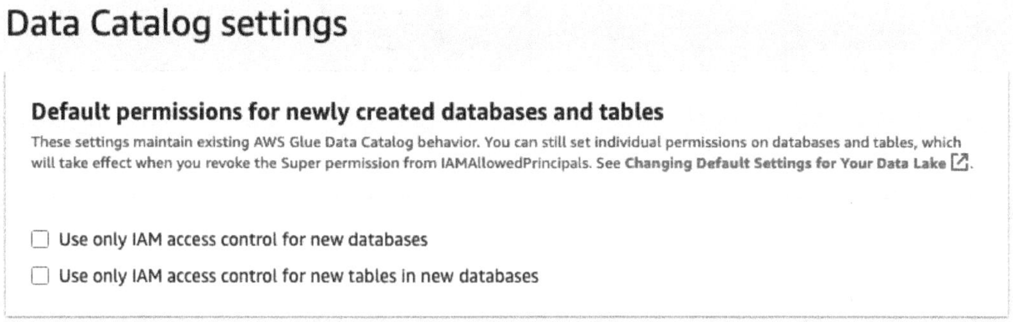

Data Catalog settings

Default permissions for newly created databases and tables

These settings maintain existing AWS Glue Data Catalog behavior. You can still set individual permissions on databases and tables, which will take effect when you revoke the Super permission from IAMAllowedPrincipals. See **Changing Default Settings for Your Data Lake** [↗].

☐ Use only IAM access control for new databases

☐ Use only IAM access control for new tables in new databases

Figure 2.9 – Disabling Data Catalog's default settings

3. Stop new tables created in the Data Catalog from having a default **super** permission to **IAMAllowedPrincipals**:

    I.   Choose **Data Catalog settings** from **Administration** in the Lake Formation navigation pane.

    II.   Clear both **Use only IAM access control for new databases** and **Use only IAM access control for new tables in this database** under the **Default permissions for newly created databases and tables** section.

    III.   Click on **Save**.

4. Register the S3 location for the Glue database or tables that you will share with Lake Formation:

    I.   In the Lake Formation navigation pane, go to **Data Lake locations** under **Register and Ingest**.

    II.   Choose **Register location**, and then choose **Browse** to select an Amazon S3 path for the database or tables you intend to share.

    III.   Repeat these steps for the different S3 locations you have for the database or tables you intend to share.

How to do it...

Let's consider that the account hosting the Glue database is referred to as *Account A*, and the account with which we intend to share the database is identified as *Account B*.

In Account A, follow these steps:

1. Log in to the AWS console (`https://console.aws.amazon.com/console/home?nc2=h_ct&src=header-signin`) and navigate to the **AWS Glue** service.

2. From the navigation pane on the left, go to **Catalog settings** under **Data Catalog service.**

3. Add the following policy statement to **Permissions** and save it (make sure to replace `aws_region` and `aws_account_id` with your specific values):

```
{
    "Version" : "2012-10-17",
    "Statement" : [ {
        "Effect" : "Allow",
        "Principal" : {
        "Service" : "ram.amazonaws.com"
        },
        "Action" : "glue:ShareResource",
        "Resource" : [
        "arn:aws:glue:<aws_region>:<aws_account_id>:table/*",

        "arn:aws:glue:<aws_region>:<aws_account_id>:database/*",

        "arn:aws:glue:<aws_region>:<aws_account_id>:catalog"
        ]
    }
    ]
}
```

4. Navigate to the **Lake Formation** service. From the navigation pane on the left, choose **Data lake permissions** under **Permissions**.

5. In **Data permissions**, click on **Grant**.

6. For principles, choose **External accounts**, then type Account B's ID and press *Enter*.

7. Under **LF-Tags or catalog resources**, choose **Named Data Catalog resources**. Choose all the databases you want to share under **Databases**:

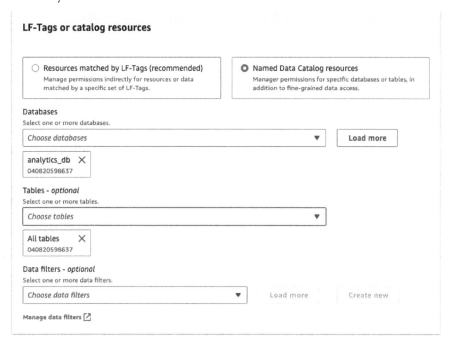

Figure 2.10 – Defining the resources to be shared

I. To grant access to the tables within those databases, choose the list of tables under **Tables**. You can optionally add **Data filters**. In the **Table permissions** section, under **Table permissions**, choose **Super** to grant all permissions or choose specific permissions from the list. You can optionally choose **Grantable permissions**, which will allow Account B to grant permissions to the shared resources.

II. To grant access to the database itself, do not select any tables. In the **Database permissions** section, choosing the permissions is like setting the table permissions – you have to choose the level of permissions and optionally grantable permissions.

III.   Click on **Grant**.

Table permissions

Table permissions
Choose specific access permissions to grant.

☐ Select   ☐ Insert   ☐ Delete   ☑ Super

This permission is the union of all the individual permissions to the left, and supersedes them.

☐ Describe   ☐ Alter   ☐ Drop

Grantable permissions
Choose the permission that may be granted to others.

☐ Select   ☐ Insert   ☐ Delete   ☐ Super

This permission allows the principal to grant any of the permissions to the left, and supersedes those grantable permissions.

☐ Describe   ☐ Alter   ☐ Drop

Figure 2.11 – Defining the table permission to be granted

In Account B, follow these steps:

1. Log in to the AWS console and navigate to the **AWS RAM** service.

2. In the left navigation pane, choose **Resource shares** under **Shared with me**.

3. You should see a resource with a pending status from Account A. Click on it and choose **Accept resource share**.

Figure 2.12 – Accepting a resource share

You should be able to see the database and tables in Account B lake formation as if they were created in Account A.

How it works...

To begin the sharing process, you created a resource share specifying the account to share with and the level of permissions that will be granted to the recipient account. This triggers the sharing workflow and sends a sharing request to the recipient account. Next, in the recipient account, you accepted the invitation for the shared Glue databases. The recipient account can access the shared Glue database using their own Lake Formation console.

There's more...

To be able to query the shared catalog tables in Account B, you have to create resource links that point to shared databases and tables. If you are sharing all tables within a database, you can create a resource link for the database, which will allow you to query all the tables within it. However, if only specific tables from a database are shared, you must create resource links for each of those individual tables. Follow the next steps to create a resource link for the tables. If you need to create a resource link for the database, you can follow a similar process by navigating to the Databases section instead of Tables.

1. Navigate to Lake Formation with a user that has the **glue:CreateTable** permission.

2. In the navigation pane, choose **Tables**, and then choose **Create table**.

3. In the **Table details** section, choose **Resource link**.

4. Give a name under **Resource link name** and then choose the local database that the table will be added to. Choose the shared table that you are creating the link for. Finally, for the table's region under **Shared table's region**, choose the same region of the shared table and click on **Create**.

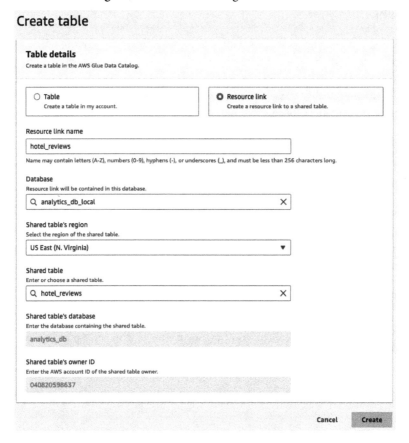

Figure 2.13 – Creating a table from a resource link

See also

- Update a resource share in AWS RAM: https://docs.aws.amazon.com/ram/latest/userguide/working-with-sharing-update.html

- Granting permissions on a database or table shared with your account: https://docs.aws.amazon.com/lake-formation/latest/dg/regranting-shared-resources.html

Enforcing fine-grained permissions on S3 data sharing using Lake Formation

AWS Lake Formation simplifies the process of setting up and managing a data lake by providing capabilities for data ingestion, organization, and access control. You can arrange your S3 data into tables within the Lake Formation Data Catalog and implement fine-grained access control on them. This means you can enforce precise access permissions at both the table and column levels, ensuring that sensitive data is safeguarded. Fine-grained access control empowers you to exercise precise control over who can view or modify specific tables and columns within the Lake Formation Data Catalog. Consequently, you can provide varying levels of access to different tables or columns, aligning with the specific requirements of different users or roles.

In this recipe, we will use Lake Formation to establish a table for an S3 dataset and provide granular access controls for it.

Getting ready

To follow this recipe, you should have an S3 dataset that you want to grant access to, as well as an IAM role to grant access for. It's also preferable that this role has access to Amazon Athena to query the table.

The recipe must be implemented with a user or role that has admin access to Lake Formation.

How to do it...

1. **Create a table for your S3 location**:

    I.   Log in to the AWS Management Console (`https://console.aws.amazon.com/console/home?nc2=h_ct&src=header-signin`) and navigate to the **Lake Formation** service.

    II.   In the **Lake Formation** console, select **Databases** under **Data Catalog** from the navigation pane on the left.

    III. Click on **Create database**.

    IV.   Give a name and description for the database under the respective fields and optionally add an S3 location if all the tables in your database will fall into a common location.

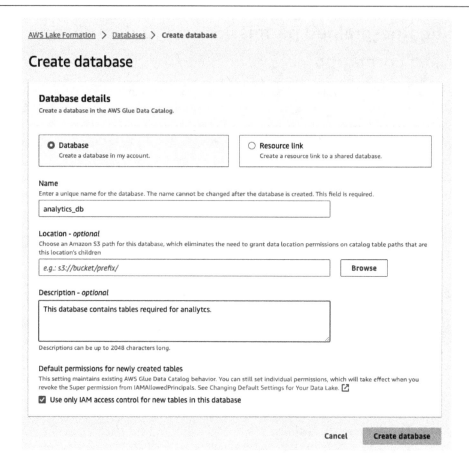

Figure 2.14 – Creating a Lake Formation database

V.   Click on **Create database**.

VI.   To create the table automatically using a crawler, go to **Crawlers** under **Data catalog** from the navigation pane on the left.

VII.   You will be directed to the Glue crawler; click on **Create crawler**.

VIII.   Give a name and description for your crawler and click on **Next**.

IX.   Choose **Not yet** for **Is your data already mapped to Glue tables?** under the **Data source** configuration.

X.   Under **Data sources**, click on **Add a data source**. For the location of S3 data, choose **In this account** and add the S3 path. Click on **Add an S3 location**.

Figure 2.15 – Adding an S3 data source

XI.   Optionally, you can add a classifier for your data under the **Custom classifiers** section.

XII. Click on **Next**.

XIII. Click on **Create new IAM role**, give a name to the role, and click on **Create | Next**.

XIV. Under **Target database**, choose the database you have created.

XV.   Optionally, provide a **Table name prefix**.

XVI. You can schedule your crawler or keep it on demand.

XVII. Click on **Next** and then click on **Create crawler**.

XVIII. Now click on **Run crawler**.

XIX.   Once the run is done, you should see that one table was changed:

Figure 2.16 – Crawler completed a run

XX.   You can view the table by going to **Lake Formation** and clicking on **Tables** under **Data catalog** from the navigation pane on the left.

2. **Create an LF-Tag for data classification**:

I.   Navigate to **LF-Tags and permissions** under **Permissions** in the left navigation pane and click on **Add LF-Tag**.

II. Give a key and list of comma-separated values or a single value to the tag, click on **Add**, and then click on **Add LF-Tag**.

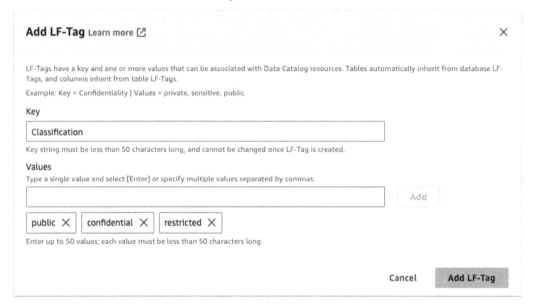

Figure 2.17 – Adding an LF-Tag

3. **Assign the tag to your table**:

   I.   Navigate to **Tables** under **Data Catalog** in the left navigation pane and open your table.

   II. Under the **Schema** section, click on **Edit schema**.

   III. Select the columns you want to give access to, click on **Edit LF-Tags**, then click on **Assign new LF-Tag**. Choose the LF-Tag you created under **Assigned keys**, select one of the values of the tag, and click on **Save**. You can assign different values to the tag for the remaining columns in your table.

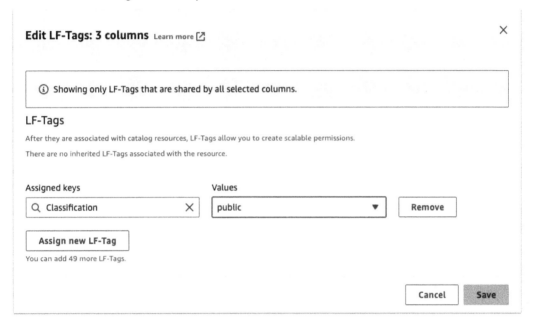

Figure 2.18 – Assigning tags to columns

   IV. Click on **Save as new version**.

4. **Give permission to the IAM role**:

   I.   Navigate to **Data lake permissions** under **Permissions** in the navigation pane on the left.

   II. Click on **Grant**.

   III. Under the **Principals** section, choose **IAM users and roles** and choose the IAM users and roles you want to give access to.

   IV. Under **LF-Tags or catalog resources**, choose **Resources matched by LF-Tags (recommended)**. Choose the key to the tag you created and the values that you are allowing this user access to.

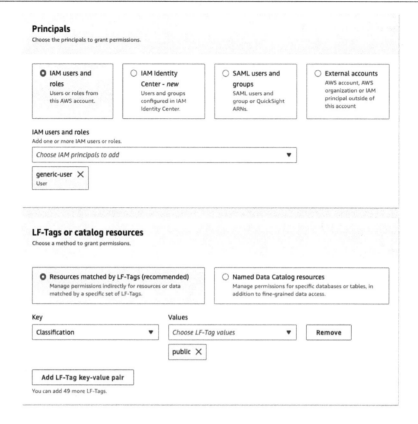

Figure 2.19 – Defining the resource to set up access for

V.   Under **Table permissions**, choose **Select** and click on **Grant**.

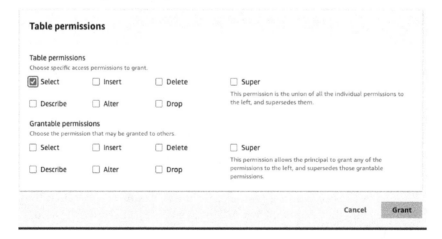

Figure 2.20 – Defining table permissions

5. **Revoke IAMAllowedPrincipals access**:

I. Navigate to Lake Formation and select **Data lake permissions** under **Permissions** in the navigation pane on the left.

II. Search for the permission for **IAMAllowedPrincipals** principle on your table.

III. Select the permission and click on **Revoke**.

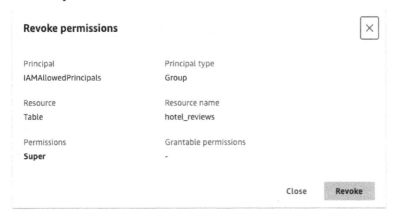

Figure 2.21 – Revoking IAMAllowedPrincipals access

How it works...

We initiated this recipe by registering our S3 location with the data lake, followed by the creation of a database in AWS Lake Formation to host our table. To automate the table creation process, we employed AWS Glue Crawler, configuring it to scan a designated S3 location. The information gathered by the crawler was utilized to define the table's structure, and added to the Lake Formation Data Catalog, which serves as a logical representation of the data. Moving forward, we created an LF-Tag for tagging specific columns with predefined values. Subsequently, we granted select access to a particular IAM user or role based on a specific tag value, providing access to a defined set of columns. To align with these access permissions, it was essential to revoke super access from the **IAMAllowedPrincipals** principle, an automatically generated entity that includes IAM users and roles with access to Data Catalog resources as per IAM policies. With these access controls, when users attempt to access the table, Lake Formation will grant temporary credentials, granting them access to the specified data.

There's more...

You can choose to set up tags at the database level, and all underlying tables will inherit these same tags. Similarly, tags can be defined at the table level, and all associated columns will also inherit these tags. Additionally, when granting access to users, you have the flexibility to utilize a combination of various tags. Leveraging LF-Tags provides an efficient means of managing permissions at scale. Alternatively, if you prefer using named resources rather than LF-Tags-based resources, you can implement data filters on your tables. This allows you to precisely specify the rows and columns that can be accessed following the outlined steps:

1. Navigate to **Data filters** under **Data Catalog** in the left navigation pane and click on **Create new filter**.

2. Give a name for your data filter and choose the target database and table to which the filter will be applied.

3. Under **Column-level access**, choose **Include columns**, then list the columns you want to give access to. Alternatively, you can choose to exclude columns and list the columns you don't want to give access to.

4. Under **Row-level access**, choose **Filter rows**, and then write down your condition the same way as you would add it in a `where` clause in SQL query and click on **Create data filter**.

Figure 2.22 – Defining data filter constraints

Now you can use this filter when you define the table you are giving access to under **Named Data Catalog resource**.

See also

- Data filtering and cell-level security in Lake Formation: `https://docs.aws.amazon.com/lake-formation/latest/dg/data-filtering.html`

- Viewing the resources that an LF-Tag is assigned to: `https://docs.aws.amazon.com/lake-formation/latest/dg/TBAC-view-tag-resources.html`

Sharing your S3 data temporarily using a presigned URL

S3 presigned URLs provide a secure and controlled way to grant time-limited access to S3 objects. You can specify an expiration time when generating the URL, after which the URL becomes invalid. Presigned URLs can have specific permissions associated with them; you can generate URLs that allow only specific actions (such as read, write, or both) on individual objects. This provides granular control over which operations can be performed on the object while minimizing the exposure of your AWS credentials, as you can avoid the need to embed AWS access credentials directly into your application or to share your AWS access keys.

In this recipe, we will learn how to create a presigned URL to download an S3 object.

Getting ready

This recipe assumes that you have an S3 bucket with an object that you will be creating a presigned URL for. Also, you need to perform these steps using a user who has access to download the object as the URL will inherit the same credentials.

How to do it...

1. Log in to the AWS console (`https://console.aws.amazon.com/console/home?nc2=h_ct&src=header-signin`) and navigate to **S3**.

2. Navigate to the bucket that has the object you want to create a presigned URL for. Select the object to be shared from the **Objects** list.

3. From the **Object actions** menu, choose **Share with a presigned URL**.

4. Specify the **Time interval until the presigned URL expires** value and then **Number of hours** or **Number of minutes** for the URL to be valid. The interval must be up to seven hours only; however, if you are doing the setup through the CLI, you can set the time interval up to 12 hours.

5. Choose **Create presigned URL**.

Figure 2.23 – A presigned URL's expiration time

6. You will get confirmation that the URL was created and it will automatically be copied to your clipboard. You can copy it again using the **Copy the presigned URL** button.

How it works...

By clicking on **Create presigned URL**, you have initiated the process of creating a time-limited URL that allows temporary access to the selected object. You can share the presigned URL with the intended users, applications, or services. They can use this URL to access the S3 object for the duration specified in the expiration time you set.

There's more...

Anyone in possession of the generated presigned URL will have the capability to download your S3 object. However, you can control who can utilize this URL for downloading by limiting access based on IP address ranges. To implement this access restriction, presigned URLs inherit the access permissions of the user or role that created them. Therefore, you can achieve this by attaching an IAM policy to the user or role responsible for generating the presigned URL, restricting its access to a specific IP or IP range. Consequently, the presigned URL will only be accessible within that defined IP range.

Here's an illustrative IAM policy. Ensure that you substitute `IP-address` and `s3-bucket-name` with the actual IP address you intend to use and the S3 bucket containing the files to be shared, respectively. Also, ensure that there are no conflicting policies in place that grant broader access:

```
{
    "Version": "2012-10-17",
    "Statement": [
      {
          "Sid": "IPAllow",
          "Effect": "Allow",
          "Action": "s3:*",
          "Resource": "arn:aws:s3:::<s3-bucket-name>/*",
          "Condition": {
              "IpAddress": {
                  "aws:SourceIp": "IP-address"
              }
          }
      }
    ]
}
```

See also

- Uploading objects with presigned URLs: `https://docs.aws.amazon.com/AmazonS3/latest/userguide/PresignedUrlUploadObject.html`

Real-time sharing of S3 data

Real-time sharing of S3 data ensures immediate responsiveness to changes in data, facilitating seamless communication between various components of a system. By setting up an S3 bucket to invoke a Lambda function upon the occurrence of any new event, performing the required processing, and publishing the data through notifications to an SNS topic, this method provides an effective and scalable means for broadcasting events to multiple subscribers. It offers a dynamic and responsive approach to sharing information in real time across different systems or applications. This approach proves particularly advantageous in scenarios where the timely distribution of information is crucial for maintaining up-to-date and synchronized systems.

In this recipe, we will learn the process of triggering a Lambda function when new files are created in an S3 bucket, subsequently processing the files and broadcasting them as SNS notifications.

Getting ready

For this recipe, you need to have an S3 bucket whose content will be shared.

How to do it...

1. **Create an SNS topic**:

    I. Log in to the AWS Management Console (`https://console.aws.amazon.com/console/home?nc2=h_ct&src=header-signin`), navigate to **SNS service**, and click on **Create topic**.

    II. Under **Details**, choose **Standard type**, give a name and optionally a display name for the topic, and click on **Create topic**.

    III. In the **Subscription** tab, click on **Create subscription**.

    IV. Select **Email** under **Protocol**, provide your email under **Endpoint**, and click on **Create subscription**.

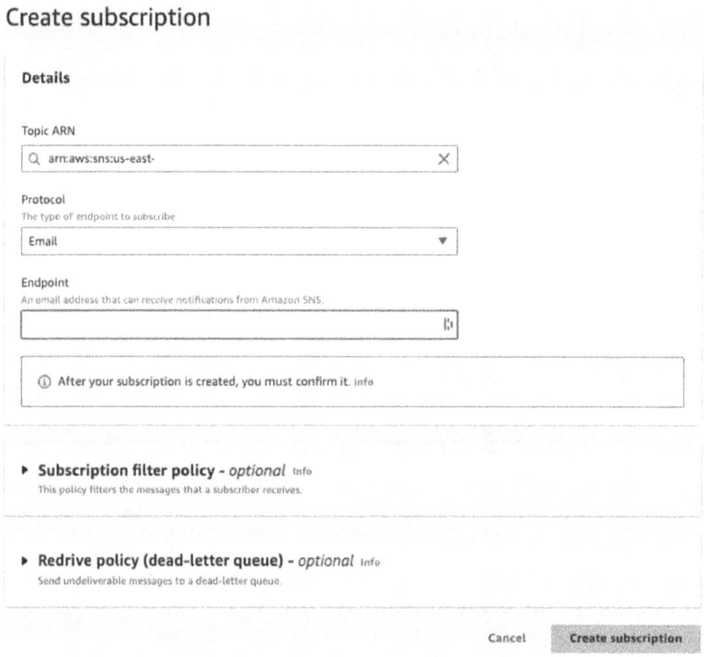

Figure 2.24 – Setting up an email subscription to SNS

You should receive an email asking you to confirm the subscription.

2. **Create an IAM role for the Lambda function**:

    I. Create a new IAM policy using the following permissions (make sure to replace `sns-topic-arn` and `s3-bucket-arn` with your specific values):

```
{
    "Version": "2012-10-17",
    "Statement": [
```

```
        {
            "Sid": "VisualEditor0",
            "Effect": "Allow",
            "Action": [
                "s3:GetObject",
                "sns:Publish"
            ],
            "Resource": [
                "<sns-topic-arn>",
                "<s3-bucket-arn>/*"
            ]
        }
    ]
}
```

II.   Create a role for Lambda service, attaching the policy from the previous step and the AWSLambdaBasicExecutionRole managed policy.

3. **Create a Lambda function**:

I.   Navigate to the Lambda service.

II.   Click on **Create function**, select **Author from scratch**, and write a function name. Under **Runtime**, choose **Python 3.11**, and under **Permissions**, expand **Change default execution role**, select **Use an existing role**, and choose the role you created in the previous step. Finally, click on **Create function**.

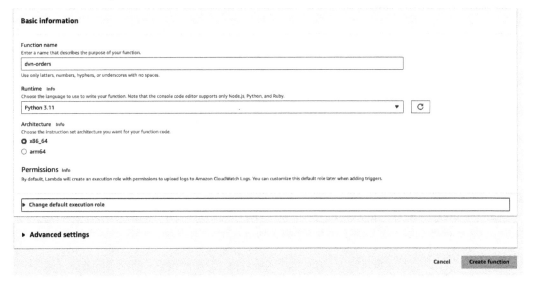

Figure 2.25 – Creating a Python Lambda function

4. **Create an S3 event**:

    I.   Navigate to the S3 service.

    II.   Open the S3 bucket whose files you need to publish and go to the **Properties** tab.

    III.   Under the **Event notification** section, click on **Create event notification**.

    IV.   Give an event name and optionally add a prefix (for example, a folder name) for the files to be shared.

    V.   Under **Event types**, select **All object create events** under **Object creation**.

    VI.   Under the **Destination** section, choose **Lambda function**, select the Lambda function created in the previous step, and click on **Save changes**.

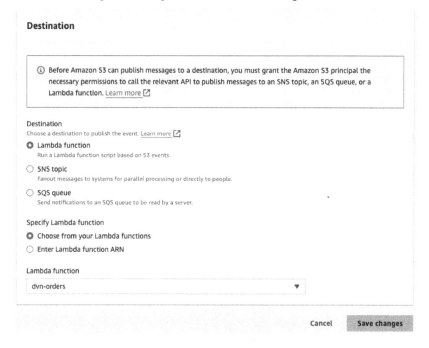

Figure 2.26 – Configuring the S3 event for Lambda

5. **Update the Lambda code**:

    I.   Navigate back to your Lambda function; you should see an S3 trigger added to your function.

    II.   In the **Code** tab, update the code with the following script. Make sure to update the `TopicArn` and `Subject` fields with your own values:

```
import json
import boto3
```

```
import logging

logger = logging.getLogger()
logger.setLevel(logging.INFO)

def publish_to_sns(message):
    try:
        sns_client = boto3.client('sns')
        # add any processing required on the file or extract
certain fields before publishing the message.
        sns_client.publish(
            TopicArn = 'yourSNSTopicArn',
            Message = message,
            Subject = 'emailSubject'
        )
        logging.info('Published SNS message successfully')
    except ClientError:
        logging.info('Failed to publish SNS due to: {}'.
format(str(ClientError)))
        return False
    return True

def lambda_handler(event, context):
    # retrieve S3 file name
    s3_event = event['Records'][0]['s3']
    object_key = s3_event['object']['key']
    bucket_name = s3_event['bucket']['name']
    s3_client = boto3.client('s3')

    # retrieve file from S3 by its object key
    response = s3_client.get_object(Bucket=bucket_
name,Key=object_key)
    file_content = response['Body'].read().decode('utf-8')
    # publish S3 file content to SNS
    publish_to_sns(file_content)
    return {
        'statusCode': 200
    }
```

III.   Click on **Deploy**.

6. **Test**: Test the process by uploading a file to your S3 bucket. This should trigger your Lambda function and you should receive an email with the notification from SNS.

How it works...

The IAM role we established provides Lambda with the necessary permissions to access and read S3 files within our designated bucket. It also allows Lambda to publish messages to our SNS topic and generate log files in CloudWatch. An S3 event has been configured to trigger the Lambda function whenever an object is added to the bucket, covering various methods such as object putting or copying. Upon triggering, the Lambda function retrieves the S3 file's location, reads its content, processes the information, and subsequently publishes it to the specified SNS topic. All subscribers connected to the SNS topic are then notified of the published message.

There's more...

You have the flexibility to incorporate multiple subscribers into your SNS topic, including AWS services within your AWS account, or extending to various AWS accounts. Moreover, if your intention is to directly share S3 events with consumers, you have the option to configure the S3 event notification destination to be an SNS topic. This allows for the subscription of multiple consumers to the topic, providing them with the choice to utilize it as is or integrate it into AWS services for customized processing.

See also

- Event notification types and destinations: `https://docs.aws.amazon.com/AmazonS3/latest/userguide/notification-how-to-event-types-and-destinations.html`

Sharing read-only access to your CloudWatch data with another AWS account

CloudWatch cross-account sharing in AWS is valuable for centralizing monitoring and management. It enables organizations to share monitoring data, including alarms, dashboards, and logs, across AWS accounts while maintaining resource isolation. This facilitates consolidated reporting, collaboration, troubleshooting, and cost management. Cross-account sharing allows for customized access control and fine-grained permissions, and is scalable for growing organizations. It simplifies the sharing of critical monitoring data, enhancing operational efficiency and visibility without compromising security.

In this recipe, we will learn how to share read-only access to CloudWatch data with another AWS account.

Getting ready

To follow this recipe, you need to have two AWS accounts, one of which must have the CloudWatch log group to be shared. The other account will share it and use it for monitoring.

How to do it...

In the source AWS account, follow these steps:

1. Log in to the AWS Management Console (`https://console.aws.amazon.com/console/home?nc2=h_ct&src=header-signin`) and navigate to the **CloudWatch** service.

2. From the navigation pane on the left, choose **Settings**.

3. Go to the **Enable account switching** section then select **Configure** for the **Share your CloudWatch data** option.

Figure 2.27 – Enabling the sharing of CloudWatch data

4. In the **Sharing** section, click on **Add account**, then type the monitoring account ID and hit *Enter*.

Figure 2.28 – Sharing to a specific AWS account

5. For **Permissions**, choose **Provide read-only access to your CloudWatch metrics, dashboards, and alarms** and check both the **Include CloudWatch automatic dashboards** and **Include X-Ray read-only access for ServiceLens** checkboxes.

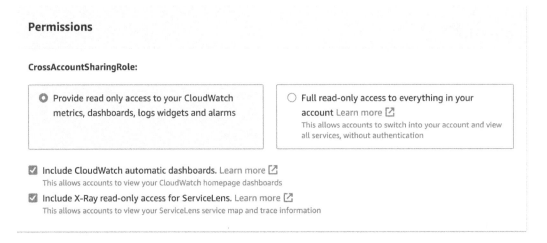

Figure 2.29 – Configuring permissions

6. Click on **Launch CloudFormation template** from the **Create CloudFormation stack** section.

7. Type `Confirm` on the confirmation screen and click on **Launch Template**.

8. On the **Quick create stack** page, select **I acknowledge that AWS CloudFormation might create IAM resources with custom names**. Then click on **Create stack**.

In the destination AWS account, follow these steps:

1. Log in to the AWS Management Console and navigate to the **CloudWatch** service.

2. From the navigation pane on the left, choose **Settings**.

3. Go to the **Enable account switching** section, then select **Configure** for the **View cross-account cross-region** option.

4. In the **Enable account selector** section, select **Show selector** in the console.

5. Select **Custom account selector**, and then enter the owner account ID and a label to identify it.

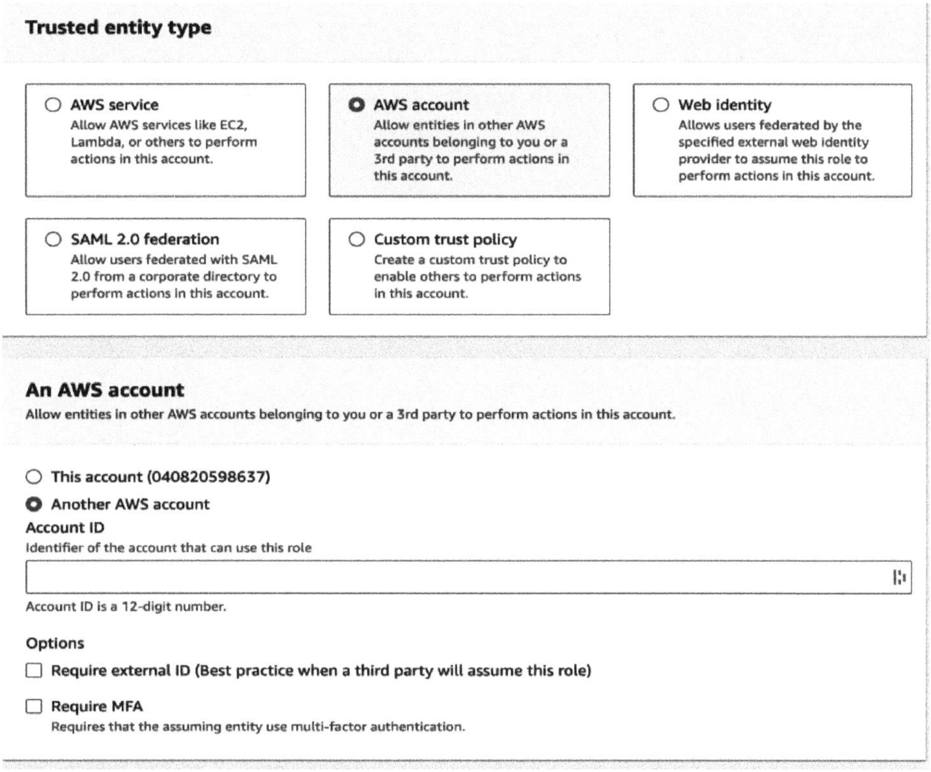

Trusted entity type

○ **AWS service**
Allow AWS services like EC2,
Lambda, or others to perform
actions in this account.

◉ **AWS account**
Allow entities in other AWS
accounts belonging to you or a
3rd party to perform actions in
this account.

○ **Web identity**
Allows users federated by the
specified external web identity
provider to assume this role to
perform actions in this account.

○ **SAML 2.0 federation**
Allow users federated with SAML
2.0 from a corporate directory to
perform actions in this account.

○ **Custom trust policy**
Create a custom trust policy to
enable others to perform actions
in this account.

An AWS account

Allow entities in other AWS accounts belonging to you or a 3rd party to perform actions in this account.

○ **This account (040820598637)**
◉ **Another AWS account**
Account ID
Identifier of the account that can use this role

Account ID is a 12-digit number.

Options
☐ **Require external ID (Best practice when a third party will assume this role)**

☐ **Require MFA**
Requires that the assuming entity use multi-factor authentication.

Figure 2.30 – Configuring the source AWS account

6. Click on **Enable**.

How it works...

In the source AWS account, we have granted read-only access to the destination AWS account (monitoring account), enabling it to create dashboards using widgets from the source account, view automatically generated dashboards, and access X-Ray data with read-only permissions for ServiceLens in the source account. The Lake Formation template will create an IAM role with policies that provide the specified access levels and establish a trust relationship with the destination AWS account.

In the destination AWS account, we have enabled the ability to view shared CloudWatch data from the source AWS account. This access is configured through a custom account selector, which offers a drop-down list of accounts to choose from based on the labels assigned to these accounts. This streamlined access management ensures a seamless experience when logging into the destination AWS account.

There's more...

In the destination AWS account, upon accessing CloudWatch, you will encounter a **View data** feature that presents a drop-down menu. This menu facilitates seamless switching between viewing data from your own account and the source account. By selecting the source account based on its designated label, you can proceed to create dashboards or set up alerts using metrics generated in the source account, just as you would for your own account.

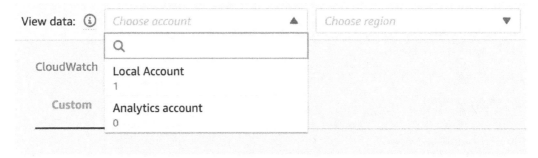

Figure 2.31 – Selecting the account of CloudWatch data

See also

- Cross-account cross-region dashboards: https://docs.aws.amazon.com/AmazonCloudWatch/latest/monitoring/cloudwatch_xaxr_dashboard.html

3

Ingesting and Transforming Your Data with AWS Glue

In data engineering, implementing integration between systems to **extract, transform, and load** (ETL) data is frequently what consumes the most time and cost. AWS Glue is a serverless data integration service that provides different engines and tools to build ETL jobs in a simple and scalable way, paying for what you use.

Glue is comprised of many components and features to serve many kinds of data products and users, including a Hive-compatible metastore and multiple engines for different needs, from single-node Glue Python shell jobs to distributed clusters that auto-scale using Glue for Spark, and the latest addition: Glue for Ray. Each of those engines has connectors for common data storage systems.

Glue also offers on-demand clusters via interactive sessions, which can be used for interactive development and analysis via Jupyter notebooks (either provided by Glue or your own). Finally, Glue Studio offers a visual environment where you can create visual jobs and build ETL pipelines with little or no coding.

This chapter includes the following recipes:

- Creating ETL jobs visually using AWS Glue Studio
- Parameterizing jobs to make them more flexible and reusable
- Handling job failures and reruns for partial results
- Processing data incrementally using bookmarks and bounded execution
- Handling a high quantity of small files in your job
- Reusing libraries in your Glue job
- Using data lake formats to store your data
- Optimizing your catalog data retrieval using pushdown filters and indexes
- Running `pandas` code using AWS Glue for Ray

Technical requirements

Many recipes in this chapter require you to have a `bash` shell or equivalent available with the AWS CLI installed with access to AWS. Check the AWS documentation installation guide: `https://docs.aws.amazon.com/cli/latest/userguide/getting-started-install.html`.

If using Microsoft Windows, you can enable **Windows Subsystem for Linux** (**WSL**) to get a `bash` shell. The instructions will also assume you have configured default credentials and the default region as the one you intend to use, using `aws configure` or an AWS CLI profile.

The easier way to experience the recipes is using a test account, on which you can use an admin user to access the console and the command line. Otherwise, you'll have to add the required permissions to create and use Glue components.

A Glue job requires a **Simple Storage Service** (**S3**) bucket to store scripts, temporary files, and, potentially, data; you will create a bucket for this purpose. None of the recipes use or generate large amounts of data that would incur high usage costs in the short term; it is up to you to clean up the files or even delete the bucket once you no longer want to keep the example data.

Define some variables in the shell, making sure the region configured on the AWS CLI is the region you intend to use later on the console:

```
ACCOUNT_ID="$(aws sts get-caller-identity --query \
  'Account' --output text)"
AWS_REGION="$(aws configure get region)"
GLUE_BUCKET ="glue-recipes-$ACCOUNT_ID"
GLUE_ROLE=AWSGlueServiceRole-Recipe
GLUE_ROLE_ARN=arn:aws:iam::${ACCOUNT_ID}:role/$GLUE_ROLE
```

The variables will be lost once you close the shell, but you can rerun the previous code definition block if a recipe requires one or more of them (this will be specified in the *Getting ready* section of each recipe that requires it).

Create a bucket in the region using the default configuration:

```
aws s3api create-bucket --create-bucket-configuration \
  LocationConstraint=$AWS_REGION --bucket $GLUE_BUCKET
```

The default permissions block public access but allow reading/writing to roles in the same account, which have permission to use S3.

After each recipe, you can wipe the bucket objects if you no longer need them.

In addition, all Glue jobs require a role that Glue can assume at runtime. By following the convention of naming it starting with AWSGlueServiceRole, it doesn't require an explicit iam:passRole permission. Create it using the following command lines:

```
aws iam create-role --role-name $GLUE_ROLE --assume-\
role-policy-document '{"Version": "2012-10-17","Statement":
[{ "Effect": "Allow", "Principal": {"Service":
"glue.amazonaws.com"}, "Action": "sts:AssumeRole"}]}'
aws iam attach-role-policy --policy-arn \
  arn:aws:iam::aws:policy/service-role/AWSGlueServiceRole \
  --role-name $GLUE_ROLE
aws iam put-role-policy --role-name $GLUE_ROLE \
--policy-name S3Access --policy-document  '{"Version": "2012-10-17",
"Statement": [{"Effect": "Allow", "Action": ["s3:*"], "Resource":
["arn:aws:s3:::'$GLUE_BUCKET'", "arn:aws:s3:::'$GLUE_BUCKET'/*",
"arn:aws:s3:::aws-glue-*","arn:aws:s3:::aws-glue-*/*"]}]}'
```

The wildcard in the bucket name allows it to accommodate a variable bucket name.

You can find the script and code files on GitHub: https://github.com/PacktPublishing/Data-Engineering-with-AWS-Cookbook/tree/main/Chapter03.

Creating ETL jobs visually using AWS Glue Studio

Typical ETL tasks consist of moving data from one data storage to another, with some simple transformations in the process. For such cases, building a job using a visual data diagram allows users without coding skills to develop such a pipeline, using their knowledge of the business and the data. These kinds of jobs are also easier to maintain and update, thus reducing the **total cost of ownership** (**TCO**).

One of the multiple types of jobs that AWS Glue Studio allows for creating is a visual job. This allows the user to define the pipeline as a graph of nodes, and then the code is generated automatically so that it runs like a regular script job.

Getting ready

Create a bucket and a role as indicated in the *Technical requirements* section.

To follow this recipe, you require a CSV file with headers and some data, which can be your own data or the sales_sample.csv file provided on the code repository: https://github.com/PacktPublishing/Data-Engineering-with-AWS-Cookbook/blob/main/Chapter03/Recipe1/sales_sample.csv.

Upload the sample CSV file to the bucket created and copy the S3 uploaded file URL; you will need it in *step 2* of the *How to do it...* section of this recipe. Here's an example:

```
aws s3 cp Recipe1/sales_sample.csv s3://$GLUE_BUCKET/
```

How to do it...

1. In the AWS console, navigate to Glue, select **Visual ETL** in the left menu, and create a job using the button with the same name. It will open Glue Studio where you can start building your visual job. As you add components, the one selected automatically becomes the parent of the one added. You can also set the parent manually by selecting it and choosing a parent in the properties.

2. Add an S3 source node. Then, on the right-hand side, navigate to **Data source properties - S3**. In the **S3 URL** field, enter the path to the input CSV file you loaded in the *Getting ready* section (*Note*: Normally, you would enter an input directory so that it picks up many files recursively.):

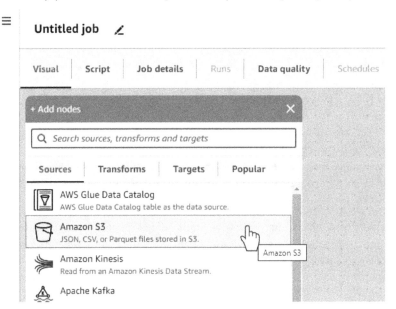

Figure 3.1 – New visual job prompting to add the source node

3. Select **CSV** for **Data format** and then choose **Infer schema**.

4. Change the tab on the bottom panel to **Output schema**. The CSV headers have been used to define the fields; however, all of them are of type `String`.

5. Using the top left + button, add a **Change Schema** transform as a child of the S3 source. On the **Transform** tab, change the type of some of the columns as in the example (the conversion has to be valid; otherwise, the data will be lost):

Change Schema (Apply mapping)

Source key	Target key	Data type	Drop
sale_id	sale_id	string ▼	☐
product_id	product_id	string ▼	☐
date	date	string ▼	☐
quantity	quantity	long ▼	☐

Figure 3.2 – Change Schema sample configuration

6. Add an S3 target node (in the **Add node** menu, select the **Target** tab) as a child of the **Change Schema** node. Select **parquet** for **Format** and enter a target S3 location under the bucket created for recipes; for instance,

```
s3:// glue-recipes-<account id>/visual_recipe/table/.
```

7. Select the **Create a table** option in **Data Catalog**, and on subsequent runs, update the schema and add new partitions, choose a database, and enter a table name of your choice as shown:

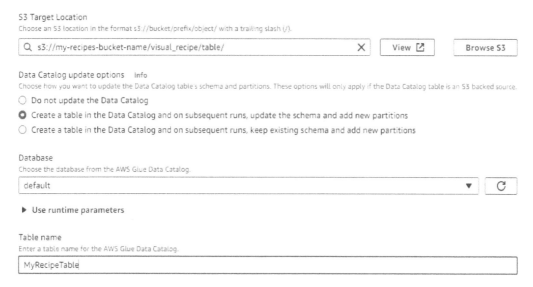

Figure 3.3 – S3 target node properties

8. Edit the job name at the top, and in the **Job details** tab, select `AWSGlueServiceRole-Recipe` as the **IAM Role** value. Save and run the job; you can monitor it on the **Runs** tab. It should only take about a minute for such a small file.

9. Navigate to Amazon Athena in the console and select the database you stored the table in. The table should be listed and using the action menu for the table. Select **Preview table**, which will create a SQL `SELECT` statement you can run to visualize the table content. If you haven't previously used Athena on this account and region, it will ask you to select a temporary path S3 to store the query results.

How it works...

As you define your nodes and their configuration, the editor builds the job pipeline with the components, from which it will generate the code that the job uses at runtime. You can view the generated code in the **Script** tab:

Figure 3.4 – Script tab with the autogenerated code

If you export the job definition (for instance, using the AWS CLI), you will notice that the JSON produced is like a script Glue job, but it has a property that stores the diagram and the node properties: `codeGenConfigurationNodes`. When it gets saved, it generates a Python script on the specified destination. Visual jobs only generate PySpark code, not Scala. If you edit the code in the **Script** tab, you get a warning that the job will become a script job, which means you can no longer edit it visually.

There's more...

Check the AWS Glue documentation for a list of visual components, available at `https://docs.aws.amazon.com/glue/latest/ug/edit-jobs-transforms.html`.

Data Preview is a key element of data pipeline development; it shows you how the sample data looks after each node of the pipeline has processed it. That way, you can gain confidence and make sure you are handling it as intended. Bear in mind that behind the scenes it uses a tiny interactive session Glue cluster, which has a cost. You can stop the preview and then disable the automatic start if you are not going to use it (for instance, if you don't have data yet) to save on cost.

If you need to troubleshoot the job and **Data Preview** is not enough, you can always copy the script code and use it in a separate script job to troubleshoot, comment out some parts, print partial results and schemas, and so on. That way, you can make code changes that help you troubleshoot without losing the visual job if you convert it.

See also

- You can create your own visual components for Visual Studio. See how at `https://docs.aws.amazon.com/glue/latest/ug/custom-visual-transform.html`.

- The catalog source node can use parameters. See more about job parameters in general in the *Parameterizing jobs to make them more flexible and reusable* recipe.

- Visual jobs can also benefit from Glue bookmarks; see how they work in the *Processing data incrementally using bookmarks and bounded execution* recipe.

Parameterizing jobs to make them more flexible and reusable

A job without any parameters normally does the same task in each run, with specific data sources and destinations. Using parameters, you can reuse the same job on different data sources or destinations, both to run recurring jobs on new data or to reuse the same logic for different purposes, such as data transformation or cleaning.

For instance, the same type of data comes from various sources but needs the same processing in a centralized data store.

Glue allows you to define your parameters for your own purposes, which you then can use in your script. You can set default values on the job and then override them as needed for each run when starting a job run manually using the console, the AWS CLI, or an API such as `boto3` or the Java SDK.

Getting ready

For this recipe, you need to follow the instructions in the *Technical requirements* section at the beginning of the chapter to create a role for Glue.

How to do it...

1. Create a new job in the Glue console by selecting **ETL jobs** in the left menu and then, using the **Script Editor** button, give it a name and select the `AWSGlueServiceRole-Recipe` role in the **Job details** tab. Leave Spark as the engine and Python as the language (see the *There's more...* section later to learn about how it could be done in Scala).

2. First, we will define the parameter and assign it a default value. In the **Job detail** tab, scroll down and open the **Advanced properties** section, add a new parameter, and enter `--DATE` as the key and `TODAY` as the value. This is a special placeholder value; otherwise, we expect on that parameter a date string in the format `yyyy-mm-dd`, such as `2000-01-01`. The key name is arbitrary but needs to start with a double hyphen:

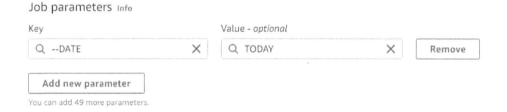

Figure 3.5 – Job parameters job configuration

3. Replace the sample code provided with the following code block:

```python
import sys
from datetime import datetime
from awsglue.utils import getResolvedOptions
from pyspark.context import SparkContext
from awsglue.context import GlueContext
from awsglue.job import Job

args = getResolvedOptions(sys.argv,
            ['JOB_NAME', 'DATE'])
data_date = args["DATE"]
date_format = '%Y-%m-%d'
if data_date == 'TODAY':
    data_date = datetime.today().strftime(date_format)

print(f"Running job for date: '{data_date}'")
```

4. Run the job using the **Run** button and then switch to the **Runs** tab to see your run progress. Select the job run and wait for it to finish. In the job run details, use the **Output logs** link to view the output logs on Amazon CloudWatch. There should be a message indicating that the job is using the current date.

5. Now, instead of using the **Run** button, select in the **Job actions** dropdown the **Run with parameters** option. Open the job parameters, enter a date in the format we have defined for the parameter (yyyy-mm-dd), and run the job:

Key	Value	
--DATE	2023-01-01	Remove

Add new parameter

Cancel   **Run job**

Figure 3.6 – Overriding the parameter for this run

Notice the value you pass when running this way is used just for this run and does not alter the default parameter value (TODAY) on the job.

In the output log for this run, you should now see a message referencing the date entered. If the parameter does not follow the format, the job run will fail, and you will get a corresponding error.

How it works...

In the script you pasted in *step 3* when creating the job, it uses the Glue getResolvedOptions API to extract the specified parameters from the command-line arguments Glue used to run the script. You could use another library or your own code to parse the parameters.

The script then checks for the value of the DATE argument. For the special value of TODAY, it replaces this keyword with the current system date.

In *step 5*, you triggered the run overriding the DATE argument with a specific date instead of the default value. This is useful if you need to rerun a specific date or you need to backload if the job hasn't been running in a timely manner each day.

Once the parameter value is stored in a variable in the script, you can use it as needed. For instance, you could use it to build a pushdown predicate when reading a table or an S3 path or the name of the table to write into.

There's more...

If Scala is the language you use to script your Glue jobs, then the code to extract the parameter is slightly different in syntax but works the same way. The main method receives a string array (named `sysArgs` in this example), which is then parsed by `getResolvedOptions`:

```
import com.amazonaws.services.glue.util.GlueArgParser
val args = GlueArgParser.getResolvedOptions(sysArgs,
        Seq("JOB_NAME"," DATE ").toArray)
val region = args("DATE")
```

You can trigger multiple instances of a job, each using different parameters. In the job run history, you can review the specific parameters that were used for each run.

If you need to trigger a job with multiple parameters, you might want to trigger them in parallel since they are doing different things, such as handling different data. The total run cost would be the same, but you can get them all completed faster that way. To allow that parallelism, make sure you have enough concurrent runs allowed on the **Job details** configuration:

Maximum concurrency

Sets the maximum number of concurrent runs that are allowed for this job. An error is returned when this threshold is reached.

1

Figure 3.7 – Concurrency config in the Job details tab

See also

- If you invoke a job by other mechanisms such as `boto3` (for instance, from AWS Lambda) or the AWS CLI, you also have the option to use specific parameters that override the default for that run. Refer to the API documentation:

  - `https://docs.aws.amazon.com/cli/latest/reference/glue/start-job-run.html`

  - `https://boto3.amazonaws.com/v1/documentation/api/latest/reference/services/glue/client/start_job_run.html`

Handling job failures and reruns for partial results

When building ETL pipelines (with Glue or in general), it's important to consider different scenarios that could make the job fail and how to deal with it. Ideally, we want the recovery to be automatic, at least for transient issues, but regardless of the recovery method, the most important aspect is that the jobs don't result in permanent data loss or duplication due to the issue. In traditional databases, this is solved using transactions, but in the case of big data ETL, that is rarely an option or would cause too much overhead.

In this recipe, you will see how to deal with job failures and resulting partial results.

Getting ready

This recipe requires a bash shell with the AWS CLI installed and configured and the GLUE_ROLE_ARN and GLUE_BUCKET environment variables set, as indicated in the *Technical requirements* section at the beginning of the chapter.

How to do it...

1. Create a job script as a local file running the following multiline bash command. It will execute when you enter the line with just EOF:

```
cat <<EOF > RetryRecipeJob.py
from pyspark.context import SparkContext
from pyspark.sql.functions import *
from awsglue.context import GlueContext
from awsglue.dynamicframe import DynamicFrame

glueContext = GlueContext(SparkContext())
spark = glueContext.spark_session

s3_output_path = "s3://${GLUE_BUCKET}/retry_recipe"
# The first attempt is just an hexadecimal number
# the retries have a suffix with the retry number
is_retry = "_attempt_" in \
  spark.conf.get("spark.glue.JOB_RUN_ID")

df = spark.range(1 << 10, numPartitions=4)
# Simulate the retry works ok
if is_retry:
    df = df.withColumn("fail", lit(False))
else:
    df = df.withColumn("fail", expr(
        f"case when id=10 then true else false end"))

# Introduce a failure when flagged
fail_udf = udf(lambda fail: 1/0 if fail else fail)
df = df.withColumn("fail", fail_udf(df.fail))
failDf = DynamicFrame.fromDF(df, glueContext, "")
glueContext.write_dynamic_frame.from_options(
    frame=failDf,
    connection_type='s3',
    format='csv',
```

```
        connection_options={"path": s3_output_path}
    )
    EOF
```

2. Upload the script to the S3 bucket and delete the local copy:

```
aws s3 cp RetryRecipeJob.py s3://$GLUE_BUCKET
rm RetryRecipeJob.py
```

3. Create a job, making sure to use \ only at the lines indicated:

```
aws glue create-job --name RetryRecipe --role \
    $GLUE_ROLE_ARN --number-of-workers 2 --worker-type \
    "G.1X" --glue-version 4.0 --command '{"Name":
    "glueetl", "ScriptLocation":
    "s3://'$GLUE_BUCKET'/RetryRecipeJob.py"}' \
    --max-retries 1 --default-arguments \
    '{"--job-language":"python", "--TempDir":
    "s3://'$GLUE_BUCKET'/tmp/"}'
```

4. Run the job:

```
aws glue start-job-run --job-name RetryRecipe
```

5. Open the AWS console, navigate to Glue, and select **ETL Jobs** on the left menu. On the table of jobs, it should list the RetryRecipe job (if not, make sure you are in the same region where you are using the command line). Select the job name to view the details and then select the **Runs** tab. Refresh until you see two attempts complete, the first one failing and the second succeeding:

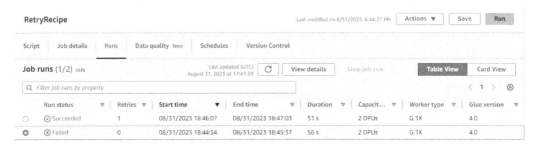

Figure 3.8 – Runs tab showing both attempts

6. Check the output folder by running the following in the command line:

```
aws s3 ls s3://${GLUE_BUCKET}/retry_recipe/
```

7. Examine the files listed. Notice there are two sets of four files; one of them has one file that is much smaller. This is the set from the first run that produced incomplete results due to an error. The retry worked but didn't clean the previous run's partial results. Let's solve that.

8. Back in the console, open the **Script** tab for the job. Near the end of the script, locate the `glueContext.write_dynamic_frame.from_options` line. Just before it, add a new line with the following content:

    ```
    glueContext.purge_s3_path(s3_output_path,
                      {"retentionPeriod": 0})
    ```

9. Save the script change and run the job. Open the **Runs** tab again and refresh to see the new job run, again with an error and a retry.

10. Once the retry has succeeded, in the command line, again list the files as done in *step 6*. Now, there is only one step of four files with similar sizes.

11. Clean the files and the job if no longer needed:

    ```
    aws glue delete-job --job-name RetryRecipe
    aws s3 rm --recursive \
     s3://${GLUE_BUCKET}/retry_recipe/
    ```

How it works...

In this recipe, you created a Glue job with one retry configured, which means that if the job fails (regardless of the reason), then it will immediately retry the same run, using the same configuration and parameters. In most cases, the job fails because an uncaught exception was raised from the script.

The script creates a dataset of 1,024 rows divided into four partitions (Spark memory partitions, not to be confused with table partitions). Then, it checks the job ID to detect whether this is the first run or a retry. The retry uses the name job ID but appends the number of retry attempts starting with 1. If it's the first run, it marks the row with the ID 10 so that the **user-defined function** (**UDF**) that is applied later simulates a runtime error.

This artificial failure simulates a transient error that can succeed on retry; for instance, a call to an external system or some missing external data that was delayed and the job cannot be completed without.

The exception the UDF caused is propagated, and after Spark exhausts the task retries, it causes the script action to raise an exception and the job fails. Unfortunately, the failed job has left incomplete results; most tasks succeeded, and the one that failed produced a partial output file (up to row 10, which the UDF failed).

Then, the job run retry succeeded, but because of the file naming based on timestamp, it didn't override the previous attempt's files, resulting in most of the data being duplicated.

In *step 8*, you added a line that purges the output path, to make sure previous attempts are cleaned before writing the data. This is useful not only on retries but also on reruns. For instance, imagine the code has a mistake doing calculations and has to be run again to correct the resulting data.

There's more...

In a Glue job, there are multiple levels of retries. First, Spark will retry a failed task three times by default. It will also retry stages if the cause is missing partial data; once Spark gives up and fails the Spark job because the exception is not captured and handled in your script, then Glue will check whether retries are configured on the job.

The issue experienced with the partial writes occurs because Glue by default uses a direct committer, which means the files are written directly into the destination folder. This is not the default on Apache Spark in general, which historically assumes it runs on **Hadoop Distributed File System** (**HDFS**), where it can write onto a temporary folder and then do atomic renames once all files are generated correctly; this is not possible on S3, where renaming a file means copying it and thus has a high cost. Glue writes files directly on the destination path, with the drawback that it can leave partial results.

If using a DataFrame to write files, instead of calling `purge`, you can just call: `.mode("overwrite")`; in fact, by default, it will refuse to write if it detects the output folder exists to prevent this situation. If you want it to behave like the DynamicFrame write, you can specify the `"append"` mode.

When writing partitioned data, things get more complicated since you just want to overwrite the partitions the job is writing and not the ones created by previous runs (or other jobs). Fortunately, Spark provides an option to do exactly that:

```
spark.conf.set("spark.sql.sources.partitionOverwriteMode",
                "dynamic")
```

See also

- In this example, you have seen how to do simple automatic retries. For more sophisticated retries such as exponential backoff, you can invoke the Glue job as part of an AWS Step Functions or an **Amazon Managed Workflows for Apache Airflow** (**MWAA**) workflow: `https://docs.aws.amazon.com/mwaa/latest/userguide/what-is-mwaa.html`.

- The open source `s3a` implementation provides a more sophisticated way of writing files that provides efficient writing without partial writes; it's called the Magic Committer. Previously, it had the drawback that it required an external system for consistency, but now that S3 is consistent, that dependency is no longer needed: `https://hadoop.apache.org/docs/r3.1.1/hadoop-aws/tools/hadoop-aws/committers.html#The_Magic_Committer`.

Processing data incrementally using bookmarks and bounded execution

Data pipelines often need to process data as it gets continuously generated and the ETL pipelines have to run on a regular basis. For such cases where the use (and extra cost) of streaming is not justified (for instance, if the data is uploaded once a day), using bookmarks is a simple way of keeping track of which files are already processed and which are new since the last run. With bookmarks, you can run a scheduled job on a regular basis and process only new data added since the last run.

In addition, Glue provides an optional feature called `bounded execution`; with it, a limited amount of data (size or files) is handled in each bookmarked run. This allows the job to run in a timely fashion and predictably with a volume of data that has been tested and not run into issues with memory, disk, or latency. This can be useful if you are backloading a large amount of data or new data arrives in bursts.

Getting ready

This recipe requires a `bash` shell with the AWS CLI installed and configured and the `GLUE_ROLE_ARN` and `GLUE_BUCKET` environment variables set, as indicated in the *Technical requirements* section at the beginning of the chapter.

In addition, run the following code to set up two environment variables, pointing each to a S3 URL for the job source and destination, on the bucket for recipes:

```
RECIPE_S3_SRC=s3://$GLUE_BUCKET/bookmarkrecipe/input/
RECIPE_S3_DST=s3://$GLUE_BUCKET/bookmarkrecipe/output/
```

For this recipe, you'll generate synthetic data. To do so, run the following `bash` commands in the shell, from a directory where you can write. The script will create 10 tiny JSON files locally; update them to the path specified by the `RECIPE_S3_SRC` variable you just set, and finally delete the local files:

```
mkdir ./bookmark_recipe_data/
for i in 1 2 3 4 5 6 7 8 9 10;do
  echo '{"file_number": '$i'}' > \
        ./bookmark_recipe_data/$i.json
done
aws s3 sync ./bookmark_recipe_data/ $RECIPE_S3_SRC
rm ./bookmark_recipe_data/*.json
rmdir ./bookmark_recipe_data/
```

How to do it...

1. Create a job script as a local file running the following multiline shell command. It will execute when you enter the line with just EOF:

```
cat <<EOF > BookmarksRecipeJob.py
import sys
from awsglue.utils import getResolvedOptions
from pyspark.context import SparkContext
from awsglue.context import GlueContext
from awsglue.job import Job

args = getResolvedOptions(sys.argv, ['JOB_NAME'])
sc = SparkContext()
glueContext = GlueContext(sc)
spark = glueContext.spark_session
job = Job(glueContext)
job.init(args['JOB_NAME'], args)

dynf = glueContext.create_dynamic_frame_from_options(
    connection_type = "s3",
    connection_options = {
        "paths": ["$RECIPE_S3_SRC"],
        "boundedFiles": 5
    },
    transformation_ctx="json_source",
    format = "json"
)

glueContext.write_dynamic_frame.from_options(
    frame=dynf.repartition(1),
    connection_type='s3',
    format='json',
    transformation_ctx="csv_dst",
    connection_options={"path": "$RECIPE_S3_DST"
    }
)

job.commit()
EOF
```

2. Upload the script to the S3 bucket and delete the local copy:

```
aws s3 cp BookmarksRecipeJob.py s3://$GLUE_BUCKET
rm BookmarksRecipeJob.py
```

3. Create a job with the following command:

```
aws glue create-job --name BookmarksRecipe --role \
 $GLUE_ROLE_ARN --glue-version 4.0 --command\
 '{"Name": "glueetl", "ScriptLocation":
 "s3://'$GLUE_BUCKET'/BookmarksRecipeJob.py"}'\
 --default-arguments '{"--job-language":"python",
 "--TempDir": "s3://'$GLUE_BUCKET'/tmp/",
 "--job-bookmark-option": "job-bookmark-enable"}'
```

4. Run the job:

```
JOB_RUN_ID=$(aws glue start-job-run --job-name \
 BookmarksRecipe --output text)
```

5. Get the run details; in the response, check the JobRunState job run to check the progress until it completes in a couple of minutes. If the final state is ERROR or FAILED, it will give you an error message with the cause:

```
aws glue get-job-run --job-name BookmarksRecipe \
  --run-id $JOB_RUN_ID
```

6. Retrieve the output files to a local directory and visualize the content. It should have 5 rows taken from 5 of the 10 files in the source:

```
aws s3 sync $RECIPE_S3_DST bookmarks_recipe_result
cut -b 1- bookmarks_recipe_result/* | sort
```

7. Repeat *steps 4, 5,* and *6*. If you try to start the job too soon after the previous run is finished, you might get a concurrency error. Wait for a few seconds and retry. Now, *step 6* prints the lines from the 10 files; this shows each input file was processed once, either on the first or the second run.

8. Wipe the local directory and delete the job if no longer needed:

```
rm bookmarks_recipe_result/*
rmdir bookmarks_recipe_result
aws glue delete-job --job-name BookmarksRecipe
```

How it works...

The first step created the code to use Glue S3 bookmarks. Bookmarks need three things:

- The `init()` method called on the job object.

- The read operation to specify a `transformation_ctx` parameter. Each source must have a different one to track the bookmarks separately.

- Calling `commit()` on the job when the data is processed successfully. Until that method is called, the bookmarks are not updated; there are no partial bookmark updates if some of the files are processed correctly but some fail.

The code uses `repartition(1)` so that the output is generated onto a single output file for convenience. On a real job with large data, such a low number could cause a bottleneck.

Then, the script is uploaded and used on a job where bookmarks are enabled; note the job argument configured on creation: `--job-bookmark-option`.

When you run the job again, it will pick up the remaining files not already bookmarked. So, after two runs, there are two output files, each with the rows of five input files (in this example, each source file has just one row, for easier traceability).

You can run the job once again and see if it produces an empty file because there are no new files. You could avoid generating empty files by checking in the code for `dynf.count() > 0`, before writing the output.

In this example, the script used bounding based on files, but you can do bounding based on data size. This is a better option if the data files have very different sizes.

When using bookmarks, don't change the default job currency of 1. In a job using bookmarks, if there are multiple concurrent runs, at commit time only one can succeed, and the others will fail because they will detect the concurrent bookmark update.

There's more...

As listed before, a `transformation_ctx` parameter on the source is a requirement for bookmarks to work. You can use this to your advantage and combine the same job sources that need bookmarking and sources that do not. For instance, if you process orders and need to join with customers, you want to bookmark the orders but not the customers, so each run can join with all existing customers and not just new ones. You can also use bookmarks on catalog tables, which will keep track of all files on all partitions.

The `--job-bookmark-option` argument has a third option other than `enable` or `disable`. You can pause the bookmark, which means the job will use the bookmark to read but won't update it. You can use this option for testing/troubleshooting.

It is possible to reset the bookmarks of a job or rewind to the state in which a previous job run left it. This is useful if you need to reprocess data due to some issue.

Glue also has bookmarks for **Java Database Connectivity** (**JDBC**) sources. For that kind of bookmark, it needs one or more numerically monotonic (always increasing or decreasing on updates) columns to be specified so that Glue can remember the last number processed and in the next run take the records with a larger number (or smaller, if configured that way).

For instance, if you have a `sequence` column, you can use that to bookmark records already processed versus new ones created, but it won't detect updates unless the `bookmark` column is updated as well, using a `timestamp` column as bookmark.

You can read more about Glue bookmarks at `https://docs.aws.amazon.com/glue/latest/dg/monitor-continuations.html`.

See also

- You can also use bookmarks in a visual job. See how to create one in the *Creating ETL jobs visually using AWS Glue Studio* recipe.

- Bookmarks control the input data being processed at least once, but you also must make sure the job output is tolerant to errors and doesn't generate duplicates. Learn more in the *Handling job failures and reruns for partial results* recipe.

Handling a high quantity of small files in your job

Frequently, when ingesting data, the source data is not optimized and it comes in tiny files, maybe because it was produced at short intervals or by many sources such as different sensors sending their individual reports. Apache Spark was designed as a big data tool and it struggles when handling such cases, causing inefficiency when processing too many partitions and also causing memory issues on the driver when building a plan.

To handle data efficiently, we want to consolidate small files to make the reading more efficient, especially if using a columnar format such as Parquet; as a rule of thumb, at least 100 MB bytes on each file. The simple way to control that is to repartition/coalesce the data to the target number of output files, but that often requires a costly shuffle operation.

In this recipe, you will see a simple and effective way provided by Glue to group small files at reading time.

Getting ready

This recipe requires a `bash` shell with the AWS CLI installed and configured. The GLUE_ROLE_ARN and GLUE_BUCKET environment variables need to be set, as per the *Technical requirements* section at the beginning of the chapter.

To demonstrate this feature, we need data with lots of small files. To prepare such an input dataset, execute the following `bash` commands. Make sure you have at least 50 MB of disk space free in the local directory. It will generate 10k tiny CSV gzipped files. It will take a few minutes, printing a dot each time it has completed 10k files:

```bash
echo placeholder > template_file.csv
gzip template_file.csv
mkdir smallfiles_recipe_input
# Generate the files locally
for i in {1..10000}; do
  cp template_file.csv.gz smallfiles_recipe_input/$i.csv.gz;
  if (($i % 1000 == 0)); then echo -n .; fi
done
echo
# Upload to s3 and cleanup the local files
rm template_file.csv.gz
S3_INPUT_URL=s3://$GLUE_BUCKET/smallfiles_input/
aws s3 sync smallfiles_recipe_input $S3_INPUT_URL
rm smallfiles_recipe_input/*.csv.gz
rmdir smallfiles_recipe_input
S3_OUTPUT_URL=s3://$GLUE_BUCKET/smallfiles_output/
```

How to do it...

1. Create a job script as a local file by running the following `bash` command:

```python
cat <<EOF > GroupingFilesRecipeJob.py
from pyspark.context import SparkContext
from awsglue.context import GlueContext
glueContext = GlueContext(SparkContext())
dynf = glueContext.create_dynamic_frame.from_options(
    connection_type="s3",
    connection_options={
        "paths": ["$S3_INPUT_URL"],
        "useS3ListImplementation": True,
        "groupFiles": "inPartition",
        "groupSize": "100000"
    },
```

```
        format="csv"
    )
    writer=dynf.toDF().write.format("parquet")
    writer.mode("overwrite").save("$S3_OUTPUT_URL")
    EOF
```

2. Upload the job Python script to S3:

    ```
    aws s3 cp GroupingFilesRecipeJob.py s3://$GLUE_BUCKET
    rm GroupingFilesRecipeJob.py
    ```

3. Create a job:

    ```
    aws glue create-job --name GroupingFilesRecipe \
      --role $GLUE_ROLE_ARN --number-of-workers 2 \
      --glue-version 4.0 --command '{"Name":
      "gluestreaming", "ScriptLocation":
      "s3://'$GLUE_BUCKET'/GroupingFilesRecipeJob.py"}'\
      --worker-type "G.025X" --default-arguments \
      '{"--job-language":"python", "--TempDir":
      "s3://'$GLUE_BUCKET'/tmp/"}'
    ```

4. Run the job:

    ```
    JOB_RUN_ID=$(aws glue start-job-run --job-name \
        GroupingFilesRecipe --output text)
    ```

5. Check the JobRunState job run until it completes (should be less than 10 minutes). If the job fails for some reason, it will give you an error message:

    ```
    aws glue get-job-run --job-name GroupingFilesRecipe \
        --run-id $JOB_RUN_ID
    ```

6. List the output files; the 10K tiny files have been consolidated into just 5:

    ```
    aws s3 ls $S3_OUTPUT_URL
    ```

7. Remove the job and the S3 files, if no longer needed:

    ```
    aws glue delete-job --job-name GroupingFilesRecipe
    aws s3 rm --recursive $S3_INPUT_URL
    aws s3 rm --recursive $S3_OUTPUT_URL
    ```

How it works...

The script specifies in the source that the files should be grouped into 100k bytes on each partition. In this case, there is only one partition, which is the input directory. Each gzipped file is 50 bytes (notice the data is so small that it takes more space compressed, but that is what Glue sees at planning time).

Therefore, we have 100k / 50 = 2000 files per group, so 10k files grouped result in 5 Spark partitions overall (not to be confused with the S3 partitions, which the grouping parameter refers to).

Then, you created a Glue job that uses the script uploaded to S3. Notice something peculiar? It is defined as Spark Streaming so that the job can use the smallest node size G.025X (a quarter of a DPU). This kind of node is intended for streaming because it is normally too small for ETL jobs in terms of memory, but in this case, it is enough because this job script requires minimum memory for planning.

The outcome was five Parquet files since the files were grouped into five Spark partitions. Notice here the number of partitions is maintained when converting from DynamicFrame to DataFrame.

Unfortunately, the conversion caused a small delay since DynamicFrame couldn't figure out the schema on the fly and had to do a two-pass to first determine it for DataFrame and then the actual processing. This wouldn't have been the case if writing directly from DynamicFrame.

There's more...

If you configure the Spark UI logs and visualize the execution, you will see there is no shuffle of data. The files are assigned to different partitions from the source.

With such small files, most of the time is not spent doing the actual reading and writing but listing the files from s3, which is not reflected in the Spark tasks. The job uses the `useS3ListImplementation` option, which lists files directly using the AWS Java SDK instead of the generic HDFS that Spark uses.

This grouping feature will kick in automatically if `GlueContext` detects the source has more than 50k files. If you do not want this optimization for some reason, you can disable the setting as `"groupFiles": "none"`.

Not all formats support this feature; check the documentation for further details: `https://docs.aws.amazon.com/glue/latest/dg/aws-glue-programming-etl-format.html`.

See also

- For DataFrame, in Spark 3.5 or later, you can specify the desired partitions with `spark.sql.files.maxPartitionNum` and it will try to group files to honor that. On older versions, there are very limited capabilities to group files.

- Glue for Ray is better suited for handling small files; learn more about this engine in the *Running pandas code using AWS Glue for Ray* recipe.

Reusing libraries in your Glue job

Spark provides a rich data framework that can be extended with additional plugins, libraries, and Python modules. As you build more jobs, you would likely reuse your own code, whether it's UDFs to process data when it's not possible to do the same using the Spark functions or some pipeline code you want to reuse; for instance, a function with some transformations that you do regularly.

In this recipe, you will see how you can reuse Python code on Glue for Spark jobs.

Getting ready

This recipe requires a bash shell with the AWS CLI installed and configured and the GLUE_ROLE_ARN and GLUE_BUCKET environment variables set, as indicated in the *Technical requirements* section at the beginning of the chapter.

How to do it...

1. The following bash commands will create a Python module and config file:

```
mkdir my_module
cat <<EOF > my_module/__init__.py
from random import randint

def do_some_calculation(a):
  return randint(1, 10) + a

def get_config_value():
  with open('/tmp/my_config') as f:
    lines = f.readlines()
    return lines[0].strip()
EOF
zip -r my_module my_module
echo "recipe_example_value" > my_config
```

2. Upload both files to S3 and clean up:

```
RECIPE_S3_PATH=s3://$GLUE_BUCKET/reuse_module
aws s3 cp my_module.zip $RECIPE_S3_PATH/
rm my_module.zip
rm my_module/__init__.py
rmdir my_module
aws s3 cp my_config $RECIPE_S3_PATH/
rm my_config
```

3. Prepare the job script:

```
cat <<EOF > ReuseLibrariesRecipe.py
from pyspark.sql import SparkSession
from pyspark.sql.functions import lit, udf
from my_module import do_some_calculation, get_config_value
spark = SparkSession.builder.getOrCreate()
df = spark.range(1 << 4).toDF("id")
df = df.withColumn("config_val",
        lit(get_config_value()))
calc_udf = udf(do_some_calculation)
df = df.withColumn("calc", calc_udf(df["id"]))
df.repartition(1).write.csv("$RECIPE_S3_PATH/out")
EOF
aws s3 cp ReuseLibrariesRecipe.py $RECIPE_S3_PATH/
rm ReuseLibrariesRecipe.py
```

4. Create a Glue job; be careful to use \ only in the lines indicated:

```
aws glue create-job --name ReuseLibraryRecipe --role\
  $GLUE_ROLE_ARN --glue-version 4.0 --command \
  '{"Name": "glueetl", "ScriptLocation":
  "'$RECIPE_S3_PATH'/ReuseLibrariesRecipe.py"}' \
  --number-of-workers 2 --worker-type "G.1X" \
  --default-arguments '{"--job-language":"python",
  "--extra-py-files":"'$RECIPE_S3_PATH'/my_module.
zip",   "--extra-files": "'$RECIPE_S3_PATH'/my_config",
  "--TempDir": "'$RECIPE_S3_PATH'/tmp/"}'
```

5. Run the job:

```
JOB_RUN_ID=$(aws glue start-job-run --job-name \
    ReuseLibraryRecipe --output text)
```

6. Check the `JobRunState` job run until it completes (should be less than 10 minutes). If the job fails for some reason, it will give you an error message:

```
aws glue get-job-run --job-name ReuseLibraryRecipe \
    --run-id $JOB_RUN_ID
```

7. Show the result file content:

```
aws s3 sync $RECIPE_S3_PATH/out/ .
cat part-*-c000.csv
rm part-*-c000.csv
```

8.   Remove the job if no longer needed:

```
aws glue delete-job --job-name ReuseLibraryRecipe
```

How it works...

In the first step, you created a basic Python module. The module has two functions: one that simulates a calculation providing a random value and another that reads a local text file and provides the context.

When a job is created, it is configured with the S3 path to the text file as an extra file and the module as an extra Python file. At runtime, the text file will be deployed on the /tmp directory of each node of the cluster, and each node will download, extract the ZIP file with the module, and make it available in the Python runtime.

When the job runs, the code creates a DataFrame with the help of the Python functions and stores it on S3 as a CSV file, with three columns:

- A sequential number for each row

- A column containing the text provided by the function – this function is called on the driver and then used as a constant for all rows

- The third column is the result of calling the do_some_calculation module on the row ID using a UDF, and the result is a random number between 1 and 10 added to the row ID

There's more...

In this recipe, the code was reused directly on the driver and distributed in the form of UDFs. Notice this is for demonstration purposes, but using Python UDFs results in performance degradation when running with a significant amount of data.

Instead of a ZIP file, you can use --extra-py-files with individual Python files, which then you can import into the code using the filename. If you want to use wheel files (created by you or from a Python repository), then you instead need to use the --additional-python-modules parameter.

Reusing Java/Scala code is similar, but instead, you specify the location of the JAR files using the --extra-jars parameter.

See also

- You can reuse code in the same way on Glue Studio visual jobs when you use the custom code node to extend the capabilities. See an example of a visual job in the *Creating ETL jobs visually using AWS Glue Studio* recipe.

- Visual jobs also allow creating your own components to reuse code; see an example at https://aws.amazon.com/es/blogs/big-data/create-your-own-reusable-visual-transforms-for-aws-glue-studio/.

Using data lake formats to store your data

Historically, big data technologies on the Hadoop ecosystem have taken some trade-offs to scale to volumes that traditional databases cannot handle. In the case of Apache Hive, which became the standard Hadoop SQL database, the external tables just point to files on some object storage such as HDFS or S3, and then jobs access those files without a central system coordinating access or transactions. This is still how the standard tables work on the Glue catalog.

As a result, the **atomicity, consistency, isolation, and durability (ACID)** properties of RDBMSs were relaxed to allow for scalability in use cases where write concurrency or the lack of transactions is not an issue, such as historical append-only tables.

In recent years, the desire has been to bring back those ACID properties while keeping the data on a scalable object store for cheap and virtually infinite scalability, with many clients and engines using the data in a distributed way.

The result is the growing popularity of so-called "data lake table formats," which define tables that are no longer just plain data files but have a structure of metadata that can handle transactions, keep track of changes and versions, and allow time travel.

The most popular formats are Apache Iceberg, Apache Hudi, and Delta Lake.

Iceberg's differentiating feature is that it allows dynamic partitioning. In this recipe, you will see an example of how to easily enable and use Iceberg in your Glue jobs.

Getting ready

This recipe requires a `bash` shell with the AWS CLI installed and configured. The `GLUE_ROLE_ARN`, `ACCOUNT_ID`, and `GLUE_BUCKET` environment variables must be set, as indicated in the *Technical requirements* section.

How to do it...

1. Create a job script as a local file running the following multiline `bash` command. It will execute when you enter the line with just EOF:

    ```
    cat <<EOF > IcebergRecipe.scala
    import org.apache.spark.sql.SparkSession
    import org.apache.spark.sql.functions.{col,lit,rand}

    object GlueApp {
      def main(sysArgs: Array[String]) {
        val spark = (SparkSession.builder
          .config("spark.sql.extensions",
                "org.apache.iceberg.spark.\
    ```

```
extensions.IcebergSparkSessionExtensions")
        .config("spark.sql.catalog.iceberg",
            "org.apache.iceberg.spark.SparkCatalog")
        .config("spark.sql.catalog.iceberg.warehouse",
            "s3://$GLUE_BUCKET/iceberg")
        .config("spark.sql.catalog.iceberg.catalog-\
impl", "org.apache.iceberg.aws.glue.GlueCatalog")
        .config("spark.sql.catalog.iceberg.io-impl",
            "org.apache.iceberg.aws.s3.S3FileIO")
        .getOrCreate())

    val db = "iceberg_recipe_db"
    spark.sql("CREATE DATABASE IF NOT EXISTS iceberg." + db)
    val df = (spark.range(1 << 10).toDF("id")
                .withColumn("value1", rand())
                .withColumn("region", lit("region1"))
            )
    df.writeTo("iceberg." + db + ".icetable").
        partitionedBy(col("region")).createOrReplace()
    }
}
EOF
```

2. Upload the script to S3:

```
aws s3 cp IcebergRecipe.scala s3://$GLUE_BUCKET
rm IcebergRecipe.scala
```

3. Create a job:

```
aws glue create-job --name IcebergRecipe --role \
 $GLUE_ROLE_ARN --glue-version 4.0 --worker-type \
 "G.1X" --number-of-workers 2 --default-arguments \
 '{"--job-language":"scala", "--class": "GlueApp",
 "--datalake-formats": "iceberg", "--TempDir":
 "s3://'$GLUE_BUCKET'/tmp/"}' --command '{"Name":
 "glueetl","ScriptLocation":
 "s3://'$GLUE_BUCKET'/IcebergRecipe.scala"}'
```

4. Run the job:

```
JOB_RUN_ID=$(aws glue start-job-run --job-name \
   IcebergRecipe --output text)
```

5. Check the job status until it succeeds:

```
aws glue get-job-run --job-name IcebergRecipe \
  --run-id $JOB_RUN_ID
```

6. On the AWS console, navigate to Athena and run the following query to show the table content (it might ask to select an S3 output location first):

```
SELECT * FROM iceberg_recipe_db.icetable LIMIT 20;
```

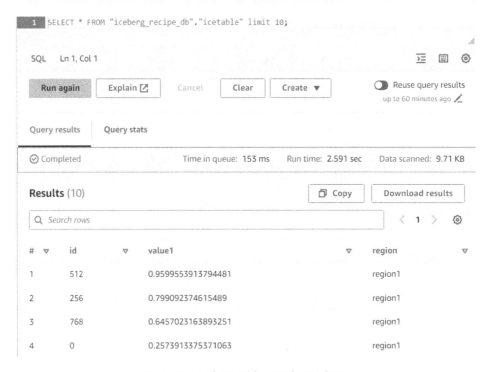

Figure 3.9 – Iceberg table sample in Athena

7. If you no longer need them, delete the table created on the catalog (this doesn't delete the data on S3), the catalog policy, and the job:

```
aws glue delete-database --name iceberg_recipe_db
aws glue delete-job --job-name IcebergRecipe
```

How it works...

In the first step, you defined the script (in this case, using Scala) and then created a job to run it. Notice the code in this recipe is pure Spark; there is no Glue API-specific code. This shows you can bring standard Spark code and run in Glue without changes.

The job code defines a special Spark catalog configuration, using the properties starting with `spark.sql.catalog`, and then an arbitrary name for the catalog. In this example named `iceberg`, you can use your own name as long as when you reference the table in the code, you use the same catalog prefix so that Spark knows it needs to use that configuration for that table.

Later in the script, it inserts the sample DataFrame created into a table named `iceberg.iceberg_recipe_db.icetable`. The first part is the catalog to match the configuration name with the database and table names. Iceberg will create a directory using the database and table names (adding the `.db` suffix for the database). Then, it will create a matching database and table in the catalog with the S3 location.

To enable the Iceberg framework, all that was required was to add the `--datalake-formats=iceberg` argument. You can enable Hudi or Delta in the same way, specifying `hudi` or `delta` respectively or multiple of them separated by a comma.

Each framework has its own particularities, configuration, and way to expose the table in the Glue catalog, but the way to enable them in a Glue job is the same.

There's more...

The updates on the table are transactional, so you will always see a consistent view. When there are multiple writers on the same table, things get more complicated; therefore, it's best to have a single job writing on the table.

Iceberg and the other frameworks have optimistic lock detection, which means they can detect when another writer has made changes concurrently and rectify it. In addition, the frameworks have the option to configure an external system such as DynamoDB to hold locks and do pessimistic locking so that other writers wait. This makes sense when the optimistic locking has many collisions and must constantly redo work due to conflicts.

Before Glue supported these frameworks natively, the way to add them was by subscribing to a marketplace connector. This option is not only more laborious but adds a runtime dependency to the `us-east-1` **Elastic Container Registry** (**ECR**) repository to download the container at runtime. Therefore, this option is no longer recommended for data lake formats.

You also have the option to add the JARs yourself to the job as extra libraries, which allows you to use the latest version of the framework.

See also

- When creating a visual job in Glue Studio, the support for data lake formats is added automatically as needed. See how to create a visual job in the *Creating ETL jobs visually using AWS Glue Studio* recipe.

- If you want to read incrementally from one of these formats, you need to track the snapshots read or use a streaming source. If this incremental data doesn't need to be updated or queried, it might be more effective for doing incremental ingestion to use Glue bookmarks; see how in the *Processing data incrementally using bookmarks and bounded execution* recipe.

Optimizing your catalog data retrieval using pushdown filters and indexes

The AWS Glue Data Catalog is a key component in a big data cloud architecture. It doesn't hold data but acts as an Apache Hive-compatible metastore, defining table metadata that acts as a layer of abstraction. It shows clients how to locate and interpret the data stored in a system such as Amazon S3.

In the traditional Hive-compatible catalog tables, the catalog doesn't keep track of data files. It just points to a directory prefix, and then the client will list the prefix to get a list of files currently present there. The way these kinds of tables can scale is by using partitions, each one corresponding to a prefix on S3, to avoid listing all files for any query. A partitioned table defines one or more columns as partition columns.

For instance, if you have a table with the `year`, `month`, and `day` partition columns as strings, then the data for each day will be placed under a specific prefix, which when following the Hive conventions contains the partition values; for instance:

`s3://mybucket/mytable/year=2023/month=08/day=01/.`

This way, a client tool that only needs data from that day just needs to read that specific data and not all the files from the table, which might potentially have years of data.

In this recipe, you will see how to benefit from partitions, even in cases where a table contains a massive number of them.

Getting ready

This recipe assumes you have created the `AWSGlueServiceRole-Recipe` role and an S3 bucket with your account ID of `glue-recipes-<your accountid>`, as indicated in the *Technical requirements* section at the beginning of this chapter.

In the `bash` shell, add a policy to allow the role to run notebooks:

```
GLUE_ROLE=AWSGlueServiceRole-Recipe
aws iam put-role-policy --role-name $GLUE_ROLE \
--policy-name GlueSessions --policy-document '{"Version":
"2012-10-17", "Statement": [{"Effect": "Allow",
"Action":["glue:*Session", "glue:RunStatement",
"iam:PassRole"], "Resource":["*"]}]}'
```

How to do it...

1. Log in to the AWS console, navigate the Glue service, and on the left menu select **Notebooks**. Then, choose the **Notebook** button to create it.

2. On the popup, leave the **Spark (Python)** and **Start fresh** default options in the `AWSGlueServiceRole-Recipe` IAM role and complete the creation.

3. After a few seconds, it will load Jupyter and open a notebook with some sample cells already filled in. The sample code might change, but there should always be a cell to help you set up your job. This setup cell does some configuration (the lines starting with %) and then sets up the `GlueContext` and `SparkSession` objects with the names `glueContext` and `spark`, respectively (should the name of these objects change, you need to update the code provided accordingly). To reduce costs, you can change `%number_of_workers` down to 2.

    Then, run this cell by selecting it and using the **Run** button on the toolbar or by using the *Shift + Enter* shortcut. Below the cell, you will see the initialization progress until it confirms the session has been created:

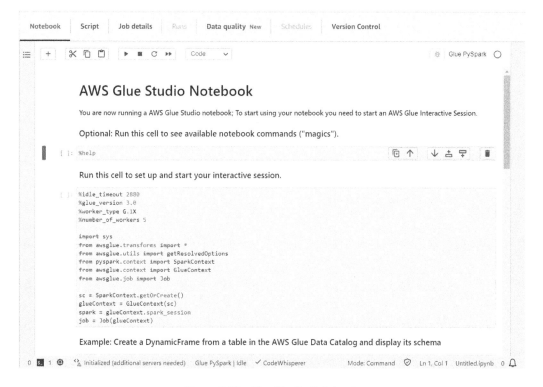

Figure 3.10 – Glue Studio Notebook

4. Now, add a new cell to the notebook; you can use the plus (+) button on the top left. Inside, enter the following code and replace the bucket name on the first line with your own. Then, run the cell and wait until it completes:

```
import boto3
from pyspark.sql.functions import rand
from awsglue.dynamicframe import DynamicFrame

account_id = boto3.client('sts').get_caller_identity(
                                    )["Account"]
s3_bucket = f"glue-recipes-{account_id}"
s3_path = f"s3://{s3_bucket}/pushdown_recipe_table/"
database = "default"
table_name = "pushdown_recipe_table"

df = spark.range(1 << 8).withColumn("value", rand())
sink = glueContext.getSink(connection_type="s3",
            path=s3_path, enableUpdateCatalog=True,
            updateBehavior="UPDATE_IN_DATABASE",
            partitionKeys=["id"])
sink.setFormat("avro")
sink.setCatalogInfo(catalogDatabase=database,
                    catalogTableName=table_name)
sink.writeFrame(DynamicFrame.fromDF(df, glueContext))
```

5. Now, add a cell with the following code and run it. It will take about a minute to print the result with the count:

```
glueContext.create_dynamic_frame.from_catalog(
    database=database,
    table_name=table_name).count()
```

6. Now, add a cell with this alternative code and run it. It should just take a few seconds to complete:

```
glueContext.create_dynamic_frame.from_catalog(
    database=database, table_name=table_name,
    push_down_predicate="id in (3, 6, 9)",
    additional_options={
        "catalogPartitionPredicate":"id < 10"
    }
).count()
```

7. Add another cell to run this code to get the same result:

```
spark.sql(f"SELECT * FROM {database}.{table_name} "
        "WHERE id IN (3, 6, 9)").count()
```

8. Extend the left menu, select **Tables**, and then search for pushdown_recipe_table. Observe the schema with the ID as the partition column and switch to the **Partitions** tab. It contains 32 partitions. Switch to the **Indexes** tab and select **Add index**. Give it the name main and select the id column. If *steps 6* and *7* had been slow due to the high number of partitions, adding this index would significantly speed up queries on the id column:

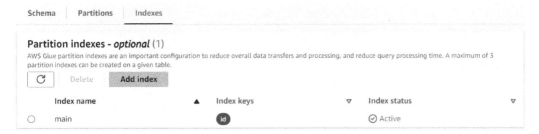

Figure 3.11 – Table indexes with the one index added

9. Add a new cell and run it with this code to remove the table:

```
boto3.client('glue').delete_table(
    DatabaseName=database,Name=table_name)
```

How it works...

In this recipe, instead of using a Glue job, you used a Glue interactive session. This is an equivalent of a cluster, which you can configure using magics instead of the **Job details** page. The magics can be line magics (starting with %), where the magic parameters are set on the same line (for instance, %glue_version 3.0), and cell magics (starting with %%), where the whole cell is used for the magic (for instance, %%configure to add arguments). You can run the %help cell to get a listing of magics and their uses.

For simplicity, it's common to use the same role for the notebook and the session, but it's possible to use a different use using the %iam_role magic.

The interactive session starts the moment you run any code (any line that is not a magic or a comment). Changes in the session configuration (for example, number of workers) won't take effect until the kernel or the notebook is restarted.

Notice that while a cell is running, it has an asterisk at the left of it. Only one cell can run at a time; you can queue multiple cell executions.

Then, you ran a cell that created simple data with a sequential ID and a random value, which was stored as a partitioned table with the ID as the partition column. Therefore, the table has just 32 partitions; a real table could have thousands of them.

Then, you ran multiple queries on the table. In the first case, any operation will have to read all the files, even if you filter out the data later because you only need a subset. The other queries leveraged the table partition. The performance difference should be noticeable even for such a small table (note the cluster is also tiny with just two nodes, one of which is the driver).

Finally, you added an index to the table. In such a small table, it will be added instantly but won't make a real difference at query time since most of the time is spent checking files. In a table with tens of thousands of partitions or more, the index will make a big difference in retrieving partitions quickly.

There's more...

In this example, the sample DataFrame created with sample data is converted to a DynamicFrame to create a table. It's also possible to create a table directly from the DataFrame using `saveAsTable()`, but using the DynamicFrame writer is more efficient in adding partitions and can update the schema automatically as needed.

Notice the difference in the usage of `DynamicFrame` and SparkSQL APIs to read the partitioned table. In the case of SparkSQL, the engine uses the query criteria to push down the filters applicable to the partitions and only read the related data. On the other hand, in the case of `create_dynamic_frame.from_catalog`, the partition filters are explicitly indicated in two possible ways:

- `push_down_predicate` allows using rich SQL filters, which are applied on the partitions retrieved from the server. So, it avoids checking files not needed but can still spend a lot of time listing table partitions if the table has many.

- `catalogPartitionPredicate` applies the filter on the server side, so it scales better, but it's more limited in the kind of filters you can build – basically, equals and greater/smaller than conditions, of which you can add multiple but not make them optional (no OR clause). In addition, this predicate can use a table index like the one you created in the recipe , if it filters on the index columns.

Both types of predicates can be combined to benefit from the efficiency of `catalogPartitionPredicate` and then apply richer filters to narrow further the results by using `push_down_predicate`.

Check the Glue documentation for details and examples about syntax and usage:

`https://docs.aws.amazon.com/glue/latest/dg/aws-glue-programming-etl-partitions.html`.

See also

- This recipe is based on traditional Hive-style tables (using directories). While this is a simple and efficient way to partition data, it has a big limitation: changing the partitioning means you need to rebuild the table. Instead, you could use Apache Iceberg, which provides dynamic partitions that allow evolving the partitioning without rebuilding the table or impacting the users. See the *Using data lake formats to store your data* recipe for the recommended way to use Apache Iceberg in Glue.

- When adding partitions using the DynamicFrame writer, at the time of this writing, it updates the main table schema to make sure it matches (whether is needed or not); this can keep increasing the number of table versions and eventually hit the account limit.

  There is an open-source tool provided by AWS to clean up old versions:

  `https://github.com/aws-samples/aws-glue-table-versions-cleanup-utility`.

Running pandas code using AWS Glue for Ray

The `pandas` library is a highly popular Python library for data manipulation and analysis, based on the well-established `numpy` library, handling data in a table-like format. It is so well established among Python analysts and data scientists, that it has become a de facto standard to the point that other libraries implement their interfaces so that they can run existing `pandas` code. This is often done to overcome `pandas`' limitations, namely being a single process memory-based library, which limits scalability.

One such `pandas`-compatible library is Modin. It can run `pandas` code by just changing the imports while being able to scale by using an engine such as Dask or Ray. In this recipe, you will see how to run `pandas` code on Glue for Ray using Modin.

Getting ready

This recipe requires a `bash` shell with the AWS CLI installed and configured. The `GLUE_ROLE_ARN` and `GLUE_BUCKET` environment variables need to be set, as indicated in the *Technical requirements* section at the beginning of the chapter.

Make sure that Glue for Ray is available in the AWS region you are using.

How to do it...

1. Create a job script as a local file running the following multiline bash command. It will execute when you enter the line with just EOF:

```
cat <<EOF > RayModinRecipeJob.py
import ray
import modin.pandas as pd
import numpy as np

ray.init()
s3_path = "s3://$GLUE_BUCKET/ray_recipe_output"
num_samples = 10000
df = pd.DataFrame({
        'id' : range(num_samples),
        'value' : np.random.randn(num_samples),
    })
result = df[df['value'] > 0]
print(f"Std Dev: {df['value'].std()}")
print(f"Out of {num_samples}, {result.shape[0]} will\
 be saved after filtering")
result.to_parquet(s3_path)
EOF
```

2. Upload the script to S3:

```
aws s3 cp RayModinRecipeJob.py s3://$GLUE_BUCKET
rm RayModinRecipeJob.py
```

3. Create a job, making sure to only use \ in the lines indicated:

```
aws glue create-job --name RayModinRecipeJob \
--role $GLUE_ROLE_ARN --command '{"Name": "glueray",
 "Runtime": "Ray2.4", "ScriptLocation":
 "s3://'$GLUE_BUCKET'/RayModinRecipeJob.py"}'\
 --number-of-workers 1 --worker-type "Z.2X"\
 --glue-version 4.0 --default-arguments\
 '{"--pip-install": "modin,s3fs"}'
```

4. Run the job:

```
JOB_RUN_ID=$(aws glue start-job-run --job-name \
        RayModinRecipeJob --output text)
```

5. Check the `JobRunState` job run until it completes (should be less than 10 minutes). If the job fails for some reason, it will give you an error message:

```
aws glue get-job-run --job-name RayModinRecipeJob \
    --run-id $JOB_RUN_ID
```

6. Print the job using `stdout` and observe the messages printed from the code:

```
aws s3 cp \
s3://$GLUE_BUCKET/jobs/RayModinRecipeJob/\
$JOB_RUN_ID/job-result/stdout /tmp
cat /tmp/stdout && echo -e "\n" && rm /tmp/stdout
```

7. List the files generated by the job:

```
aws s3 ls s3://$GLUE_BUCKET/ray_recipe_output/
```

8. Remove the job if no longer needed:

```
aws glue delete-job --job-name RayModinRecipeJob
```

How it works...

If you have run other Glue recipes that use script jobs, you might have noticed the process is almost identical to Glue for Ray. First, you created a Python script, uploaded it to S3, then created a job specifying `glueray` as the command. This command requires the version of Ray to be specified and uses different instance types as other types of Glue jobs. The job specifies that the Modin and `s3fs` libraries need to be installed.

Running the job and checking the status was identical to other kinds of Glue jobs.

Glue stores the job logs, including `stdout` and `stderr`, under the path where the script was located. This allowed you to easily check the result of the `print` statements.

Logs are also available on CloudWatch, as usual.

The script generated a DataFrame with 1,000 rows and a column of values based on a standard normal distribution, then the code filtered only the positive values and saved them as Parquet files on S3.

There's more...

When using Glue for Ray, the smallest instance at the time of this writing is two DPUs. However, unlike Glue for Spark, you can use a single node since the Ray driver can also do work. Modin automatically detected the Ray framework. If the job had more nodes assigned, it would have used them for distributing the data and processing.

In this example, we assumed that a requirement was to be able to reuse existing `pandas` code (or existing `pandas` skills); if you want more control of the distributed processing, you can use the Ray framework directly with the APIs it provides.

As Ray gains popularity, other libraries are adding support for it, such as the popular **machine learning** (**ML**) library `pytorch` or the versatile AWS library `awsdatawrangler`, which provides integration to AWS data services such as S3, DynamoDB, Redshift, the Glue Catalog, or OpenSearch.

See also

- If you want to avoid doing any coding at all, you can refer to the *Creating ETL jobs visually using AWS Glue Studio* recipe. At the time of this writing, visual jobs are only able to generate Glue for Spark code.

- To reuse existing `pandas` code, you also have the option to use the recently added Pandas API on Spark. Unfortunately, this API has more limitations and discrepancies in terms of `pandas` compatibility. See details on the Apache site: `https://spark.apache.org/docs/latest/api/python/user_guide/pandas_on_spark/index.html`.

4

A Deep Dive into AWS Orchestration Frameworks

Welcome to this chapter on orchestration frameworks in the AWS ecosystem. In this chapter, you will gain valuable insights into setting up and managing orchestration frameworks using AWS services. Specifically, we will explore four key AWS offerings: AWS Glue workflows, **Amazon Managed Workflows for Apache Airflow** (**MWAA**), AWS Step Functions, and Amazon EventBridge.

In distributed systems, coordinating tasks across different services and components can be challenging. Building complex architecture and implementing workflows with code can lead to spaghetti code that's hard to maintain and debug. Throughout this chapter, you will find practical examples and guidance on creating, updating, and implementing rollback strategies based on metrics emitted by your workflows. By the end of this chapter, you'll be well-equipped to harness the power of these AWS orchestration tools to optimize your workflow management and automation processes.

The following recipes will be covered in this chapter:

- Defining a simple workflow using AWS Glue workflows
- Setting up event-driven orchestration with Amazon EventBridge
- Creating a data workflow using AWS Step Functions
- Managing data pipelines with MWAA
- Monitoring your pipeline's health
- Setting up a pipeline using AWS Glue to ingest data from a JDBC database into a catalog table

Technical requirements

For the recipes in this chapter, you'll need an active AWS account with appropriate permissions to create and manage the following:

- Glue workflows

- Glue triggers

- Step Functions

- MWAA environment

- AWS roles and permissions

- CloudWatch

You can find the code files for this chapter on GitHub at `https://github.com/PacktPublishing/Data-Engineering-with-AWS-Cookbook/tree/main/Chapter04`.

Defining a simple workflow using AWS Glue workflows

In this recipe, we will explore the world of AWS Glue workflows. AWS Glue workflows are powerful tool for orchestrating complex **extract**, **transform**, and **load** (**ETL**) processes on AWS. By combining jobs, crawlers, and triggers, workflows can automate data pipelines, ensuring data is consistently processed and delivered to its intended destinations. This makes them ideal for a variety of data-driven applications, from data warehousing and analytics to machine learning and real-time data processing.

Imagine a retail company that needs to load sales data from multiple sources into its data lake on Amazon S3. The data comes in raw CSV format, and the goal is to transform this data into a structured format that can be used for reporting and analysis. The data is updated daily, and you need an automated pipeline to clean, transform, and store this data.

In this scenario, you can leverage an AWS Glue workflow, crawler, job, and trigger to automate the process.

Getting ready

To build a simple AWS Glue workflow that reads CSV data from S3 and writes it out as partitioned Parquet, you need an S3 bucket with source CSV data.

If you have an existing S3 bucket, then use it, or else create a new S3 bucket. To create an S3 bucket, first log in to the AWS Management Console and navigate to the S3 service. Click on **Create bucket** and choose a unique name using lowercase letters, numbers, periods, and hyphens. Select your desired name. Finally, after reviewing your configuration, click on **Create bucket** to finalize.

How to do it...

The high-level steps to build a simple AWS Glue workflow for CSV-to-Parquet conversion are as follows:

1. **Prepare your data**: Store your CSV data in an S3 bucket. This will be the input source for your Glue job.

2. **Create an IAM role**: Create an IAM role with the necessary permissions for your Glue crawler and Glue job to access S3 and the Glue Data Catalog.

3. **Create a Glue database**: A Glue database is like a container that holds tables and their schemas within the AWS Glue Data Catalog. It helps organize your metadata, making it easier to manage and query your data.

   In the AWS Glue Data Catalog, create a database to organize the tables that your crawler will generate.

4. **Create an AWS Glue crawler**: Glue crawlers automatically discover and catalog your data stored in various locations such as S3, databases, and data lakes. They extract schema information and create table definitions, making your data readily available for analysis and querying.

   Configure the crawler to read the CSV data from your S3 bucket. The crawler will infer the schema of your data and create corresponding tables in the AWS Glue Data Catalog, within the database you created.

5. **Create an AWS Glue job**: Glue jobs are the heart of your ETL processes in AWS Glue. They run your scripts (written in Python or Scala) to extract data from sources, transform it according to your needs, and load it into target destinations:

   I.   Use a Python or Scala script to define your ETL logic.

   II.   In the script, read the CSV data from the S3 bucket using the table created by the crawler.

   III. Perform any necessary transformations on the data.

   IV.   Write the transformed data to a new location in the S3 bucket in Parquet format, partitioning the data as needed.

6. **Create an AWS Glue workflow**: Workflows provide a visual representation of your data processing steps in AWS Glue. They allow you to connect various components, define dependencies, and monitor the execution flow, making it easier to manage complex ETL processes:

   I.   Create a Glue workflow to orchestrate the crawler and job.

   II.   Add the crawler and job as nodes in the workflow.

   III. Define the dependencies between the nodes (for example, the job should run after the crawler has finished).

7. **Create an AWS Glue trigger**: Triggers act like "event listeners" for your Glue workflows. They initiate your data processing tasks based on schedules, data arrival events, or completion of other jobs, making your data pipelines dynamic and responsive:

    I.   Configure the trigger to start the Glue workflow based on a schedule or an event (for example, new files arriving in the S3 bucket).

Now let's go through the detailed explanation of each step:

1. **Prepare CSV data in S3**:

    I.   Ensure you have CSV data stored in an S3 bucket. For this example, let's assume your CSV files are stored in s3://your-folder-name/data/.

    II. Download the sample_data.csv file from GitHub (https://github.com/PacktPublishing/Data-Engineering-with-AWS-Cookbook/blob/main/Chapter04/Recipe1/glue-workflow/sample.csv).

    III. Upload it to the S3 bucket (s3://your-folder-name/data/). It should present the following content:

```
id,name,age,city,state,country,zip,email,phone,registered_date
1,John Doe,30,New York,NY,USA,10001,john.doe@example.com,212-555-1234,2020-01-01
2,Jane Smith,25,San Francisco,CA,USA,94105,jane.smith@example.com,415-555-7890,2020-02-01
3,Bob Johnson,40,Chicago,IL,USA,60601,bob.johnson@example.com,312-555-5678,2020-03-01
4,Alice Williams,35,Seattle,WA,USA,98101,alice.williams@example.com,206-555-4321,2020-04-01
5,Tom Brown,28,Boston,MA,USA,02101,tom.brown@example.com,617-555-8765,2020-05-01
6,Emily Davis,32,Denver,CO,USA,80201,emily.davis@example.com,303-555-3456,2020-06-01
7,Michael Miller,45,Austin,TX,USA,73301,michael.miller@example.com,512-555-6543,2020-07-01
8,Sarah Wilson,29,Portland,OR,USA,97201,sarah.wilson@example.com,503-555-3210,2020-08-01
9,David Lee,38,Miami,FL,USA,33101,david.lee@example.com,305-555-7894,2020-09-01
10,Laura Kim,27,Atlanta,GA,USA,30301,laura.kim@example.com,404-555-6789,2020-10-01
11,Peter Jackson,34,Dallas,TX,USA,75201,peter.jackson@example.com,214-555-1234,2020-11-01
12,Olivia Harris,31,Houston,TX,USA,77001,olivia.harris@example.com,713-555-4567,2020-12-01
13,James Clark,39,Las Vegas,NV,USA,88901,james.clark@example.com,702-555-7890,2021-01-01
14,Emma Martinez,26,San Diego,CA,USA,92101,emma.martinez@example.com,619-555-2345,2021-02-01
15,William Roberts,41,Phoenix,AZ,USA,85001,william.roberts@example.com,602-555-6789,2021-03-01
16,Isabella Lewis,30,Philadelphia,PA,USA,19101,isabella.lewis@example.com,215-555-3456,2021-04-01
17,Benjamin Walker,35,Charlotte,NC,USA,28201,benjamin.walker@example.com,704-555-1234,2021-05-01
18,Sophia Hall,28,Columbus,OH,USA,43001,sophia.hall@example.com,614-555-5678,2021-06-01
19,Alexander Young,33,Indianapolis,IN,USA,46201,alexander.young@example.com,317-555-7890,2021-07-01
20,Mia King,29,San Antonio,TX,USA,78201,mia.king@example.com,210-555-4321,2021-08-01
21,Daniel Allen,37,Nashville,TN,USA,37201,daniel.allen@example.com,615-555-8765,2021-09-01
22,Charlotte Scott,32,Jacksonville,FL,USA,32099,charlotte.scott@example.com,904-555-3456,2021-10-01
23,Matthew Green,45,Fort Worth,TX,USA,76101,matthew.green@example.com,817-555-6543,2021-11-01
24,Amelia Baker,31,Memphis,TN,USA,37501,amelia.baker@example.com,901-555-3210,2021-12-01
25,Lucas Nelson,38,El Paso,TX,USA,79901,lucas.nelson@example.com,915-555-7894,2022-01-01
26,Evelyn Carter,26,Detroit,MI,USA,48201,evelyn.carter@example.com,313-555-6789,2022-02-01
27,Henry Mitchell,34,Washington,DC,USA,20001,henry.mitchell@example.com,202-555-1234,2022-03-01
28,Abigail Perez,29,Louisville,KY,USA,40201,abigail.perez@example.com,502-555-4567,2022-04-01
29,Mason White,40,Baltimore,MD,USA,21201,mason.white@example.com,410-555-7890,2022-05-01
30,Sofia Thompson,27,Milwaukee,WI,USA,53201,sofia.thompson@example.com,414-555-2345,2022-06-01
31,Logan Adams,36,Albuquerque,NM,USA,87101,logan.adams@example.com,505-555-6789,2022-07-01
32,Charlotte Collins,33,Fresno,CA,USA,93650,charlotte.collins@example.com,559-555-3456,2022-08-01
33,James Alexander,39,Tucson,AZ,USA,85701,james.alexander@example.com,520-555-7890,2022-09-01
34,Victoria Evans,31,New Orleans,LA,USA,70112,victoria.evans@example.com,504-555-1234,2022-10-01
35,Elijah Turner,45,Cleveland,OH,USA,44101,elijah.turner@example.com,216-555-5678,2022-11-01
36,Madison Roberts,27,Sacramento,CA,USA,94203,madison.roberts@example.com,916-555-7890,2022-12-01
37,Aiden Bennett,38,Kansas City,MO,USA,64101,aiden.bennett@example.com,816-555-4321,2023-01-01
38,Harper Murphy,26,Virginia Beach,VA,USA,23450,harper.murphy@example.com,757-555-8765,2023-02-01
39,David Richardson,34,Atlanta,GA,USA,30301,david.richardson@example.com,404-555-3456,2023-03-01
40,Isabella Bell,30,Colorado Springs,CO,USA,80901,isabella.bell@example.com,719-555-6543,2023-04-01
```

Figure 4.1 – The content of the sample_data.csv file

2. **Create an IAM role for AWS Glue:**

   I.   Choose an IAM role or create one that allows AWS Glue to access your S3 data. To create an IAM role and policy in AWS, first, navigate to the IAM service in the AWS Management Console and open the IAM console at `https://console.aws.amazon.com/iam/`.

   II.   Create the policy with the following steps:

     i.   In the navigation pane, click on **Policies**.

     ii.   Click on the **Create policy** button.

     iii. Select the **JSON** tab and paste the policy document (available at `https://github.com/PacktPublishing/Data-Engineering-with-AWS-Cookbook/blob/main/Chapter04/Recipe1/glue-workflow/glue-policy.json`).

     iv.   Then, create the `AWSGlueServiceRole` role. Select **Role | Glue** and then select the policy name you created in the previous steps.

> **Note**
>
> Don't forget to update your S3 bucket name and create the same folder structure under your S3 bucket.

In this policy, we have given the following permissions:

- `GlueAccess:`

  - Allows most common Glue actions. You can further refine this list based on your specific needs.

  - The policy allows these actions on all Glue resources (`"Resource": "*"`).

- `S3Access:`

  - Grants permissions to get, put, and delete objects, list the bucket, and get the location for the specified bucket (`my-data-bucket`).

  - Replace `my-data-bucket` with your actual bucket name.

  - You can extend the `"Resource"` list to include other buckets or paths within buckets as necessary.

3. **Create an AWS Glue database**:

   I.   Open the AWS Glue console at `https://console.aws.amazon.com/glue/`.

   II.   Create a database with the help of the following steps:

      i.   Navigate to the **Databases** section.

      ii.   Click on **Add database**.

      iii. Name your database `glue_workshop`.

4. **Create an AWS Glue crawler**:

   I.   Open the AWS Glue console at `https://console.aws.amazon.com/glue/`.

   II.   Create a Glue crawler with the help of the following steps:

      i.   Navigate to the **Crawlers** section.

      ii.   Click on **Add crawler**.

      iii. Name your crawler `csvCrawler`.

      iv.   Set the data store to **S3** and specify the S3 path (`s3://your-folder/data/`).

      v.   Configure the IAM role to use `AWSGlueServiceRole`.

      vi.   Set the frequency to **Run on demand**.

      vii. Create or select a database (created in *step 3*) where the crawler results will be stored, for example, `aws_workshop`.

   Review and create the crawler.

> **Note**
>
> Please ensure that the IAM role used by the AWS Glue crawler is the same as the one that has the necessary permissions for creating tables in the database. Use an already created role with these permissions.

5. **Run the crawler**:

I.   Select the crawler, `csvCrawler`, and click on **Run crawler**.

II.   Wait for the crawler to complete. It will create a table in the specified database.

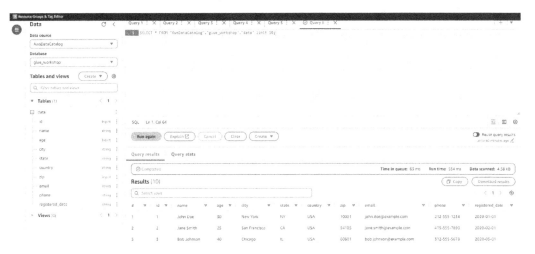

Figure 4.2 – Crawler output

6. **Create an AWS Glue job**:

I.   Open the AWS Glue console at `https://console.aws.amazon.com/glue/`.

II.   Create a Glue job as follows:

i.   Download the `csv_to_parquet.py` file from your GitHub for upload. You can find scripts on following GitHub path `https://github.com/PacktPublishing/Data-Engineering-with-AWS-Cookbook/blob/main/Chapter04/Recipe1/glue-workflow/csv_to_parquet.py`.

ii.   Go to the **Jobs** section in AWS Glue.

iii. Click on **Author code with a script editor** and then on **Script editor**.

iv.   Then select **Spark**, choose **Upload scripts**, and add the `csv_to_parquet.py` script you downloaded from GitHub.

v.   Click on **Create script** to complete the setup.

```
import sys
from awsglue.transforms import *
from awsglue.utils import getResolvedOptions
from pyspark.context import SparkContext
from awsglue.context import GlueContext
from awsglue.job import Job

args = getResolvedOptions(sys.argv, ['JOB_NAME'])
sc = SparkContext()
glueContext = GlueContext(sc)
spark = glueContext.spark_session
job = Job(glueContext)
job.init(args['JOB_NAME'], args)

# Read the data from the Glue catalog
database = "glue-workshop"
table_name = "tbl_data"

# Create a DynamicFrame from the table
dynamic_frame = glueContext.create_dynamic_frame.from_catalog(
    database=database,
    table_name=table_name
)

# Convert DynamicFrame to DataFrame
data_frame = dynamic_frame.toDF()

# Write DataFrame to S3 as partitioned Parquet files
output_s3_path = "s3://aws-data-eng-recipes-bucket/parquet-data/"
data_frame.write.mode("overwrite").partitionBy( "country", "state","city").parquet(output_s3_path)

# Commit job
job.commit()
```

Figure 4.3 – Glue ETL job

vi.   Fill in the following job properties:

- **Name**: CsvToParquetJob (double-click, update the Glue Job name, and save)

- Now go to **Job detail** tab then click **IAM Role: AWSGlueServiceRole** (select from dropdown)

- **Type: Spark** (select from dropdown)

- **Script file name**: s3://aws-glue-assets-accountNo-us-east-1/scripts/

- **Spark UI logs path**: s3://aws-glue-assets-accountNo-us-east-1/sparkHistoryLogs/

- **Temporary directory**: s3://your-folder/temp/

vii. After starting the job, monitor its progress in the AWS Glue console. Upon completion, your Parquet data should be available in the specified S3 location, partitioned according to the keys you specified.

Note

1. In the previous script `csv_to_parquet.py` code don't forget to update following:

- `database` with the Glue database name you created as `glue_workshop`.

- `table_name` with the table your crawler created (since we are using `sample.csv` for crawling data, your table name will be `sample`).

- `output_s3_path` with your S3 bucket and folder locations. The script path is created by the job by default if you want to change and add your S3 bucket and path update it.

2. Always monitor AWS costs associated with running AWS Glue crawlers and jobs and storing data in S3. For large datasets, converting data from CSV to Parquet can be cost-effective, as Parquet is a columnar storage format optimized for analytics. Ensure you've set up appropriate data retention, backup, and lifecycle policies for your S3 buckets.

---

7. **Configure the Glue job**:

I. To set the Glue job script path, click on **Job details**, and in the dropdown, choose your S3 path: `s3://your-folder_name/scripts/csv_to_parquet.py`.

II. Enable job bookmarking if you want to keep track of processed data.

III. Choose an appropriate worker type and number of workers based on your data size and complexity.

Note

Standard workers are general-purpose workers suitable for a wide range of ETL tasks; we will use the same for our recipe.

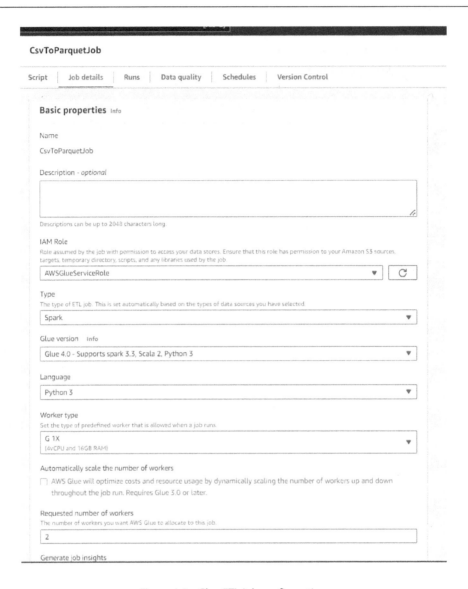

Figure 4.4 – Glue ETL job configuration

IV.   In **Advanced property**, select **Job parameters** and add the following values:

- `--enable-metrics true`

- `--enable-job-insights =true`

- `--enable-observability-metrics =true`

8. **Create and configure the Glue workflow**:

   I. **Workflow creation**:

      i.   Navigate to the AWS Glue console.

      ii. Click on the **Workflows** menu on the left and then click on the **Add workflow** button.

      iii. On the next screen, type in glueworkflow as the workflow name and click on the **Add workflow** button. The workflow is created.

   II. **Trigger configuration**:

      i.   Select the workflow and click on the **Add trigger** link.

      ii. On the **Add trigger** popup, select the **Add new** tab. Type in startcrawler as the trigger name and select **On demand** for the trigger type. (You select the **On demand** trigger type because you will start the workflow manually in this workshop.) Click on the **Add** button. The trigger is added to the workflow.

   III. **Node addition**:

      i.   Click on the **Add node** link to configure what you want to run after the trigger.

      ii. On the pop-up screen, select the **Crawlers** tab. Select **csvCrawler** (created in *step 4 Create an AWS Glue crawler*) and click on the **Add** button.

      iii. The crawler node is added as the next step to the trigger. Next, select the **Add trigger** option under the **Action** menu to add another trigger.

      iv. On the pop-up screen, select the **Add new** tab. Type in startjob as the name. Select **Event** for the trigger type. Select **Start after ANY watched event** for the trigger logic. Finally, click on the **Add** button. The trigger is added.

   IV. **Trigger and job configuration**:

      i.   Select the **Start job** trigger and select **Add jobs crawlers** to the **Watch** option under the **Action** menu.

      ii. On the pop-up screen, select the **Crawlers** tab. Select **SUCCEEDED** for the **Crawler event to watch** field. Finally, click on the **Add** button.

      iii. The startjob trigger is now configured to run when the crawler finishes execution successfully. Click on the **Add node** icon next to **startjob** to configure what job or crawler the statjob trigger will invoke.

      iv. On the pop-up screen, select the **Jobs** tab. Select **csvtoparquetjob** and click on the **Add** button. The workflow is now configured end to end. It will first run the crawler and then the job.

V.   **Workflow execution**:

i.   Select **glueworkflow** and click on the **Run** option under the **Action** menu.

ii. The workflow execution will start with the status of **Running**. Wait till the status changes to **Completed**. You can see the crawler run in the workflow has added the table customers under the `glue-workshop` database.

VI. **Verification**:

i.   Confirm that the crawler has successfully created the table definition for the data in the `glue-workshop` database.

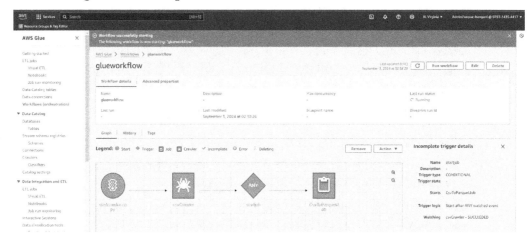

Figure 4.5 – Glue workflow UI

That's a high-level overview of creating a simple AWS Glue workflow. Adjust the steps based on the specifics of your use case, such as the complexity of the transformations needed or additional sources and sinks.

Once the Glue workflow is finished, you can see the job's status is shown as successful:

Figure 4.6 – Glue workflow status

You can also check the results in the S3 bucket; go to the S3 bucket and explore the data:

Figure 4.7 – Results of the ETL job in the S3 bucket

When you open the folder, you will see the Parquet file:

Figure 4.8 – Parquet file in the S3 bucket

See also

- *Overview of workflows in AWS Glue*: https://docs.aws.amazon.com/glue/latest/dg/workflows_overview.html

- *AWS Glue Immersion day – Introduction*: https://catalog.us-east-1.prod.workshops.aws/workshops/ee59d21b-4cb8-4b3d-a629-24537cf37bb5/en-US/intro

Setting up event-driven orchestration with Amazon EventBridge

Creating a workflow that reacts to specific events to trigger AWS Glue components on-demand typically requires the combination of AWS Glue with other AWS services, such as AWS Lambda and Amazon EventBridge (formerly known as Amazon CloudWatch Events).

Now, we want our workflow to react to the addition of a new file in an S3 bucket to trigger an AWS Glue job.

Let's understand when this is required. A media company processes daily uploads of large CSV files to an S3 bucket. Each time a new CSV file is uploaded, the company needs to trigger an ETL pipeline that extracts data from the CSV, processes it, and stores the information in a data lake for analysis. This workflow is fully automated and event-driven, ensuring immediate processing upon file arrival.

Getting ready

Before proceeding with this recipe, ensure you have completed the steps in the *Defining a simple workflow using AWS Glue workflows* recipe earlier in this chapter.

How to do it...

1. **Set up an S3 bucket event notification**: Ensure you have an S3 bucket where files will be uploaded. Please use the same bucket we created in the *Defining a simple workflow using AWS Glue workflows* recipe.

2. **Create an IAM role and policy for Glue**: Please use the same policy and role we created in the *Defining a simple workflow using AWS Glue workflows* recipe.

3. **Create an AWS Glue database and crawler**:

    I.   **Create Glue database**:

        i.   Go to the AWS Glue console at `https://console.aws.amazon.com/glue/`.

        ii.   Navigate to **Databases** and click on **Add database**.

        iii. Name your database `csv_database`.

        iv.   Click on **Create**.

> **Note**
>
> Please use the same database we created in the *Defining a simple workflow using AWS Glue workflows* recipe.

II.   **Create Glue crawler**:

    i.   Go to the **Crawlers** section in the Glue console.

    ii.   Click on **Add crawler**.

    iii. Name your crawler `csvCrawler`.

    iv.   For the data store, choose **S3** and specify the S3 path (`s3://your-bucket/csv-data/`).

    v.   Configure the IAM role to use `glue-sample-role`.

    vi.   Set the frequency to **Run on demand**.

    vii. Create or select the `csv_database` database where the crawler results will be stored.

    viii. Review and create the crawler.

    ix.   Do not run the crawler yet.

> **Note**
>
> Please use the same crawler we created in the *Defining a simple workflow using AWS Glue workflows* recipe.

4. **Create AWS Glue job**:

    I.   **Prepare the Glue ETL script**: Save the script presented in the *Defining a simple workflow using AWS Glue workflows* recipe as `csv_to_parquet.py` and upload it to your S3 bucket.

    II.   **Create the Glue job**:

        i.   Go to the **Jobs** section in the Glue console.

        ii.   Click on **Add Job**.

        iii. Name the job `CsvToParquetJob`.

        iv.   Choose the **glue-sample-role** IAM role.

        v.   For the ETL language, select **Python**.

        vi.   For the script, choose **A new script to be authored by you**.

        vii. Set the script filename to `s3://your-bucket/scripts/csv_to_parquet.py`.

        viii. Set the temporary directory to `s3://your-bucket/temp/`.

        ix.   Click on **Next** and configure the job properties as needed.

        x.   Click on **Save**.

> **Note**
>
> Please use the same job we created in the *Defining a simple workflow using AWS Glue workflows* recipe.

5. **Create and configure the Glue workflow:**

    I.   **Create workflow:**

        i.   Go to the **Workflows** section in the Glue console.

        ii. Click on **Add workflow**.

        iii. Name your workflow glueWorkflow.

        iv. Click on **Create**.

    II. **Add crawler to the workflow:**

        i.   In the Glue console, select your glueWorkflow workflow.

        ii. Click on **Add trigger** and choose **Add new**.

        iii. Name the trigger csvCrawlerTrigger.

        iv. Set the type to **On-demand**.

        v.   In the **Actions** section, add an action to start the csvCrawler crawler.

        vi. Save the trigger.

    III. **Add job to the workflow:**

        i.   In the Glue console, select your glueWorkflow workflow.

        ii. Click on **Add trigger** and choose **Add new**.

        iii. Name the trigger csvToParquetJobTrigger.

        iv. Set the type to **Event-based**.

        v.   Under **Conditions**, add a condition:

            • Logical operator: **EQUALS**

            • State: **SUCCEEDED**

            • Crawler name: **csvCrawler**

        vi. In the **Actions** section, add an action to start the job, CsvToParquetJob.

        vii. Save the trigger.

> **Note**
>
> Please use the same workflow we created in the *Defining a simple workflow using AWS Glue workflows* recipe.

6. **Create an EventBridge rule to trigger the workflow:**

    I. **Create EventBridge rule:**

        i.   Go to the Amazon EventBridge console at `https://console.aws.amazon.com/events/`.

        ii. Click on **Create rule**.

        iii. Name your rule `S3FileUploadRule]`.

        iv. Select **Rule with an event pattern** and click on the **Next** button.

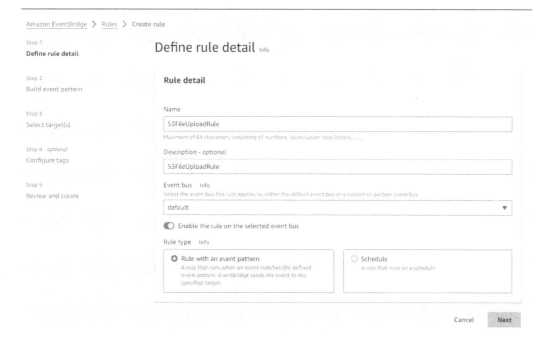

Figure 4.9 – EventBridge rule

v.   Choose **Event source** on next page and select **AWS events or EventBridge partner events**.

Build event pattern Info

Event source

Event source
Select the event source from which events are sent.

◉ AWS events or EventBridge partner events
Events sent from AWS services or EventBridge partners.

○ Other
Custom events or events sent from more than one source, e.g. events from AWS services and partners.

○ All events
All events sent to your account.

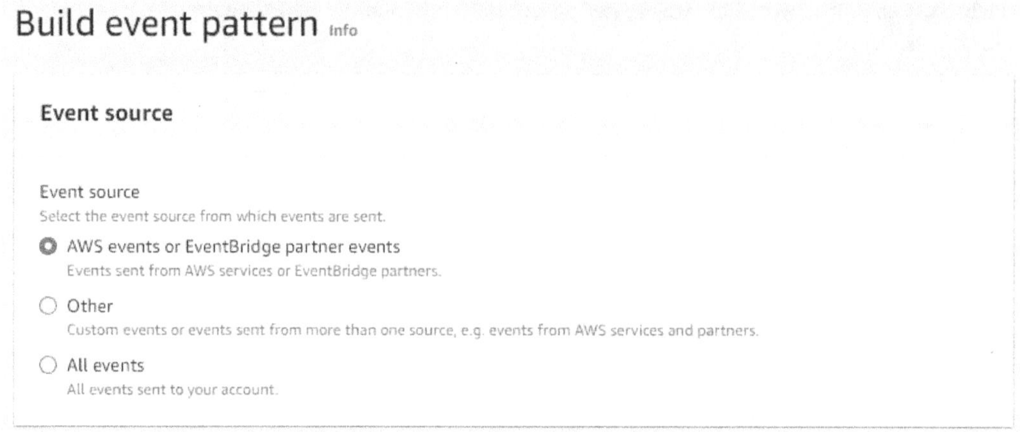

Figure 4.10 – EventBridge rule event source

vi.   Now choose **Custom pattern JSON editor**:

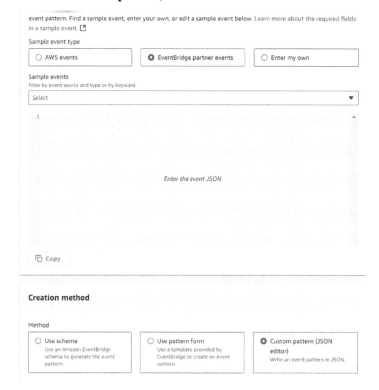

Figure 4.11 – EventBridge patterns

vii. Then, on the next page, set **Creation method** as **Custom pattern (JSON editor)** and add the following JSON in the input box:

```
{   "source": ["aws.s3"],
    "detail-type": ["Object Created"],
    "detail": {
      "bucket": {
        "name": ["your-bucket"]
      },    "object": {
        "key": [{"prefix": "data/"}]
      }  }}
```

> **Note**
> Replace your-bucket with your bucket name.

viii. Click on **Save**.

ix. In the **Targets** section, click on **Add target** and select **AWS Lambda function**.

x. Choose the Lambda function you created to start the Glue workflow (**TriggerGlueWorkflow**).

xi. Click on **Create**.

7. **Create AWS Lambda function to start Glue workflow**:

I. **Create Lambda function**:

i. Go to the AWS Lambda console at https://console.aws.amazon.com/lambda/.

ii. Click on **Create function**.

iii. Choose **Author from scratch**.

iv. Enter a function name, for example, TriggerGlueWorkflow-lambda.

v. Choose the **Python 3.x** runtime.

vi. Choose or create an execution role and ensure it uses `lambda-glue-trigger-role` (choose **Use an existing role in lambda**). You need to go to IAM, open `lambda-glue-trigger-role`, and add **AWSGlueConsoleFullAccess**.

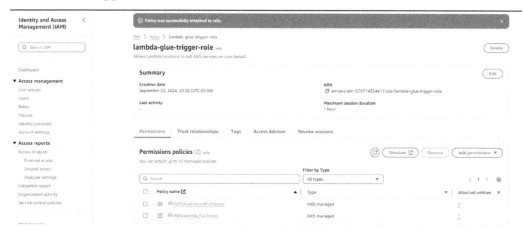

Figure 4.12 – Lambda Glue trigger role

vii. Click on **Create function**.

II.   **Add Lambda function code**: Replace the default Lambda function code with the following:

```python
import json
import boto3

def lambda_handler(event, context):
    glue = boto3.client('glue')
    # Start the Glue workflow
    workflow_name = 'glueworkflow'
    response = glue.start_workflow_run(Name=workflow_name)

    return {
        'statusCode': 200,
        'body': json.dumps('Workflow started: {}'.
format(response['RunId']))
    }
```

III. **Add S3 trigger to Lambda function**:

i.   In the Lambda function, go to the **Configuration** tab.

ii. Click on **Add trigger**.

iii.   Select **S3**.

iv.   Choose the bucket name (`your-bucket`).

v.   Set the event type to **All object create events**.

vi.   Set the prefix to `csv-data/` (or whatever path you want to monitor).

vii. Click on **Add**.

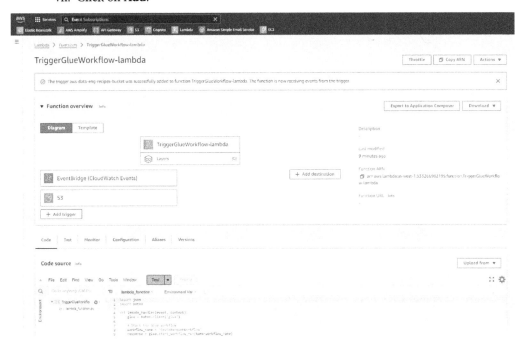

Figure 4.13 – Lambda function

Upon uploading the file to the S3 bucket, a Lambda function will be triggered to initiate a Glue workflow using Boto3.

Creating a data workflow using AWS Step Functions

AWS Step Functions is a serverless workflow orchestration service that enables you to connect and manage various AWS services within a streamlined workflow. This allows for the rapid development and updating of applications.

Step Functions employs a state machine approach, where each step in the workflow is represented by a state. These states are defined using the Amazon States Language, a JSON-based format. By chaining together different states, you can create complex workflows that integrate services such as AWS Lambda, AWS Fargate, and Amazon SageMaker to build powerful applications.

In this recipe, we'll create a simple AWS Step Functions workflow that uses two AWS Lambda functions: one that gets data and one that processes data.

How to do it...

1. **Create two Lambda functions**:

    I.   Go to the AWS Management Console.

    II.   Navigate to **Lambda** by searching for Lambda in the services search bar.

    III. Create a new Lambda function

    IV.   In the Lambda Console, click on **Create function**.

    V.   Select **Author from scratch**.

    VI.   Enter the function name (for example, GetDataFunction).

    VII. For **Runtime**, select **Python 3.x** from the dropdown menu.

    VIII. Under **Permissions**, choose or create a role with necessary Lambda execution permissions (for example, basic permissions to write logs to CloudWatch).

    IX.   In the **Function code** section, you can edit the default Python code and replace it with following code same step repeat for ProcessDataFunction.

    X.   Once you have written the code and configured the settings, click on **Deploy** to save and activate the function.

    The first Lambda function is GetDataFunction. Use the Python 3.x runtime for this:

    ```python
    def lambda_handler(event, context):

        # Sample data fetching logic
        data = {
            "message": "Hello from GetDataFunction"
        }    return data
    ```

    The second Lambda function is ProcessDataFunction, again using the Python 3.x runtime:

    ```python
    def lambda_handler(event, context):
        # Sample data processing logic
        processed_data = event['message'].upper()
        return {
            "processedMessage": processed_data
        }
    ```

2. **Define a state machine in Amazon States Language:**

Now, create a state machine with the help of the following JSON that first invokes `GetDataFunction` and then `ProcessDataFunction`:

```
{     "Comment": "A simple AWS Step Functions state machine that
invokes two Lambda functions.",
    "StartAt": "GetDataState",
    "States": {
        "GetDataState": {
            "Type": "Task",
            "Resource": "arn:aws:lambda:REGION:ACCOUNT_
ID:function:GetDataFunction",
            "Next": "ProcessDataState"
        },          "ProcessDataState": {
            "Type": "Task",
            "Resource": "arn:aws:lambda:REGION:ACCOUNT_
ID:function:ProcessDataFunction",
            "End": true
        }     }}
```

> **Note**
>
> Replace `REGION` and `ACCOUNT_ID` with your AWS Region and account ID, respectively.

3. **Create the Step Functions state machine:**

I.   Download `stepfunction.json` from GitHub at `https://github.com/PacktPublishing/Data-Engineering-with-AWS-Cookbook/tree/main/Chapter04/Recipe3/stepfunction`.

II.   Open the AWS Management Console and navigate to step function or navigate to `https://us-east-1.console.aws.amazon.com/states`.

III. Click on the **Create state machine** button and select **Blank**.

IV.   Click on the **Action** dropdown and select **Import definition**. Select your file and upload it. Your step function will be ready after upload.

V.   The other option is to navigate to the AWS Step Functions dashboard and create a new state machine. Paste the Amazon States Language JSON from the previous step into the state machine definition. Ensure your AWS Step Functions role has permission to invoke Lambda functions.

You can view the JSON data and its corresponding visual representation in the Workflow studio.

Figure 4.14 – Step function with JSON code

4. **Execute the state machine**: Once the state machine is created, you can start executing the step function (click on **Start execution**, enter { }, and then click on the **Start execution** button).

Figure 4.15 – Start step function execution

This will run `GetDataFunction`, take its output, and pass it as an input to `ProcessDataFunction`.

AWS Step Functions offers a clear visual representation of your workflow, making it easier to follow the execution process. As the state machine runs, each step is highlighted in real time, allowing you to track progress visually. Additionally, for each step in the workflow, you can view detailed input and output data, helping you understand how the information flows and changes throughout the execution.

5. **Workflow Studio in AWS Step Functions**: Workflow Studio is a visual tool that allows users to design workflows by dragging and dropping AWS services onto a canvas. This graphical interface provides an easy way for those who might not be familiar with **Amazon States Language (ASL)** or prefer visual design overwriting code.

Here's a basic overview of how to use Workflow Studio:

I.   **Navigate to AWS Step Functions**:

   i.   Open the AWS Management Console.

   ii.   Navigate to the Step Functions service.

II.   **Create a new state machine**:

   i.   Click on **Create state machine**, select **Blank**, and click on **Create**.

   ii.   Choose **Design with Workflow Studio**.

III. **Design your workflow**:

   i.   You'll be provided with a blank canvas. On the right side of the interface, there's a palette of AWS services and flow control elements.

   ii.   Simply drag and drop services or elements onto the canvas.

   iii. For each service or element, configure its parameters using the properties panel on the right side.

IV.   **Connect the elements**:

   i.   Click on the output connector (a small circle) of an element and drag it to the input connector of another element to establish a transition between them.

   ii.   You can also add error catchers, retries, and other flow control mechanisms.

V.   **Deploy the workflow**:

   i.   Once you've designed the workflow visually, click on **Next**. Workflow Studio will automatically generate the ASL definition for your workflow.

   ii.   Provide a name for the state machine.

   iii. Define or select an IAM role for the state machine to use.

   iv.   Click on **Create state machine**.

VI.   **Execute and monitor the workflow**: After creating your workflow, you can initiate its execution to test its functionality. AWS Step Functions offers a visual representation of the workflow, allowing you to observe the progress as each step is executed. While we've previously shown how to create a Step Function using JSON, you can also utilize a more intuitive drag-and-drop interface for a simplified workflow creation process.

See also

- *AWS Step Functions*: https://aws.amazon.com/step-functions/
- *Using Amazon States Language to define Step Functions workflows*: https://docs.aws.amazon.com/step-functions/latest/dg/concepts-amazon-states-language.html
- *Creating a workflow with Workflow Studio in Step Functions*: https://docs.aws.amazon.com/step-functions/latest/dg/workflow-studio-use.html
- *New – AWS Step Functions Workflow Studio – A Low-Code Visual Tool for Building State Machines*: https://aws.amazon.com/blogs/aws/new-aws-step-functions-workflow-studio-a-low-code-visual-tool-for-building-state-machines/

Managing data pipelines with MWAA

MWAA is a fully managed service provided by AWS that simplifies the deployment and operation of Apache Airflow, an open source workflow automation platform. Apache Airflow is widely used for orchestrating complex data workflows, scheduling batch jobs, and managing data pipelines. MWAA takes the power of Apache Airflow and makes it easier to use, maintain, and scale in the AWS cloud environment.

MWAA is commonly used to orchestrate complex data pipelines. Users can define and schedule tasks to transform, process, and move data between various AWS services, databases, and external systems. Another good use case is that MWAA simplifies the management of ETL workflows. Users can easily schedule and automate data extraction, transformation, and loading tasks, ensuring data accuracy and consistency. MWAA also supports batch processing and batch jobs, such as data aggregation, report generation, and data synchronization, which can be efficiently managed and scheduled using MWAA.

How to do it...

1.   **Setting up your MWAA environment**: Before you can start orchestrating data pipelines with MWAA, you need to set up your environment:

   I.   **Create IAM roles and policies**:

      i.   Go to the IAM console at https://console.aws.amazon.com/iam/.

ii.   Navigate to **Policies** and click on **Create policy**.

iii.   Select the **JSON** tab and paste the following policy document:

```
{     "Version": "2012-10-17",
      "Statement": [
          {             "Effect": "Allow",
              "Action": [
                  "glue:*",
                  "s3:GetObject",
                  "s3:PutObject",
                  "s3:ListBucket",
                  "logs:CreateLogGroup",
                  "logs:CreateLogStream",
                  "logs:PutLogEvents",
                  "cloudwatch:PutMetricData"
              ],            "Resource": [
                  "arn:aws:s3::your-mwaa-bucket",
                  "arn:aws:s3::your-mwaa-bucket/*",
                  "arn:aws:logs:*:*:*",
                  "arn:aws:glue:*:*:catalog",
                  "arn:aws:glue:*:*:database/*",
                  "arn:aws:glue:*:*:table/*",
                  "arn:aws:glue:*:*:connection/*",
                  "arn:aws:glue:*:*:job/*",
                  "arn:aws:glue:*:*:crawler/*",
                  "arn:aws:glue:*:*:workflow/*",
                  "arn:aws:glue:*:*:trigger/*",
                  "arn:aws:glue:*:*:classifier/*",
                  "arn:aws:cloudwatch:*:*:metric/*"
              ]            }        ]}
```

> **Note**
> Update your S3 bucket.

2.   **Create an IAM role for MWAA**:

I.   Go to the IAM console, navigate to **Roles**, and click on **Create role**.

II.   Select **AWS service** and then choose **MWAA**.

III.   Click on **Next: Permissions**.

IV.   Attach the policy you just created.

V.   Click on **Next: Tags**, optionally add tags, then click on **Next: Review**.

VI.   Name your role `mwaa-service-role`.

VII. Click on **Create role**.

3. **Create MWAA environment**:

I.   Go to the MWAA console at `https://us-east-1.console.aws.amazon.com/mwaa/home?region=us-east-1#home?landingPageCheck=1`.

II.   Click on **Create environment**.

III. Enter the environment details, such as name and description.

IV.   For **DAG folder** S3 Path, enter `s3://your-mwaa-bucket/dag`.

V.   For **Plugins file** S3 Path, enter `s3://your-mwaa-bucket/plugins` (optional). You can find details about plugins on AWS Documents (`https://docs.aws.amazon.com/mwaa/latest/userguide/configuring-dag-import-plugins.html`).

Figure 4.16 – DAG configuration

VI.   For **Requirements file** S3 Path, enter `s3://your-mwaa-bucket/requirements.txt` (optional). You can find details about `reuirments.txt` in AWS documents (`https://docs.aws.amazon.com/mwaa/latest/userguide/best-practices-dependencies.html`).

VII. Select the `mwaa-service-role` you created earlier in the *Create IAM roles and policies* section at the beginning of *step 1*.

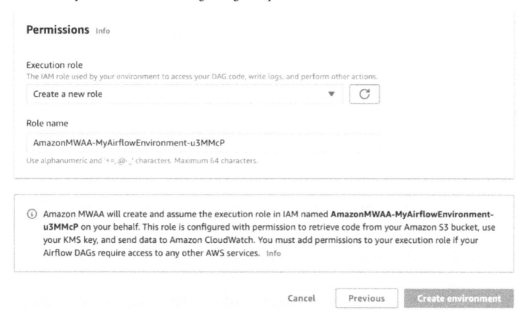

Figure 4.17 – Add role in configuration

VIII. Configure networking settings (VPC, subnets, and security groups).

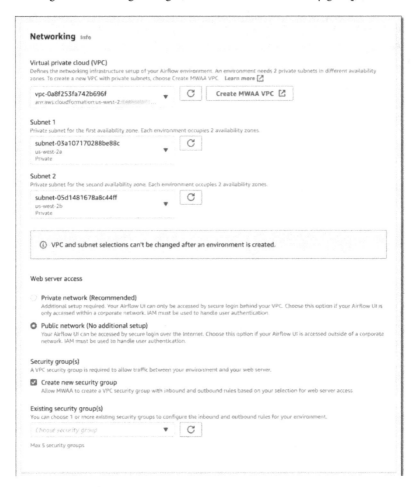

Figure 4.18 – Networking configuration

IX. Select **Create new VPC**. Wait for a few minutes. Once the VPC is created, select **New**; you will see that a subnet will be automatically assigned. Now, choose **VPC security group** and **Public network** (internet access).

> **Note**
> Before you start MWAA, create an S3 bucket that will store your **Directed Acyclic Graphs** (**DAGs**). This bucket will also store MWAA logs and metrics.

X.   **Logging configuration**: Specify the S3 prefix where you want logs to be stored in your S3 bucket and enable the checkbox in the **Amazon Managed Workflows for Apache Airflow (MWAA)** UI.

XI.   **Airflow configuration options**: If needed, override the default `airflow.cfg` settings in the CFG file.

XII.   **Permissions**: Add permissions using the following two steps:

i.   Attach an execution role that allows MWAA to access necessary resources (download the policy from GitHub and attach it to your role).

ii.   Attach a task role if your tasks need to access other AWS services.

iii.   Click on **Create environment**. This can take a few minutes.

Figure 4.19 – MWAA environment created

XIII.   **Access the Airflow UI**: Once the environment is created, you can access the Airflow UI through the link provided in the MWAA console.

Figure 4.20 – MWAA UI

XIV.   Once your UI is visible, download `dag-sample.py` from GitHub (`https://github.com/PacktPublishing/Data-Engineering-with-AWS-Cookbook/blob/main/Chapter04/Recipe4/mwaa/dag-sample.py`) and upload it into the `dags` folder in your S3 bucket. Your uploaded DAGs will be visible in a few minutes:

Figure 4.21 – MWAA sample DAG

XV. Clicking on the MWAA DAG will display a graph view. Explore the user interface by interacting with all available options.

Figure 4.22 – MWAA DAG

Now, we will look into MWAA and its integration with AWS Glue:

1. **Create an S3 bucket for MWAA**:

   I. Use the S3 bucket we created in the previous steps to upload your DAGs (please use the steps from *Create MWAA environment*).

      i. Create a folder for dags in your S3 bucket.

      ii. Upload your Airflow DAGs to the dags folder.

      iii. Upload any custom plugins to the plugins folder.

      iv. Upload the requirements.txt file to the root of the bucket.

2. **Create an Airflow DAG to trigger AWS Glue jobs**:

I. **Create a DAG**: Save the following DAG as mwaa-glue-dag.py and upload it to your S3 bucket in the dags folder. Let's discuss the code:

i. **Introduction and initial setup**: We start by importing the required libraries: boto3 for interacting with AWS services, airflow for orchestrating tasks, and time for delays between task checks. Then, we initialize an AWS Glue client using boto3, specifying the AWS Region:

```
import boto3
from airflow import DAG
from airflow.operators.python_operator import PythonOperator
from airflow.utils.dates import days_ago
from botocore.exceptions import ClientError
import time
# Initialize AWS Glue client
glue_client = boto3.client('glue', region_name='us-east-1')
```

ii. **Starting the Glue crawler**: The start_crawler function initiates the Glue crawler by calling glue_client.start_crawler(). It includes error handling using ClientError in case of failures:

```
# Function to start Glue crawler
def start_crawler(crawler_name):
    try:
        response = glue_client.start_crawler(Name=crawler_
name)
        print(f"Started crawler: {crawler_name}")
    except ClientError as e:
        print(f"Error starting crawler: {e}")
        raise as e
```

iii. **Checking Glue crawler status**: This function continuously checks the status of the crawler and waits until it completes. The time.sleep(60) function introduces a delay of 1 minute between status checks to prevent rapid polling:

```
# Function to check the status of Glue crawler
def check_crawler_status(crawler_name):
    while True:
        response = glue_client.get_crawler(Name=crawler_
name)
        crawler_status = response['Crawler']['State']
        if crawler_status == 'READY':
```

```
        print(f"Crawler {crawler_name} has completed.")
        break
    elif crawler_status == 'RUNNING':
        print(f"Crawler {crawler_name} is still
running...")
        time.sleep(60)   # Wait for 1 minute before
checking again
    else:
        raise Exception(f"Crawler {crawler_name}
encountered an error: {crawler_status}")
```

iv. **Starting the Glue ETL job**: The `start_glue_job` function triggers a Glue ETL job using the `start_job_run()` method and returns `JobRunId`, which is used later to monitor the job's progress:

```
# Function to start Glue ETL job
def start_glue_job(job_name):
    try:
        response = glue_client.start_job_run(JobName=job_
name)
        job_run_id = response['JobRunId']
        print(f"Started Glue job: {job_name} with JobRunId:
{job_run_id}")
        return job_run_id
    except ClientError as e:
        print(f"Error starting Glue job: {e}")
        raise
```

v. **Checking Glue ETL job status**: This function monitors the status of the Glue ETL job using `JobRunId`. Similar to the crawler status check, the function loops until the job either succeeds or fails:

```
# Function to check Glue job status
def check_glue_job_status(job_name, job_run_id):
    while True:
        response = glue_client.get_job_run(JobName=job_name,
RunId=job_run_id)
        job_status = response['JobRun']['JobRunState']
        if job_status == 'SUCCEEDED':
            print(f"Glue job {job_name} succeeded.")
            break
        elif job_status in ['STARTING', 'RUNNING']:
            print(f"Glue job {job_name} is still {job_
status}...")
```

```
            time.sleep(60)  # Wait for 1 minute before
checking again
        else:
            raise Exception(f"Glue job {job_name} failed
with status: {job_status}")
```

vi. **Defining the DAG**: The DAG defines the workflow using Airflow, including task dependencies and scheduling. Default arguments such as the start date and retry count are specified for the DAG:

```
# Define the DAG
default_args = {
    'owner': 'airflow',
    'start_date': days_ago(1),
    'retries': 1
}with DAG(
    dag_id='glue_crawler_etl_workflow',
    default_args=default_args,
    description='A simple MWAA DAG to trigger Glue Crawler
and ETL Job',
    schedule_interval=None,  # Manual trigger for this
workshop
    catchup=False
) as dag:
```

vii. **Adding tasks to the DAG**: Each step of the workflow is represented as a task using Airflow's PythonOperator (https://airflow.apache.org/docs/apache-airflow/stable/howto/operator/python.html). Tasks such as starting the crawler, checking its status, starting the Glue ETL job, and monitoring the job's status are added to the DAG:

```
    # Task 1: Start Glue Crawler
    start_crawler_task = PythonOperator(
        task_id='start_crawler',
        python_callable=start_crawler,
        op_kwargs={'crawler_name': 'csvToCatalogCrawler'}
    )   # Task 2: Check Glue Crawler Status
    check_crawler_task = PythonOperator(
        task_id='check_crawler_status',
        python_callable=check_crawler_status,
        op_kwargs={'crawler_name': 'csvToCatalogCrawler'}
    )   # Task 3: Start Glue ETL Job
    start_glue_job_task = PythonOperator(
        task_id='start_glue_job',
```

```
            python_callable=start_glue_job,
            op_kwargs={'job_name': 'CsvToParquetJob'}
        )    # Task 4: Check Glue ETL Job Status
        check_glue_job_task = PythonOperator(
            task_id='check_glue_job_status',
            python_callable=check_glue_job_status,
            op_kwargs={
                'job_name': 'CsvToParquetJob',
                'job_run_id': '{{ ti.xcom_pull(task_ids="start_
glue_job") }}'
            }    )
```

viii. **Defining task dependencies**: Task dependencies are defined using >>, creating a sequential flow where the tasks are executed one after the other. This ensures that the Glue crawler runs first, followed by the ETL job after the successful completion of the crawler:

```
# Task dependencies: DAG Flow
    start_crawler_task >> check_crawler_task >> start_glue_
job_task >> check_glue_job_task
```

3. **Monitor and handle failures**:

   I.   **Set up CloudWatch alarms**:

   i.   Go to the Amazon CloudWatch console at `https://console.aws.amazon.com/cloudwatch/`.

   ii. Navigate to **Alarms** and click on **Create alarm**.

   iii. Select the Glue metrics to monitor (for example, `JobRunTime`).

   iv. Set the threshold and configure notifications.

   II. **Add retries and failure handling in the DAG**: Update your Airflow DAG to include retries and failure handling

   III. **Update your existing Glue job with CloudWatch metrics and retry**:

   i.   **Importing libraries and setting up the environment**: This section imports the necessary libraries for the AWS Glue job. Key libraries include `awsglue` for working with Glue, `pyspark` for handling Spark, and `boto3` for interacting with AWS services such as CloudWatch:

```
import sys
from awsglue.transforms import *
from awsglue.utils import getResolvedOptions
from pyspark.context import SparkContext
```

```
from awsglue.context import GlueContext
from awsglue.job import Job
import boto3
```

ii. **Initializing the Spark and Glue context**: This code block sets up the Spark and Glue context, initializing the job using parameters passed from the command line. The `GlueContext` object allows access to Glue features within Spark:

```
args = getResolvedOptions(sys.argv, ['JOB_NAME'])
sc = SparkContext()
glueContext = GlueContext(sc)
spark = glueContext.spark_session
job = Job(glueContext)
job.init(args['JOB_NAME'], args)
```

iii. **Reading data from the Glue Data Catalog**: Here, we specify the source database and table in the Glue Data catalog from which the data will be read, and the destination S3 path where the transformed data will be written:

```
# Read the data from the Glue Data catalog
database_name = "glue-workshop"
table_name = "data"
output_s3_path = "s3://mwaa-env-data-bucket-vk/parquet-
data/"
```

iv. **Sending metrics to CloudWatch**: This function sends custom metrics to Amazon CloudWatch. The `put_metric_data` method allows us to track job success or failure by pushing metric values:

```
def put_metric_data(metric_name, value):
    cloudwatch = boto3.client('cloudwatch')
    cloudwatch.put_metric_data(
        Namespace='GlueIngestion',
        MetricData=[
            {                       'MetricName': metric_name,
                'Value': value,
                'Unit': 'Count'
            },          ]       )
```

v. **Creating a DynamicFrame**: This part of the code reads data from the Glue Data Catalog as a DynamicFrame. Glue uses DynamicFrames to handle semi-structured data efficiently:

```
# Create a DynamicFrame from the table
dynamic_frame = glueContext.create_dynamic_frame.from_
catalog(
    database=database_name,
    table_name=table_name
)
```

vi. **Converting to a DataFrame and writing to S3**: The DynamicFrame is converted to a Spark DataFrame, which is then written to Amazon S3 as partitioned Parquet files, using `country`, `state`, and `city` as partition keys:

```
# Convert DynamicFrame to DataFrame
data_frame = dynamic_frame.toDF()


# Write DataFrame to S3 as partitioned Parquet files
data_frame.write.mode("overwrite").partitionBy(
    "country", "state", "city").parquet(output_s3_path)
```

vii. **Adding retry logic for data processing**: A retry mechanism is added to ensure the job can retry up to three times in case of failure. If the job succeeds, it sends a `success` metric to CloudWatch:

```
# Maximum number of retries
max_retries = 3
retries = 0
success = False

while retries < max_retries and not success:
    try:
        # Create a DynamicFrame from the table
        dynamic_frame = glueContext.create_dynamic_frame.
from_catalog(
            database=database_name,
            table_name=table_name
        )        # Convert DynamicFrame to DataFrame
        data_frame = dynamic_frame.toDF()

        # Write DataFrame to S3 as partitioned Parquet files
        data_frame.write.mode("overwrite").partitionBy(
```

```
                      "country", "state", "city").parquet(output_s3_
    path)

            success = True
            put_metric_data('JobSuccess', 1)
```

viii. **Handling errors and sending failure metrics**: If an error occurs, the job will retry after a 30-second wait. If the maximum retries are reached, a `failure` metric is sent to CloudWatch, and an exception is raised:

```
    except Exception as e:
        retries += 1
        if retries < max_retries:
            time.sleep(30)  # Wait for 30 seconds before
    retrying
        else:
            put_metric_data('JobFailure', 1)
            raise e
```

ix. **Committing the job**: The `job.commit()` call finalizes the Glue job, ensuring that all changes are saved and the job is properly completed:

```
    job.commit()
```

On the MWAA UI, your MWAA DAG will be visible in a few minutes as **Auto-refresh** is enabled:

Figure 4.23 – MWAA DAGs

On clicking on **DAG**, you can see Orchestration in the Flow graph:

Figure 4.24 – MWAA DAG

See also

- To learn more about Apache Airflow and MWAA, you can refer to the following links:

    - https://airflow.apache.org/docs/apache-airflow/stable/index.
      html

    - https://docs.aws.amazon.com/mwaa/latest/userguide/what-is-
      mwaa.html

- *What Is Amazon Managed Workflows for Apache Airflow?*: https://docs.aws.amazon.
  com/mwaa/latest/userguide/what-is-mwaa.html

- *Apache Airflow tutorial*: https://airflow.apache.org/docs/apache-airflow/
  stable/tutorial/index.html

- *Installing custom plugins*: https://docs.aws.amazon.com/mwaa/latest/userguide/
  configuring-dag-import-plugins.html

- *Installing Python dependencies*: https://docs.aws.amazon.com/mwaa/latest/
  userguide/working-dags-dependencies.html

Monitoring your pipeline's health

To create a robust monitoring and alerting system for your AWS Glue pipeline, you can use Amazon CloudWatch to emit success metrics, create alarms, and monitor the health of your pipeline. We'll use CloudWatch Metrics, CloudWatch alarms, and Amazon SNS for notifications.

How to do it...

1. **Modify the AWS Glue job to emit success metrics:**

    I.   **Update the Glue ETL script**: Modify your `csv_to_parquet.py` script to include the following code to emit custom metrics:

    ```
    import sys
    import boto3
    # Emit success metric to CloudWatch
    cloudwatch = boto3.client('cloudwatch')
    cloudwatch.put_metric_data(
        Namespace='GluePipelineMetrics',
        MetricData=[
            {                   'MetricName': 'JobSuccess',
                'Dimensions': [
                    {                               'Name': 'JobName',
                        'Value': args['JOB_NAME']
                    },              ],              'Value': 1,
                'Unit': 'Count'
            },      ])
    ```

    II. Upload the updated script to your S3 bucket using the following AWS CLI command:

    ```
    aws s3 cp csv_to_parquet.py s3://your-bucket/scripts/csv_to_
    parquet.py
    ```

2. **Create CloudWatch alarms for metrics:**

    I.   **Create an SNS topic for notifications:**

    i.   Go to the Amazon SNS console at `https://console.aws.amazon.com/sns/`.

    ii. Click on **Topic** and then click on **Create topic**.

    iii. Choose **Standard** and enter a name for your topic, for example, `GluePipelineAlerts`.

    iv. Click on **Create topic**.

    v.   Copy the topic ARN (for example, `arn:aws:sns:region:account-id:GluePipelineAlerts`).

    vi. Click on **Create subscription**.

    vii. Set **Protocol** as **Email** and enter your email address.

    viii. Click on **Create subscription**.

II.   **Create CloudWatch alarms:**

i.   Go to the Amazon CloudWatch console at `https://console.aws.amazon.com/cloudwatch/`.

ii.   Navigate to **Alarms** and click on **Create alarm**.

iii. Click on **Select metric in cloudWatch**.

iv.   Choose **Glue** from the list of services.

v.   Then, select **CloudWatch Metrics** to view the available metrics for your Glue jobs.

vi.   Choose **CsvToParquetJob**.

vii. Click on **Select metric** and proceed to configure the alarm.

viii. Configure the threshold type to **Static** and set the condition to **Threshold < 1 for 1 consecutive period(s)**.

ix.   Set the period to match your Glue job's frequency (for example, 1 hour).

x.   Click on **Next**.

xi.   Configure the actions as follows:

  • Choose **In alarm** and select **Send notification to**.

  • Select your SNS topic, **GluePipelineAlerts** (created previously in *Create an SNS topic for notifications*).

xii. Click on **Next**.

xiii. Name your alarm (for example, `GlueJobSuccessAlarm`).

xiv. Click on **Create alarm**.

You can find all alerts in the CloudWatch UI:

Figure 4.25 – Cloudwatch alarms

3. **Monitor queue size metrics**: To monitor the size of queues that are part of your data pipeline, you might be using Amazon SQS or other queue services. Here, we'll focus on monitoring an SQS queue:

I. **Create SQS queue (if not already created)**:

i. Go to the Amazon SQS console at `https://console.aws.amazon.com/sqs/`.

ii. Click on **Create queue**.

iii. Enter the queue name (for example, `GluePipelineQueue`).

iv. Configure the queue settings as required.

v. Click on **Create queue**.

II. **Monitor SQS queue metrics**:

i. Go to the Amazon CloudWatch console at `https://console.aws.amazon.com/cloudwatch/`.

ii. Navigate to **Metrics**.

iii. Choose **Metrics** and select the **SQS** namespace.

iv. Select the metrics for your queue (for example, `ApproximateNumberOfMessagesVisible`).

v. Create a dashboard to monitor these metrics by clicking on **Add to dashboard**.

vi. Configure the dashboard with relevant widgets to monitor your queue size.

III. **Create CloudWatch alarms for SQS queue**:

i. Go to the Amazon CloudWatch console at `https://console.aws.amazon.com/cloudwatch/`.

ii. Navigate to **Alarms** and click on **Create alarm**.

iii. Click on **Select metric** and choose the SQS namespace.

iv. Select the `ApproximateNumberOfMessagesVisible` metric for your queue.

v. Click on **Select metric** and proceed to configure the alarm.

vi. Configure the threshold type to **Static** and set the condition (for example, **Threshold > 100 for 5 consecutive period(s)**).

vii. Set the period (for example, 1 minute).

viii. Click on **Next**.

ix. Configure the actions as follows:

- Choose **In alarm** and select **Send notification to**.

- Select your SNS topic, **GluePipelineAlerts**.

x. Click on **Next**.

xi. Name your alarm (for example, SQSQueueSizeAlarm).

xii. Click on **Create alarm**.

You can find all alerts in CloudWatch UI:

Figure 4.26 – CloudWatch alarm SQS

To monitor AWS Glue job performance effectively, setting up CloudWatch alarms with SNS for AWS Glue metrics provides a proactive way to track and manage your data pipeline. By configuring alarms for key Glue job metrics (for example, job failures or long runtimes) and integrating them with SNS, you can receive instant notifications when thresholds are exceeded. This ensures that any issues in your pipeline are addressed promptly, maintaining the reliability and health of your data processes.

Setting up a pipeline using AWS Glue to ingest data from a JDBC database into a catalog table

Creating a full pipeline using AWS Glue to ingest data from a relational database on a regular basis involves setting up the necessary components such as a Glue job, Glue crawler, and a retry mechanism to handle transient errors. In this recipe, we are going to use the AWS Glue job with EventBridge and Step Functions workflow. We will read data from a relational database and store it in an S3 bucket.

How to do it...

1. **Set up your environment**:

    I.   Use your existing S3 bucket or create a new one. (To create a new S3 bucket, navigate to the S3 service in the AWS Management Console, click on **Create bucket**, and specify a unique name. Choose the region and configure settings such as versioning or encryption as needed, then click on **Create**.)

    II.   Create an RDS MySQL instance (please use the following link and follow the given instructions: `https://aws.amazon.com/getting-started/hands-on/create-mysql-db/`).

    III. Once the database is created, download `mysql-scripts.txt`, connect it to your RDS (MySQL) instance, and run it.

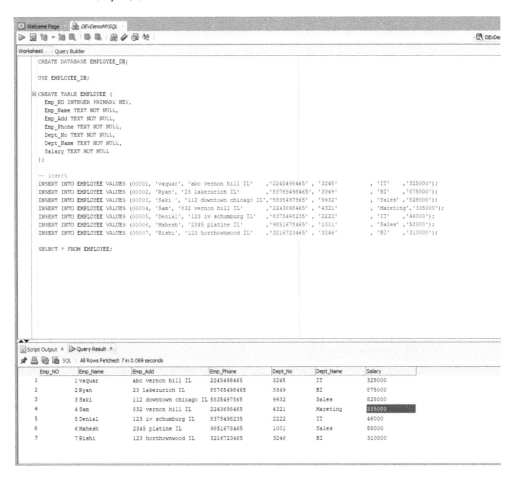

Figure 4.27 – MySQL editor with query and results

2. **Create IAM roles and policies**:

    I.   **Create an IAM policy for Glue**:

        i.   Go to the IAM console at `https://console.aws.amazon.com/iam/`.

        ii. Navigate to **Policies** and click on **Create policy**.

        iii. Select the **JSON** tab and paste the policy document. Please use the same policy and role we created in the *Defining a simple workflow using AWS Glue workflows* recipe.

    II.   **Create an IAM role for Glue**:

        i.   Go to the IAM console, navigate to **Roles**, and click on **Create role**.

        ii.   Select **AWS service** and then choose **Glue**.

        iii. Click on **Next: Permissions**.

        iv.   Attach the policy you just created (use the same policy as in *Defining a simple workflow using AWS Glue workflows*).

        v.   Click on **Next: Tags**, optionally add tags, then click on **Next: Review**.

        vi.   Name your role `AWSGlueServiceRole`. Use the same role as in *Defining a simple workflow using AWS Glue workflows*.

        vii. Click on **Create role**.

3. **Create an AWS Glue database**:

    I.   Open the AWS Glue console at `https://console.aws.amazon.com/glue/`.

    II.   Create a database with the help of the following steps:

        i.   Navigate to the **Databases** section.

        ii.   Click on **Add database**.

        iii. Name your database `relational_database`.

4. **Create an AWS Glue connection**:

    I.   **Create Glue connection**:

        i.   Go to the AWS Glue console at `https://console.aws.amazon.com/glue/`.

        ii.   Navigate to **Connections** and click on **Add connection**.

        iii. Choose the connection type as **JDBC** and click on **Next**.

        iv.   Fill in the connection details, including the JDBC URL, username, and password for your relational database.

        v.   Name your connection `glue-rds-Jdbc-connection`.

Test your connection using **Test Connection**:

Figure 4.28 – Testing your connection

When the connection is successful, you will see the following message:

Figure 4.29 – JDBC connection test

   vi. Click on **Finish**.

5. **Create AWS Glue crawler**:

   I.   **Create Glue crawler**:

     i. Go to the **Crawlers** section in the Glue console.

     ii. Click on **Add crawler**.

     iii. Name your crawler `RelationalDataCrawler`.

     iv. For the data store, choose **JDBC** and select your connection, **glue-rds-Jdbc-connection**.

     v. Include the `EMPLOYEE_DB/EMPLOYEE%` path and click on the **Add datasource** button.

     vi. Specify the database and table name to crawl.

     vii. Configure the IAM role to use `AWSGlueServiceRole`.

     viii. Set the frequency to run on demand.

ix. Create or select the `relational_database` database where the crawler results will be stored.

x.   Review and create the crawler.

6. **Create an AWS Glue job with the retry mechanism**:

I.   Download the `jdbc_to_s3.py` script from GitHub and upload it to your S3 bucket:

i.   **Importing libraries and defining the metric function**: This section imports the necessary libraries such as `awsglue`, `pyspark`, `boto3`, and `time`. The `put_metric_data` function allows us to send custom metrics (for example, job success or failure) to Amazon CloudWatch to monitor job performance:

```python
import sys
from awsglue.transforms import *
from awsglue.utils import getResolvedOptions
from pyspark.context import SparkContext
from awsglue.context import GlueContext
from awsglue.job import Job
import boto3
import time

def put_metric_data(metric_name, value):
    cloudwatch = boto3.client('cloudwatch')
    cloudwatch.put_metric_data(
        Namespace='GlueIngestion',
        MetricData=[
            {                       'MetricName': metric_name,
                'Value': value,
                'Unit': 'Count'
            },          ]     )
```

ii. **Setting up the Glue and Spark context**: This block initializes the Spark and Glue contexts. The `getResolvedOptions` function fetches the job name passed as a command-line argument, and `Job` is initialized for tracking and managing the Glue job:

```python
args = getResolvedOptions(sys.argv, ['JOB_NAME'])
sc = SparkContext()
glueContext = GlueContext(sc)
spark = glueContext.spark_session
job = Job(glueContext)
job.init(args['JOB_NAME'], args)
```

iii. **Defining the data source and destination**: `database` and `table_name` specify the Glue Data Catalog from which the data is pulled. `output_s3_path` points to the Amazon S3 bucket where the ingested data will be written in the Parquet format:

```
database = "relational_data"
table_name = "your_table_name"
output_s3_path = "s3://your-glue-data-bucket/ingested-data/"
```

iv. **Implementing retry logic for data ingestion**: The retry mechanism allows the job to retry up to three times if it encounters an issue. A DynamicFrame is created from the Glue Data Catalog, which enables handling semi-structured data:

```
max_retries = 3
retries = 0
success = False
while retries < max_retries and not success:
    try:
        dynamic_frame = glueContext.create_dynamic_frame.
from_catalog(
            database=database,
            table_name=table_name,
            transformation_ctx="dynamic_frame"
        )
```

v. **Converting data and writing to S3**: The DynamicFrame is converted into a Spark DataFrame for easier manipulation. The DataFrame is written to S3 in the Parquet format, partitioned by the specified columns. On success, a metric is sent to CloudWatch:

```
data_frame = dynamic_frame.toDF()
data_frame.write.mode("overwrite").parquet(output_s3_path)
success = True
put_metric_data('JobSuccess', 1)
```

vi. **Handling exceptions and sending failure metrics**: If an exception occurs, the retry count is incremented, and the process waits for 30 seconds before trying again. If the retries are exhausted, a failure metric is sent to CloudWatch, and an exception is raised:

```
except Exception as e:
    retries += 1
    if retries < max_retries:
            time.sleep(30)   # Wait for 30 seconds before
retrying
        else:
            put_metric_data('JobFailure', 1)
            raise e
```

vii. **Finalizing the Glue job**: The `job.commit()` method commits the job, marking its completion in the AWS Glue environment, ensuring that all job metadata is finalized:

```
job.commit()
```

II.   **Create the Glue job**:

i.   Go to the **Jobs** section in the Glue console.

ii.   Click on **Add Job**.

iii. Name the job `JdbcToS3Job`.

iv.   Choose the `GlueServiceRole` IAM role.

v.   For the ETL language, select **Python**.

vi.   For the script, choose **A new script to be authored by you**.

vii. Set the script filename to `s3://your-glue-data-bucket/scripts/jdbc_to_s3.py`.

viii. Set the temporary directory to `s3://your-glue-data-bucket/temp/`.

ix.   Click on **Next** and configure the job properties as needed.

x.   Click on **Save**.

7.   **Create an AWS Step Functions state machine**:

I.   Download and use the following state machine JSON definition to coordinate the workflow:

i.   **Initial setup and workflow start**: The state machine starts with the `StartAt` directive, which specifies the first step, `"StartCrawler"`. `"StartCrawler"` is a `Task` state that uses the `glue:startCrawler` resource to start an AWS Glue crawler named `"RelationalDataCrawler"`:

```
{   "Comment": "A description of my state machine",
  "StartAt": "StartCrawler",
    "States": {
      "StartCrawler": {
        "Type": "Task",
        "Resource": "arn:aws:states::aws-
sdk:glue:startCrawler",
        "Parameters": {
          "Name": "RelationalDataCrawler"
        },Next": "WaitForCrawler"
    },
```

ii. **Adding a wait step**: The `"WaitForCrawler"` state is a `Wait` state that pauses the workflow for 60 seconds before proceeding to the next step. This ensures the state machine allows the Glue crawler enough time to start and process the data:

```
"WaitForCrawler": {
  "Type": "Wait",
  "Seconds": 60,
  "Next": "CheckCrawlerStatus"
},
```

iii. **Checking the crawler status**: The `"CheckCrawlerStatus"` state is another `Task` state – this time, using the `glue:getCrawler` resource to retrieve the current status of the crawler. The crawler's status is critical to determine whether it has finished, is still running, or encountered an error:

```
"CheckCrawlerStatus": {
    "Type": "Task",
    "Resource": "arn:aws:states:::aws-
sdk:glue:getCrawler",
    "Parameters": {
      "Name": "RelationalDataCrawler"
    },      "Next": "CrawlerComplete?"
    },
```

iv. **Conditional check for crawler completion**: The `"CrawlerComplete?"` state is a `Choice` state that evaluates the status of the crawler. If the status is `"READY"`, it means the crawler has finished and the workflow proceeds to the next task, `"StartGlueJob"`. If not, it defaults back to the `"WaitForCrawler"` state to wait and check again:

```
"CrawlerComplete?": {
  "Type": "Choice",
  "Choices": [
    {            "Variable": "$.Crawler.State",
      "StringEquals": "READY",
      "Next": "StartGlueJob"
    }      ],      "Default": "WaitForCrawler"
  },
```

v. **Starting the Glue job**: The `"StartGlueJob"` state is the final `Task` that triggers the AWS Glue job using the `glue:startJobRun` resource. The job named `"JdbcToS3Job"` is started to process and load the data, and the state machine ends after this step (`"End":` `true`). This state machine orchestrates a simple data processing pipeline using AWS Glue. It first starts a Glue crawler to scan the data, waits for the crawler to complete, and then triggers a Glue ETL job to process the ingested data. By using `Task`, `Wait`, and `Choice` states, the workflow is controlled and monitored efficiently:

```
"StartGlueJob": {
    "Type": "Task",
    "Resource": "arn:aws:states:::aws-
sdk:glue:startJobRun",
    "Parameters": {
      "JobName": "JdbcToS3Job"
    },          "End": true
  }  }}
```

II.   **Create the state machine**:

i.   Go to the AWS Step Functions console at `https://console.aws.amazon.com/states/`.

ii.   Click on **Create state machine**.

iii.   Choose **Author with code snippets**.

iv.   Name your state machine `GlueIngestionStateMachine`.

v.   Paste the state machine definition into the editor.

vi.   Choose the `step-functions-role` execution role.

vii.   Click on **Create state machine**.

8.   **Create an EventBridge rule to trigger the state machine**:

I.   **Create an EventBridge rule**:

i.   Go to the Amazon EventBridge console at `https://console.aws.amazon.com/events/`.

ii.   Click on **Create rule**.

iii.   Name your rule `S3FileUploadRule`.

iv.   Choose **Event Source** and select **Event Pattern**.

v.   Click on **Edit** and specify the following JSON event pattern:

```
{
  "source": ["aws.s3"],
  "detail-type": ["Object Created"],
  "detail": {
    "bucket": {
      "name": ["your-bucket"]
    },
```

```
      "object": {
        "key": [{"prefix": "csv-data/"}]
      }
    }
  }
```

> **Note**
>
> Replace your-bucket with your bucket name.

    vi.   Click on **Save**.

    vii. In the **Targets** section, click on **Add target**

In AWS Step Functions Studio, you will find the following visual representations:

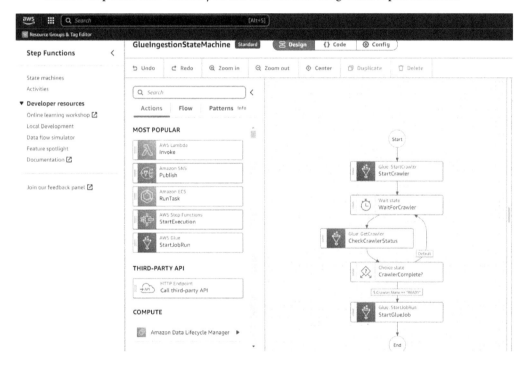

Figure 4.30 – Step Functions workflow

Once your execution completes, you can see all the steps in green, and if you click on a step and definition, you can see all inputs, outputs, definitions, events, and so on.

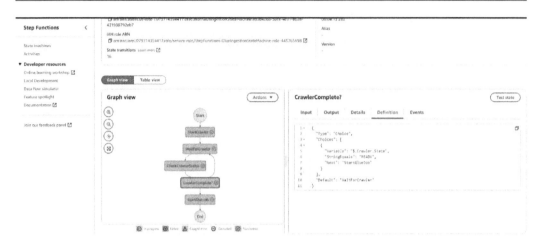

Figure 4.31 – Step Functions with Glue

Finally, you can go into your S3 bucket and validate the data created inside the `ingetion-data` folder:

Figure 4.32 – Results in the S3 bucket

In conclusion, setting up a pipeline using AWS Glue to ingest data from a JDBC database into a catalog table streamlines the process of extracting, transforming, and loading data into a centralized data catalog. This enables efficient data management, making it easier to query, analyze, and integrate with other AWS services for scalable analytics solutions.

5

Running Big Data Workloads with Amazon EMR

Amazon EMR is a managed service that allows running big data frameworks such as Apache Spark or Apache Hive on the Apache Hadoop ecosystem. It provides clusters for data applications to handle large amounts of data in a distributed and scalable way.

EMR removes the complexity of having to deploy, configure, and coordinate all these open source frameworks and tools to work together, so you can just start using them. Each version of EMR lists all the specific frameworks and the specific versions it provides.

Unlike other AWS-managed services, EMR allows you to have full control and visibility of your cluster: which hardware to run, which EC2 image to use, what to install, and even root access to the cluster (except when you run on EMR serverless mode). The recipes in this chapter will help you learn the EMR capabilities and how to make the best use of them.

This chapter includes the following recipes:

- Running jobs using AWS EMR serverless
- Running your AWS EMR cluster on EKS
- Using the AWS Glue catalog from another account
- Making your cluster highly available
- Scaling your cluster based on workload
- Customizing the cluster nodes easily using bootstrap actions
- Tuning Apache Spark resource usage
- Code development on EMR using Workspaces
- Monitoring your cluster
- Protecting your cluster from security vulnerabilities

Technical requirements

The recipes in this chapter assume that you have a Bash shell or equivalent available with the AWS CLI installed (https://docs.aws.amazon.com/cli/latest/userguide/getting-started-install.html) with access to AWS. If using Microsoft Windows, you can enable WSL (https://learn.microsoft.com/en-us/windows/wsl/install) or install Git (https://git-scm.com/downloads) to use the Bash shell that it brings.

Most commands are too long to list in the book line width. To fit them in, lines have been broken using the \ character, which tells the shell that the command continues in the next line. Be careful to respect the spaces (or lack of) on the line breaks.

It is also assumed that you have configured the default credentials and region using aws configure or using an AWS CLI profile. You need to have permission to manage EMR clusters. Also, most recipes will expect you to have a user with access to the AWS EMR console using a web browser. It's easier if you have a test account on which you can use an admin user to access the console and the command line; otherwise, you'll need to grant permissions as needed.

Most recipes create an EMR cluster that requires a subnet and an S3 directory to store logs. For simplicity, it's easier if the subnet is public and the default security groups haven't been customized, or if they have, that they are secured. If you want to run on a private subnet, you will have to meet some further requirements (see https://docs.aws.amazon.com/emr/latest/ManagementGuide/emr-clusters-in-a-vpc.html#emr-vpc-private-subnet).

For the log's path, choose a bucket in the account and region or create one.

The recipes that run AWS CLI commands will indicate the specific variables that needs to be set. You can define them as needed when you open the shell or export them on your user profile so that they are available on every shell you create.

Many recipes will ask you to define the following variables in your shell, so it's better if you have them defined beforehand. Pick or create an S3 bucket in the region you intend to use:

```
SUBNET=<enter the subnet for EMR to use here>
# Don't use a trailing / in the S3 urls
S3_SCRIPTS_URL=<enter an s3:// path where to store scripts>
S3_LOGS_URL=<enter an s3:// path where to store logs>
```

It is recommended that you set the Bash flag to warn you if a variable is missing:

```
set -u
```

The recipes try to be frugal and create small clusters that suffice for the demonstration while using commonly available nodes; if that instance type doesn't happen to be available in your region, please replace them with the most similar instance type available at the time.

Some recipes will expect you to have a keypair .pem file, created in the region, which you can do in the EC2 console (https://docs.aws.amazon.com/AWSEC2/latest/UserGuide/create-key-pairs.html#having-ec2-create-your-key-pair).

When the keypair is created, it will download a .pem file with the name of the key. Change the permissions so only your user can read it (for instance, by running cmhod 400 mykey.pem). Create a variable in the shell with the key name:

```
KEYNAME=<just the key name, no .pem extension>
```

You can find the code files for this chapter on GitHub at https://github.com/PacktPublishing/Data-Engineering-with-AWS-Cookbook/tree/main/Chapter05.

Running jobs using AWS EMR serverless

EMR serverless was created for the common case where the user just wants to run Spark and Hive jobs without having to worry about the type of nodes, capacity, and configuration.

For such cases, EMR serverless really simplifies the operation, since it does not require a cluster to be configured or maintained. You do not have to worry about which kind of EC2 is the right one or whether it is going to be present (and available in enough capacity) for your chosen region and Availability Zone. The main trade-off is that you can no longer ssh into nodes to do low-level administration and troubleshooting.

In this recipe, you will see how simple it is to run a Spark application using EMR serverless.

Getting ready

To test serverless, you will need a sample script to run. The following script is a basic example that accesses the Glue catalog from the EMR serverless job. In the shell, run the following command to create a Python file with the script that follows:

```
cat <<EOF > testEMRServerless.py
from pyspark.sql import SparkSession

spark = (SparkSession.builder.config(
        "hive.metastore.client.factory.class",
        "com.amazonaws.glue.catalog.metastore.\
AWSGlueDataCatalogHiveClientFactory")
    .enableHiveSupport()
    .getOrCreate())
spark.sql("SHOW TABLES").show()
EOF
```

Upload the file to S3 in a bucket on the same region as you intend to use Glue (otherwise, you would need a VPC for EMR serverless to download it).

How to do it...

1. Open the AWS console on EMR and select **EMR Serverless** in the left menu.

2. Use the **Get Started** button to set up EMR Studio so it logs you in and opens the application creation screen. If you had previously set EMR Studio up in the region, then select **Manage application**, and then **Create application** in the applications screen.

3. In the application creation screen, enter the name `SparkTest`. Make sure that the type is **Spark** and leave the rest as defaults.

4. Select **Create and start application** at the bottom.

5. In the **Applications** list, use the `SparkTest` link to view the details, then use the refresh button until the status becomes **Started**.

6. Now that the serverless application is started, you can submit jobs. Select **Submit job run** on the lower side of the **Application details** screen.

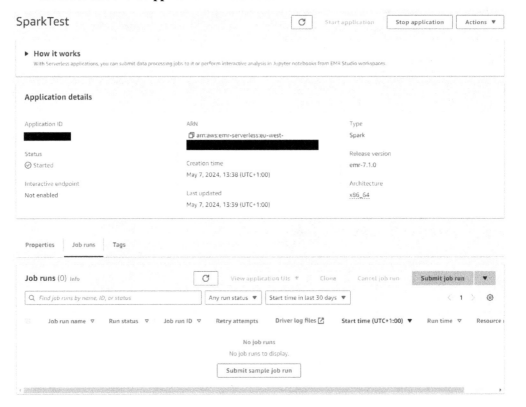

Figure 5.1 – An EMR Studio application is created

7. Select a runtime role. The dropdown will show the ones that are suitable for EMR Serverless and allow you to create a new one. Choose one of those.

At the time of this writing, the role is created without permission to use the Glue catalog. To allow this, use the link in the creation confirmation message that appears to open a new tab. From the **Add permissions** dropdown, select **Create inline policy** using this JSON policy:

```
{
    "Version": "2012-10-17",
    "Statement": [
        {
            "Effect": "Allow",
            "Action": [
                "glue:CreateDatabase",
                "glue:UpdateDatabase",
                "glue:DeleteDatabase",
                "glue:GetDatabase",
                "glue:GetDatabases",
                "glue:CreateTable",
                "glue:UpdateTable",
                "glue:DeleteTable",
                "glue:GetTable",
                "glue:GetTables",
                "glue:GetTableVersions",
                "glue:CreatePartition",
                "glue:BatchCreatePartition",
                "glue:UpdatePartition",
                "glue:DeletePartition",
                "glue:BatchDeletePartition",
                "glue:GetPartition",
                "glue:GetPartitions",
                "glue:BatchGetPartition",
                "glue:CreateUserDefinedFunction",
                "glue:UpdateUserDefinedFunction",
                "glue:DeleteUserDefinedFunction",
                "glue:GetUserDefinedFunction",
                "glue:GetUserDefinedFunctions"
            ],
            "Resource": [
                "arn:aws:glue:*:*:database/*",
                "arn:aws:glue:*:*:table/*/*",
                "arn:aws:glue:*:*:catalog"
            ]
        }
    ]
}
```

8. In the script location field, enter the S3 URL of the Python file that was uploaded earlier, as mentioned in the *Getting ready* section of this recipe.

9. Once you finish editing the field, it should detect it is a Python script and remove or empty the **Main class** field.

10. Leave the rest as default and submit it. The job will be listed in the **Job runs** tab and the status will change to **Scheduled**. After a minute or so, it should change to **Running**. You can use the refresh button to refresh the status. If you have made a mistake, such as lacking permissions, you could clone the job in the table to submit it again quickly.

11. When the job ends successfully, you can view it from the application screen listing the jobs and the logs (`stdout` and `stderr`). In this example, opening the `stdout` log should print the list of the tables in the catalog default database on the current region.

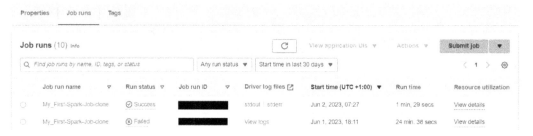

Figure 5.2 – The Spark application runs

How it works...

When the job is submitted, EMR creates a cluster taking care of all the configuration, instance selection, and capacity. This means that we only have to provide the PySpark script or the executable JAR if we are using Scala or Java.

It's possible to avoid this scheduling delay by enabling the pre-initialized capacity. With that configuration, you can set a minimum capacity to be available when the application is started, so it's ready to handle jobs promptly. Then EMR will scale the capacity as needed.

Spark can use the Glue catalog because, by default, the **AWS Glue Data Catalog as metastore** configuration setting is enabled for the jobs. This in turn sets a Spark property that tells Spark to use Glue.

As you have seen, in EMR serverless, you can submit Spark jobs in an equivalent way to how you would start it using `spark-submit` in a traditional EMR cluster or other Spark environment. You can read more about it in the Spark documentation: `https://spark.apache.org/docs/latest/submitting-applications.html`.

If instead of Python, you wrote your Spark job using Scala or Java, then instead of the path to the script on S3, you would specify the S3 path to the JAR with the job and dependencies, and then on the Spark properties tab you would specify the main class using `--class`, just like you would do if you were using `spark-submit`.

You can experiment with this by selecting the **Submit sample job** option in the application screen, which creates a job that runs a sample PI calculation using the JAR provided by EMR.

There's more...

Inside the job run, you can easily access SparkUI, both while running and after the job has completed.

You can streamline running the apps if you configure the app to auto-start when a job is submitted. You can also avoid the startup lag by adding provisioned nodes (at minimum a driver and an executor).

In addition, you can configure the application to auto-stop if it has been idle for a configured amount of time.

For more information, you can visit `https://docs.aws.amazon.com/emr/latest/EMR-Serverless-UserGuide/application-capacity.html`.

See also

- AWS Glue is an alternative solution to run Spark jobs in a managed environment, with value features such as bookmarks and incremental processing. You can experience these features with the *Processing data incrementally using bookmarks and bounded execution* recipe in *Chapter 3, Ingesting and Transforming Your Data with AWS Glue*.

Running your AWS EMR cluster on EKS

EMR offers the option of running a fully managed Spark cluster leveraging Kubernetes.

Instead of running a full Hadoop cluster with YARN and HDFS, EMR on **Elastic Kubernetes Service** (**EKS**) is a lightweight solution to run Spark applications with a reduced start time, better resource utilization, and better availability. It allows running a cluster with nodes from different Availability Zones or even AWS Outposts (on the customer data center).

Getting ready

You need to have installed both the AWS and EKS CLI client tools. You can follow the AWS instructions to install them on your machine depending on your OS:

- **AWS CLI**: `https://docs.aws.amazon.com/cli/latest/userguide/getting-started-install.html`
- **EKS CLI**: `https://github.com/weaveworks/eksctl`

You can verify that both tools are installed by running the following commands:

```
aws --version
eksctl version
```

If you haven't already set up the credentials and region to use by running `aws configure` in the commands, we'll assume that the region you want to work with is the default region set.

How to do it...

1. On the command line, run the following command:

    ```
    eksctl create cluster --name emr-recipe2 \
    --instance-types=m5.xlarge --managed
    ```

2. Wait for a few minutes while the two CloudFormation templates are deployed: one for the cluster and another for the managed node group for the cluster.

> **Tip:**
>
> You can see progress details on the AWS CloudFormation console.
>
> You might see an error such as `kubectl not found`. If the Kubernetes client is available, it will be used to check the cluster, but this is not required.

3. Check the capacity status by running the following:

    ```
    eksctl get nodegroup --cluster emr-recipe2
    ```

    Notice that these nodes are being charged, so if you are not going to complete the recipe (including the cleanup) at this time, you can manually adjust this capacity using `eksctl scale nodegroup`. See the *There's more...* section of this recipe for how to automate scaling.

4. Allow the EMR cluster to be used as a Kubernetes namespace:

    ```
    eksctl create iamidentitymapping --cluster emr-recipe2\
    --namespace default --service-name "emr-containers"
    ```

5. Create an EMR virtual cluster using the EKS cluster:

    ```
    aws emr-containers create-virtual-cluster \
    --name vc_emr_recipe2 --container-provider \
    '{"id": "emr-recipe2", "type": "EKS", "info":
    {"eksInfo": {"namespace": "default"}}}'
    ```

6. Define variables with the cluster id and role arn.

```
VIRTUAL_CLUSTER_ID=$(aws emr-containers \
list-virtual-clusters --output text --query \
"sort_by(virtualClusters, &createdAt)[-1].id")

ROLE_ARN=$(eksctl get \
iamidentitymapping --cluster emr-recipe2 | \
grep emr-containers | cut -f 1)
```

7. Prepare the job configuration (type the following without break lines inside quoted strings):

```
cat > emr-recipe2.json << EOF
{
    "name": "EmrRecipe2-Pi",
    "virtualClusterId": "$VIRTUAL_CLUSTER_ID",
    "executionRoleArn": "$ROLE_ARN",
    "releaseLabel": "emr-6.10.0-latest",
    "jobDriver": {
        "sparkSubmitJobDriver": {
            "entryPoint": "local:///usr/lib/spark/examples/jars/
spark-examples.jar",
            "sparkSubmitParameters": "--conf spark.executor.
instances=1 --conf spark.executor.memory=1G --conf spark.driver.
memory=1G --class org.apache.spark.examples.SparkPi"
        }
    }
}
EOF
```

8. Launch the job:

```
aws emr-containers start-job-run --cli-input-json \
 file://./emr-recipe2.json
```

9. Check the job run status. You can run this multiple times to get the status:

```
aws emr-containers list-job-runs \
 --virtual-cluster-id $VIRTUAL_CLUSTER_ID
```

10. Clean up to avoid incurring further costs:

```
eksctl delete cluster --name emr-recipe2
aws emr-containers delete-virtual-cluster --id \
 ${VIRTUAL_CLUSTER_ID}
```

How it works...

When you ran the `eksctl create` command, it created and deployed a CloudFormation template that creates a cluster in EKS. The `--managed` option enabled the cluster to work with managed node group, meaning that nodes of the specified type will be allocated and released as needed. It is recommended to use at least `m5.xlarge` or instances with more memory. Spark is likely to fail to run on smaller instances. This feature doesn't have an extra cost; you pay for what you use.

You can optionally download the Kubernetes client (`https://docs.aws.amazon.com/eks/latest/userguide/install-kubectl.html`). You can check the current usage by running the following:

```
kubectl get nodes
```

For simplicity, we used the `AWSServiceRoleForAmazonEMRContainers` role, which provides the minimum permissions and trusts `emr-containers` (the name of EMR on EKS) to assume it. This is enough to run a job, but in a more realistic example, you would need your own role with the same `AmazonEMRContainersServiceRolePolicy` and trust relation, but with additional permissions to write to S3, CloudWatch, and possibly others depending on the job.

When the job was submitted, it allocated nodes from the node group, retrieved the image indicated, and started one as the driver. Therefore, with this method, Spark can only run on *cluster* mode, meaning that the driver also runs in the remote cluster and not a local process (like you can do in normal EMR).

There's more...

On the AWS console, you can navigate the EMR, select virtual clusters under EKS, and view the details of the run in the SparkUI. It should look like this:

Figure 5.3 – The SparkUI job run

Notice that we didn't select the Spark version directly. This is selected by the EMR version of the selected image. This way, each job in the virtual cluster can use a different version if needed. This allows for more flexibility than using a traditional EMR cluster.

Remember that this way of running jobs allows you to ssh into the nodes (like a regular EMR cluster does). For that, you need to set up a keypair ssh keys on EC2 and then use them when creating the EKS cluster by adding the `--ssh-access` and `--ssh-public-key` parameters.

In this recipe, the provision of nodes was automated by using the default desired nodes, but you can use one of the autoscaling methods to allow the job to adapt to the demand and grow and shrink as you submit workloads (`https://docs.aws.amazon.com/eks/latest/userguide/autoscaling.html`).

See also

- Unless you have a specific reason for using EKS (such as optimizing a pool of capacity) to run a Spark job like in this recipe, it's easier to use EMR serverless and that way simplify the capacity management. See how to do this in the *Running jobs using AWS EMR serverless* recipe.

Using the AWS Glue catalog from another account

Apache Hive is the traditional SQL solution in the Hadoop ecosystem. Since its early versions, it has decoupled the catalog metastore from the query engine to store information about the tables and schemas. This means that multiple tools have added support to integrate with the Hive metastore over the years such as Spark, Presto/Trino, Impala, or the Hive server itself.

When AWS Glue was released, one of the value propositions that it could provide was a Hive-compatible store, which could massively scale and provide fault tolerance out of the box.

When you run EMR on EC2, you can run a Hive metastore or use the Glue catalog as the metastore just by checking the corresponding box in the cluster configuration screen (or setting the equivalent configuration if doing programmatically). When Glue is set as the catalog metastore, all the tools in the cluster that are compatible with Hive will use it to retrieve and store information about the databases and tables. By default, it will use the catalog on the same account (and region) as EMR is running.

In this recipe, you'll see how to enable the Glue metastore and use one from a different account in the same region, as well as how to run a PySpark script that prints the tables in the `CrossCatalogRecipeDB` database of the catalog.

Getting ready

For this recipe, you ideally need two accounts: one running EMR and the other holding the catalog to use. You can also do it on the same account if you don't have multiple accounts; in that case, you can skip the permissions step.

In the Bash shell, define the environment variables with the account IDs:

```
EMR_ACCOUNT=<id of the account running EMR>
CATALOG_ACCOUNT=<id of the account holding the catalog>
REGION=<id of the account holding the catalog>
```

The recipe assumes that the AWS CLI credentials are for the EMR account, while the catalog account will be accessed using the AWS console with a Lake Formation admin user (or a user that can create one).

This recipe assumes that you have defined the SUBNET, S3_SCRIPTS_URL, and S3_LOGS_URL variables in the shell (see the *Technical requirements* section at the beginning of the chapter for more information).

How to do it...

1. Using a text editor, create a text file named PrintTables.py with the following content:

   ```
   from pyspark.sql import SparkSession
   spark = SparkSession.builder.enableHiveSupport(
           ).getOrCreate
   spark.sql("SHOW TABLES FROM cross_catalog_recipe"
           ).show()
   ```

2. Upload the script you created to S3:

   ```
   aws s3 cp PrintTables.py $S3_SCRIPTS_URL/
   ```

3. For this recipe, we are using plain IAM permissions and not Lake Formation. Let us verify whether it is disabled while you create the test database and table.

   On the AWS console, log in using the account you specified in the CATALOG_ACCOUNT variable in *Getting ready* and go to AWS Lake Formation. If this is the first time you are using it, you will be asked to assign an administrator.

   Navigate to **Settings** on the left menu and ensure that IAM permissions are enabled for new tables and databases (you can change it back after you create a database and table in the following steps if you want to).

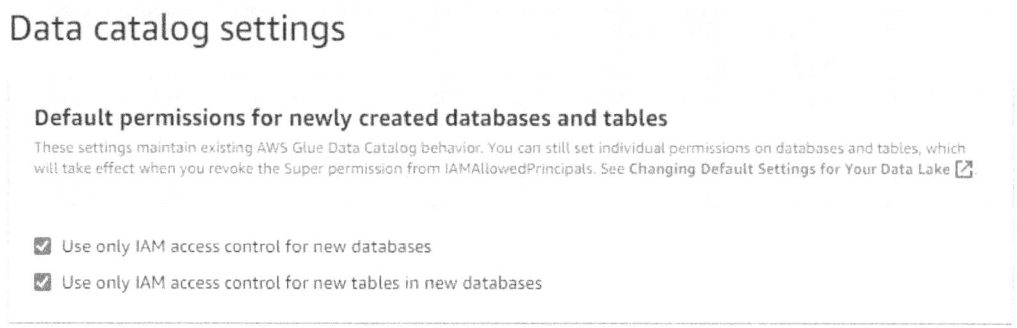

Data catalog settings

Default permissions for newly created databases and tables

These settings maintain existing AWS Glue Data Catalog behavior. You can still set individual permissions on databases and tables, which will take effect when you revoke the Super permission from IAMAllowedPrincipals. See **Changing Default Settings for Your Data Lake** ☑.

☑ Use only IAM access control for new databases

☑ Use only IAM access control for new tables in new databases

Figure 5.4 – Lake Formation settings

4. Navigate to **Glue** in the console and select **Databases** in the left menu, then choose **Create** and name it `cross_catalog_recipe`. Leave the rest as default.

5. In the left menu, select **Tables** and then **Add Table**. Enter `cross_account_table` as the name and select `cross_catalog_recipe` as the database. Enter any S3 path location; it doesn't matter, as we are not really going to read the table content. Continue to the next steps leaving everything as default until you complete the table creation.

6. Go back to the Bash shell and run the following command to generate the policy based on your variables:

```
cat << EOF
{
  "Version": "2012-10-17",
  "Statement": [ {
    "Effect": "Allow",
    "Principal": {
      "AWS":
"arn:aws:iam::${EMR_ACCOUNT}:role/EMR_EC2_DefaultRole"
    },
    "Action": [ "glue:GetDatabase",
"glue:GetPartition","glue:GetTables",
"glue:GetPartitions", "glue:BatchGetPartition",
"glue:GetDatabases", "glue:GetTable",
"glue:GetUserDefinedFunction",
"glue:GetUserDefinedFunctions"],

    "Resource" : [
"arn:aws:glue:$REGION:$CATALOG_ACCOUNT:catalog",
"arn:aws:glue:$REGION:$CATALOG_ACCOUNT:database\
/default",
"arn:aws:glue:$REGION:$CATALOG_ACCOUNT:database\
/cross_catalog_recipe",
```

```
"arn:aws:glue:$REGION:$CATALOG_ACCOUNT:table\
/cross_catalog_recipe/*",
"arn:aws:glue:$REGION:$CATALOG_ACCOUNT:database\
/global_temp"]
   } ]
}
EOF
```

7. Copy the JSON object printed by the previous step (all the lines from the one starting with { to the one ending with }). In the AWS console, under **Glue**, select **Settings** in the left menu, paste the policy in the **Permissions** textbox, and save the changes. If your policy is not empty, save the existing policy to restore it after completing the recipe, or use another account or region. If you have mixed up the account IDs, it will reject the policy.

8. Go back into the Bash shell and run the following command (note that the last seven lines don't have a \ at the end because they are inside a multiline string):

```
METASTORE_CLASS="com.amazonaws.glue.catalog.\
metastore.AWSGlueDataCatalogHiveClientFactory"
aws emr create-cluster --name CrossAccountCatalog \
--release-label emr-6.11.0 --instance-count 2 \
--log-uri $S3_LOGS_URL --instance-type=m5.xlarge \
--applications Name=Spark --use-default-roles \
--ec2-attributes SubnetId=${SUBNET} --auto-terminate\
 --steps Type=Spark,Name=SparkPi,ActionOnFailure=\
TERMINATE_CLUSTER,Args=[$S3_SCRIPTS_URL/\
PrintTables.py] --configurations '[{"Classification":
"spark-hive-site", "Properties":
{"hive.metastore.client.factory.class":
"'$METASTORE_CLASS'","hive.metastore.glue.catalogid":
"'$CATALOG_ACCOUNT'"}}]'
```

9. If all is correct, the previous command will return a JSON with a cluster ID and ARN. You can check periodically for the cluster status like this (use your ID):

```
aws emr describe-cluster --cluster-id j-XXXXXXXXX \
 | grep '"State"'
```

10. Once the cluster is in the terminated state, retrieve the stdout of the step (the Spark job) by running the following command (again, enter your cluster ID instead of j-XXXXXXXXX). It should confirm that it has downloaded a single file:

```
aws s3 cp $S3_LOGS_URL/j-XXXXXXXXX/steps/ . \
 --exclude '*' --include '*stdout.gz' --recursive
```

11. Print the contents of the zipped file (replace XXX with the path indicated by the previous command):

```
zcat s-XXXXXXXXXXXXXXXXX/stdout.gz
```

You should get confirmation that Spark was able to list the tables in the catalog from the other account:

Figure 5.5 – Job stdout listing tables

12. The next step pertains to cleanup. If you no longer need them, delete the table, catalog, and policy in the catalog account. You can also delete the logs and the script file from S3:

```
aws s3 rm --recursive $S3_LOGS_URL
aws s3 rm $S3_SCRIPTS_URL/PrintTables.py
```

How it works...

In this recipe, you first created a trivial Spark script to demonstrate that it can view the tables in a database on a different account. Then you configured the catalog account to allow EMR on the other account to use a new database that was created for this purpose.

The reason you were asked to check that only IAM permissions are set is for you to be able to use a simple catalog policy to give access with a resource rules. When a database or table is created using an **IAM only** configuration, Lake Formation adds a special permission automatically, which means that it delegates on the catalog IAM policy instead of using what Lake Formation grants.

You can go to Lake Formation on the catalog account, select **Data Lake permissions** on the left menu, and type IAMAllowedPrincipals in the table search box. This will show you the databases and tables that are secured by the catalog policy rules. You can add this special principal permission to existing databases or tables that were created without **IAM only** enabled to achieve the same result.

Review the policy that you added to the catalog. Notice that it gives typical catalog permissions to multiple resources and that the cross_catalog_recipe database is listed twice, once as a database resource and another to allow access to all tables in it using a wildcard. In addition, the policy gives access to the default and global_temp databases. This is done for historical reasons because Spark will check those databases, for instance, to check whether there are UDFs registered. This depends on the Spark version.

Then you created a small cluster. Notice that there are two properties in the configuration. The first one enables the Glue catalog client for the Spark Hive metastore; it is the equivalent of checking the option to use the Glue catalog for Spark table metadata in the UI:

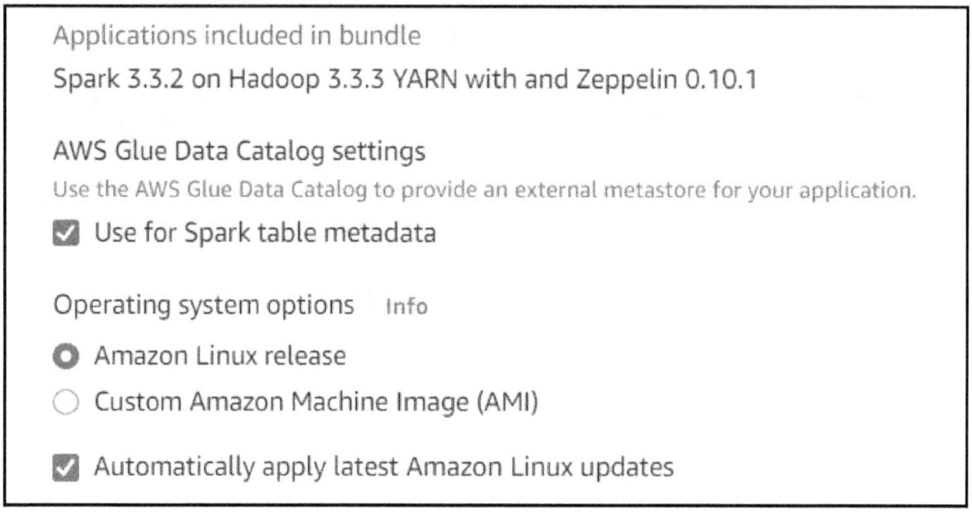

Applications included in bundle
Spark 3.3.2 on Hadoop 3.3.3 YARN with and Zeppelin 0.10.1

AWS Glue Data Catalog settings
Use the AWS Glue Data Catalog to provide an external metastore for your application.
☑ Use for Spark table metadata

Operating system options   Info
◉ Amazon Linux release
○ Custom Amazon Machine Image (AMI)

☑ Automatically apply latest Amazon Linux updates

Figure 5.6 – EMR basic configuration options

The second property, `hive.metastore.glue.catalogid`, tells the client to use another account catalog instead of using the one from the current account.

Finally, by checking the output of the step, you can verify that Spark can list the tables.

There's more...

You can go ahead and create a table with real data and use it. However, notice that the catalog policy just gives you access to the metadata. Spark will still need permission to access the actual data, normally on S3. For that, you need to create a policy in the S3 bucket granting access to the EMR role.

Lake Formation offers the option to manage S3 locations, so when access is granted to a user by Lake Formation, it also includes access to S3 (Lake Formation creates temporary credentials for S3 for that specific data). More information can be found in the official documentation: `https://docs.aws.amazon.com/lake-formation/latest/dg/access-control-underlying-data.html#data-location-permissions`.

See also

- An alternative way to access another account's catalog (while retaining access to the account's catalog) is to enable a special feature available on EMR and Glue. By configuring the `aws.glue.catalog.separator="/"` property in the `spark-hive-site` cluster configuration, you can now reference a database in another account (of course, you will still need to be granted permission on the catalog). For instance, you can use the following command:

  ```
  SELECT * FROM `111122223333/default.mytable`;
  ```

  For more information, you can refer to `https://repost.aws/knowledge-center/query-glue-data-catalog-cross-account`.

Making your cluster highly available

Historically, AWS EMR clusters were used for batch workloads and torn down afterward; errors in the worker nodes would be handled by YARN and HDFS' innate resiliency. There was still the single point of failure of the primary node (previously called **master**), in the unlikely case that the whole batch process would need to be retried.

Hadoop has added full **High Availability** (**HA**) since the early days, as it was intended to run on permanent on-premise clusters that would be shared by many users and teams.

Since EMR 5.23, it allows running with multiple primary nodes. EMR takes care of the tedious process of correctly configuring Hadoop to be HA. Over time, it has also improved the process of automatically replacing a primary node and reconfiguring the system, so the cluster can graciously survive the failure of a single node with minimal or no disruption to the cluster users.

HA is important in cases where delays have a business impact, such as real-time systems or services where the users interact directly with the cluster and an outage would directly impact productivity.

Nonetheless, this HA support has its limitations, and it is good to know them. First, the cluster can only run on a single Availability Zone (you will see later some ideas to mitigate this). Just because Hadoop is HA does not mean every service, for instance a Presto server, in the cluster is automatically HA.

In this recipe, you will see how straightforward it is to run a failure-tolerant cluster on EMR.

Getting ready

This recipe assumes that you have defined the SUBNET variable and set up KEYNAME in the shell, as well as having the `.pem` key in the current shell directory (see the *Technical requirements* section at the beginning of the chapter for more information).

How to do it...

1. Add the resource placement policy to the default EMR role. This role will have automatically been created if you have run the other recipes in this chapter in the same account and region. If the command fails because the role doesn't exist, you can execute *step 2* (which will fail) so it creates the role, then come back to execute the following command, and then continue again with *step 2*:

```
aws iam  attach-role-policy --role-name \
EMR_DefaultRole --policy-arn \
arn:aws:iam::aws:policy/\
AmazonElasticMapReducePlacementGroupPolicy
```

2. In the shell, run the following command to create an HA cluster. If everything is correct, you should just see the cluster ID printed:

```
CLUSTER_ID=$(aws emr create-cluster --release-label \
emr-6.15.0 --use-default-roles --applications \
Name=HBase --use-default-roles --ec2-attributes \
SubnetId=$SUBNET,KeyName=$KEYNAME --instance-groups \
InstanceGroupType=MASTER,InstanceCount=3,InstanceType\
=m5.xlarge InstanceGroupType=CORE,InstanceCount=1\
,InstanceType=m5.xlarge --placement-group-configs \
InstanceRole=MASTER | grep ClusterArn \
| grep -o 'j-.*[^",]');
echo $CLUSTER_ID
```

3. Check the status of the cluster by running the following command until it prints that the cluster is WAITING and both instance groups are RUNNING:

```
aws emr describe-cluster --cluster-id $CLUSTER_ID
```

If the status returned is TERMINATED_WITH_ERRORS, it's probably because the subnet and key name specified are not valid for the region.

4. By default, the security group created by EMR doesn't allow SSH access. First, you need to find the cluster ID (in case there are other clusters):

```
aws emr describe-cluster --cluster-id $CLUSTER_ID \
  | grep 'MasterSecurityGroup' | grep -o 'sg-.*[^",]'
```

5. Open the AWS console, search for the **Security Groups** feature in the top bar, and open it. In the list, find the one matching the ID from the previous step (it should be named **ElasticMapReduce-master**).

6. Open the security group using the ID link. Then, select **Edit inbound rules** | **Add rule**, and in the new rule on the dropdowns, select **SSH** and **My IP**. Save the rule changes.

Figure 5.7 – A security group's new SSH rule

7. Log in to the server using the `.pem` keypair. The command assumes that you have the `.pem` file available in the current directory and with read permission just for you:

```
aws emr ssh --cluster-id $CLUSTER_ID\
  --key-pair-file $KEYNAME.pem
```

If the command results in a **Connection timed out** message, it means that the previous step wasn't completed correctly or that the IP detected by the browser is not the one that the shell is really using to access AWS. In that case, you will need an alternative method to find the IP to allow, for instance, running the following in the shell:

```
curl ifconfig.me
```

8. In the EMR shell that you just logged in to, run the following commands to confirm that the cluster is configured in HA, with two or three servers depending on the service:

```
grep -A 1 zookeeper.quorum \
  /etc/hadoop/conf/core-site.xml
grep qjournal /etc/hadoop/conf/hdfs-site.xml
grep -A 1 yarn.resourcemanager.address \
  /etc/hadoop/conf/yarn-site.xml
```

9. Exit the EMR shell to go back to your local one:

```
exit
```

10. Remove the termination protection and terminate the cluster:

```
aws emr modify-cluster-attributes --cluster-id \
  $CLUSTER_ID --no-termination-protected
aws emr terminate-clusters --cluster-id $CLUSTER_ID
```

11. Confirm that the cluster is terminating:

```
aws emr describe-cluster --cluster-id $CLUSTER_ID \
  grep '"State"'
```

12. If you don't need to ssh to other clusters, edit the security group again to remove the **SSH** inbound rule that you added in *step 6.*

How it works...

At the beginning of the recipe, you added the `AmazonElasticMapReducePlacementGroupPolicy` policy to the default role, which was enabled using the `--placement-group-configs` option on the master node group. Currently, the only strategy is `SPREAD`, which is used by default. This means that the three nodes requested for the `MASTER` instance group will be implemented by EC2 by ensuring that the instances cannot be collocated on the same physical machine. This would reduce reliability, since Hadoop can only survive the loss of one master node without any service loss.

This is because both ZooKeeper and the HDFS journal nodes work on a quorum-based system that requires a majority to make decisions. On the other hand, other services such as the YARN ResourceManager, HBase region server, and others work on an active/passive model, but still rely on ZooKeeper to make that selection and failover. In *step 9*, you saw the configuration for these services and how they configure multiple servers, so clients can failover if one is down.

In this case, the cluster was created with a keypair, so you could ssh and check the configuration to highlight the key differences. You can compare it with the one of a single master on your own.

Finally, notice that to terminate the cluster, you first had to disable the termination protection flag despite not creating the cluster with the explicit `--termination-protected` flag. This is because when a multimaster node is created programmatically, it assumes that it is meant to be long-term and protects it from accidental deletion and potential data loss.

Since EMR 7, if the master capacity is not specified, it will use multimaster by default.

There's more...

Notice that despite spreading the master nodes across different hardware, the cluster still has a single subnet specified. This means that a single VPC and a single region and Availability Zone are used.

As a result, the cluster won't survive if the AWS AZ goes down. If you need to protect against such extreme cases, you will need to use your own measures. Examples include running multiple redundant clusters on different AZs (or even regions) and defining one as the primary and one as the backup, or another strategy that allows coordinating multiple clusters as needed.

As an exercise, you can ssh to one of the masters and issue the following command:

```
sudo shutdown now
```

You can then list the instances and observe how EMR detects that the host is no longer available and replaces it.

Often, the HA cluster will depend on having an HA catalog. The easiest way to achieve that is to configure EMR to use Glue as the Hive metastore; it provides out-of-the-box scalability and AZ fail tolerance. However, if you need to use a Hive metastore, create a database on AWS Aurora and then configure Hive to use it.

See also

- Always consider whether you really need a long-lived EMR cluster and whether there are any alternatives; for instance, instead of storing permanent data on HDFS, you could store it on S3. Instead of using a Presto or Trino server, you could use Athena. Instead of getting capacity from YARN, use serverless EMR, and so on.

  - To run workloads, see the *Running jobs using AWS EMR serverless* recipe.

  - To learn more about the Glue catalog, see the *Optimizing your catalog data retrieval using pushdown filters and indexes* recipe in *Chapter 3, Ingesting and Transforming Your Data with AWS Glue*

Scaling your cluster based on workload

The main benefit of running on the cloud compared to on-premises is the access to virtually endless capacity. When running EMR workloads, you don't want to just have resources available but also to only pay for them when needed to be cost-effective.

In this recipe, you will see how EMR can effortlessly allow you to scale your cluster capacity based on the workload.

Getting ready

This recipe assumes that you have set up the SUBNET environment variable as indicated in the *Technical requirements* section at the beginning of this chapter.

How to do it...

1. Create a cluster with autoscale and idle timeout (make sure you use \ only at the end of the lines indicated; the second command will print the cluster ID):

```
CLUSTER_ID=$(aws emr create-cluster --name AutoScale\
 --release-label emr-7.1.0 --use-default-roles \
 --ec2-attributes SubnetId=${SUBNET} \
 --auto-termination-policy IdleTimeout=900 \
 --applications Name=Spark --instance-groups \
'[{"InstanceCount":1,"InstanceGroupType":"MASTER",
"Name":"MASTER","InstanceType":"m5.xlarge"},
{"InstanceCount":1,"InstanceGroupType":"CORE",
"Name":"CORE","InstanceType":"m5.xlarge"},
{"InstanceCount":1,"InstanceGroupType":"TASK",
"Name":"TASK","InstanceType":"m5.xlarge"}]' \
 --managed-scaling-policy '{"ComputeLimits":
{"UnitType":"Instances","MinimumCapacityUnits":1,
```

```
"MaximumCapacityUnits":10,
"MaximumOnDemandCapacityUnits":10,
"MaximumCoreCapacityUnits":2}}' \
| grep ClusterArn | grep -o 'j-.*[^",]')

echo $CLUSTER_ID
```

2. Print the cluster capacity details. This will show that there is one node allocated for each group and run multiple times until it gets the RUNNING status:

    ```
    aws emr describe-cluster --cluster-id $CLUSTER_ID \
      | grep -B 3 -A 5 "RunningInstanceCount"
    ```

3. Run a Spark example app to force the cluster to scale up:

    ```
    aws emr add-steps --cluster-id $CLUSTER_ID --steps \
    Type=Spark,Name=SparkPi,ActionOnFailure=\
    TERMINATE_CLUSTER,Args=--class,org.apache.spark.\
    examples.SparkPi,/usr/lib/spark/examples/jars/\
    spark-examples.jar,10000
    ```

4. Run the following command multiple times to see how the cluster resizes until it hits the maximum capacity allowed (11 nodes in total). Notice that it first changes RequestedInstanceCount, then eventually scales up to that, and then scales down again when no longer needed. In the rare case that the step finishes before the scaling kicks in, you can repeat *step 3* and this step as needed:

    ```
    aws emr describe-cluster --cluster-id $CLUSTER_ID \
      | grep -B 2 -A 3 "RequestedInstanceCount"
    ```

5. You can wait for the step to finish and see it scale back down to the minimum capacity. Terminate the cluster when you are done observing it:

    ```
    aws emr terminate-clusters --cluster-id $CLUSTER_ID
    ```

How it works...

In the first step, you created a cluster, which included the three kinds of group types:

*   MASTER to mainly run YARN ResourceManager and the HDFS NameNode
*   CORE nodes running HDFS and YARN
*   TASK nodes that just run YARN

The distinctive feature here is that you used managed scaling, which only requires you to tell it how much to scale and lets EMR take care of applying it.

In addition, you defined an auto termination policy of 15 minutes. This means that if the cluster is idle for that period, it will be shut down automatically. This is a good practice for clusters that do not have auto-shutdown when all steps are complete, which avoids leaving clusters forgotten and building up charges over time. The criteria of what constitutes an idle cluster are more sophisticated in newer versions (for further details, see the documentation at `https://docs.aws.amazon.com/emr/latest/ManagementGuide/emr-auto-termination-policy.html`).

Then you added the SparkPI estimation but with 10,000 partitions so that it generated enough load to force the cluster to scale up. When the task completes, the cluster will scale back to the initial capacity.

There's more…

For this recipe, you defined a maximum of 10 units of on-demand capacity (you can also use cheaper spot instances for part or all), of which a maximum of two will be core nodes. That is because this example is pure computing, so it makes sense to add most of them as `TASK` nodes, so it has more resources for Spark.

Since EMR 5.10, the scaling behavior has been optimized to terminate nodes when the instance approaches the next hour to avoid incurring another hour of cost. You can change back to the old behavior by using `--scale-down-behavior "TERMINATE_AT_TASK_COMPLETION"`. See further details in the documentation: `https://docs.aws.amazon.com/emr/latest/ManagementGuide/emr-scaledown-behavior.html`.

Remember that managed scaling is based on YARN usage. With cluster services that are not based on YARN, such as Trino or HBase, it cannot work correctly. See the next section for alternatives.

See also

- In this recipe, you have seen how to use managed scaled, which is an effortless way of doing efficient scaling, just setting the limits. However, there are cases where you might want to take control and build your own scaling rules. Check the documentation for guidance on how to build such rules: `https://docs.aws.amazon.com/emr/latest/ManagementGuide/emr-automatic-scaling.html`.

Customizing the cluster nodes easily using bootstrap actions

EMR offers the freedom to control all aspects of the cluster nodes. You can provide your own **Amazon Machine Images (AMIs)** and run your own choice of OS (within reason) and configuration.

However, if you create or even branch your AMIs, you will create a maintenance burden for the future such as patching for vulnerabilities or fixes. In practice, you rarely need to deviate so much as to justify this effort. In most cases, you just need to do small customizations, such as installing a system package, changing some OS configuration, or running some special service.

For such cases, bootstrap actions provide a straightforward way of running these customizations just by running a shell script as `root` during setup. In most cases, this means that you can run the same actions on newer versions of the OS and EMR, reducing the operational burden. If new nodes are added to the cluster, they will also run the same bootstrap actions to ensure consistency.

In this recipe, let us say that we will run an application in the cluster that requires using utilities from the popular Python `scikit-learn` machine learning library. Instead of asking applications to install it or having it preinstalled in a container or image, you will use a bootstrap action to launch an EMR cluster that comes with the package installed and ready to use.

Getting ready

This recipe assumes that you have defined the SUBNET, S3_SCRIPTS_URL and S3_LOGS_URL variables in the shell (see the *Technical requirements* section at the beginning of this chapter for more information). For this recipe, it is essential that the S3 URLs point to a bucket in the same region as EMR.

How to do it...

1. Create the bootstrap script file:

```
cat > recipe_bootstrap.sh << EOF
#!/bin/bash
set -x
/usr/bin/pip3 install -U scikit-learn

EOF
```

2. Copy the file onto S3:

```
aws s3 cp recipe_bootstrap.sh $S3_SCRIPTS_URL/
```

3. Create a cluster that uses the bootstrap script you just uploaded. Take care to only use the \ character at the end of the lines indicated:

```
BOOTSTRAP_SCRIPT=$S3_SCRIPTS_URL/recipe_bootstrap.sh
CLUSTER_ID=$(aws emr create-cluster \
--name BootstrapRecipe --log-uri $S3_LOGS_URL \
--auto-terminate  --release-label "emr-6.15.0" \
--instance-type=m5.xlarge --instance-count 2 \
--applications Name=Spark --use-default-roles \
--ec2-attributes SubnetId=$SUBNET --bootstrap-action\
 Name=PythonDeps,Path=$BOOTSTRAP_SCRIPT --steps \
'[{"Name": "Validate bootstrapaction",
"ActionOnFailure":"TERMINATE_CLUSTER", "Jar":
"command-runner.jar","Properties":"","Args":
```

```
["/usr/bin/python3","-c","exec(\"import sklearn'\
'\\nprint(sklearn.__version__)\")"],
"Type":"CUSTOM_JAR"}]')
```

4. You can use the AWS console, navigate to EMR, select **Clusters** on the left menu, and then select the one named **BootstrapRecipe**. Explore the different tabs, including **Bootstrap Actions** and **Steps**. If all goes correctly, the cluster will start and then run the step configured, which will check the package version and terminate the cluster automatically.

5. Once the cluster has completed the bootstrap, run the step, and completed successfully, check the step log to verify that the bootstrap step did its job and the package was installed. It might take a few minutes after the step is executed before the log becomes available. In the **Steps** tab, open the **stdout** link. Once it becomes available, it should display the version of the library installed by the bootstrap script (for instance, 1.0.2). If something has gone wrong, the **stderr** log will show the error.

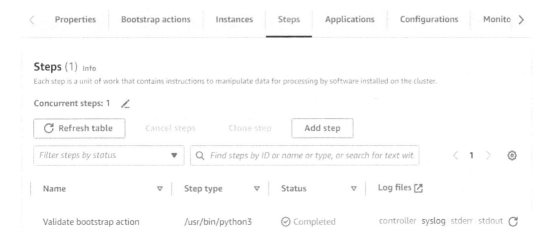

Figure 5.8 – EMR steps details

6. If you no longer need it, remove the S3 files from the shell:

```
aws s3 rm $S3_SCRIPTS_URL/recipe_bootstrap.sh
aws s3 rm --recursive $S3_LOGS_URL
```

How it works...

In *step 3*, you created a basic EMR cluster (using as little configuration as possible) with the minimum number of nodes: a primary node (formerly known as master) and a core node. The difference between a core and a task node is that the former is used for the HDFS filesystem, as well as for computing. These nodes use the modest m5.xlarge type to reduce cost.

Notice that you didn't have to specify any roles. We used the default ones provided by EMR, which serve for the common cases. Also, you didn't specify any image or OS, which means that we have let EMR use the latest version of Amazon Linux 2 and thus don't have to worry about patching it for vulnerabilities.

The subnet implicitly determines the VPC that will be used; in this case, it sets no security configuration or ssh access, since there was no need to access the cluster nodes to complete the recipe.

The cluster instructions included a bootstrap action to call the script that you put on S3 following *steps 1* and *2*. A bootstrap action is a reference to a script that gets executed as `root` once the OS is ready but before services in the cluster are started. This allows the bootstrap script to do OS configuration and tuning. Once EMR has completed the provisioning of nodes, the bootstrap, along with the installation of the configured apps (in this case just Spark), EMR starts executing the steps defined (if any).

In this recipe, the Spark step cluster runs the simple shell script that runs Python code to import the package that was installed on the OS by the bootstrap script. The step is configured to terminate the cluster if it fails; this will mark the cluster as terminated with errors.

If there is no issue, the script prints the version of the package, showing that the package is installed. Notice that the step only runs on the primary node, but the bootstrap script is applied on all nodes; therefore, we could use this package's utilities in a PySpark-distributed script.

Finally, because the cluster was started with the `--auto-terminate` flag, it shuts itself down after completing the steps. Otherwise, the cluster would be left running until it is manually shut down or hits the idle timeout (if configured).

There's more...

Remember that the intent of this feature is not to configure the EMR applications such as Spark conf, as such configuration would be overridden when the apps are installed and when using the application options provided by EMR.

In the recipe, you checked the step output log. However, there are also `stdout` and `sterr` logs for the bootstrap action. Go check them to see the bootstrap script running. You could print messages in the script and then check in these logs. Notice that there is one for each node in the cluster (two in this case). For instance, you can use: `s3://{S3_LOGS_URL}/<your cluster id>/node/<node id>/bootstrap-actions/1/`.

It's possible to pass parameters to the script. That way, you can reuse a bootstrap script and make it configurable to your needs. For instance, you could indicate the set of packages that the script needs to install based on the cluster purpose. You can pass shell arguments to the script using an `Args` parameter after the script path. For instance, here, the script will receive the `arg1` and `arg2` arguments, which the script can access using Bash syntax (`$1` and `$2`):

```
--boostrap-action \
Name=PythonDeps,Path="$BOOTRAP_SCRIPT}",Args=[args1,arg2]
```

Finally, notice that the script runs without user variables such as PATH. That's why we used full paths to run the `pip` and Python interpreter.

See also

- Bootstrap actions run in all nodes before services are started. In cases where you need to do some setup once the services are running such as uploading files to HDFS, instead of a bootstrap action, add a command type step (before any other steps), which can also run a script (but unlike bootstrap, only on the master). See how to do this in the documentation: `https://docs.aws.amazon.com/emr/latest/ReleaseGuide/emr-commandrunner.html`.

Tuning Apache Spark resource usage

Apache Spark is probably the most-used framework on EMR. It works by running YARN containers for executor instances and another one for the **Application Manager** (**AM**), which often also acts as the Spark driver.

In traditional on-premises clusters, the cluster has many resources that are shared by many cluster users, so you use as few resources as possible for the job at hand. On EMR, it is simple to run clusters on demand, which are fit for purpose, and then shut them down when the job is complete. That way, the cluster size, nodes, and configuration can be optimized for the specific job and negative interactions between users, such as resource starvation or saturation, can be avoided.

In such dedicated clusters, you want your application, such as Apache Spark, to make the most of the hardware provided since it doesn't have to share it with other users. That's why EMR added the `maximizeResourceAllocation` configuration option for Spark. It is enabled by default and indicates to EMR to automatically configure the Spark executors' resource usage based on the cluster capacity. That includes cores and memory, and, unless `spark.dynamicAllocation.enabled` is explicitly enabled, it will also calculate the number of instances to use.

However, it's possible when using EMR on EC2 that the cluster can have nodes of different sizes. That's why since EMR versions 5.32.0 and 6.5.0, a `spark.yarn.heterogeneousExecutors.enabled` flag was added and enabled by default. This feature adjusts the containers requested by Spark to run executors dynamically.

While these configuration features provide good defaults and easy configuration, they might not be optimal, and you might want to tune them yourself; you'll see how in this recipe.

Getting ready

This recipe assumes that you have defined the SUBNET and S3_LOGS_URL variables in the shell (see the *Technical requirements* section at the beginning of the chapter for more information).

How to do it...

1. In a shell, run the following command. If all is correct, you should get a JSON printed with `id` and `arn` of the cluster that was just created:

    ```
    aws emr create-cluster --name SparkResourcesRecipe \
    --release-label emr-6.15.0 --auto-terminate \
    --instance-type=m5.2xlarge --log-uri $S3_LOGS_URL \
    --instance-count 2 --applications Name=Spark \
    --use-default-roles --ec2-attributes \
    SubnetId=$SUBNET --steps Type=Spark,Name=SparkPi\
    ,ActionOnFailure=TERMINATE_CLUSTER,Args=\
    --deploy-mode,cluster,--class,org.apache.spark.\
    examples.SparkPi,/usr/lib/spark/examples/jars/\
    spark-examples.jar,12
    ```

2. Go to the AWS EMR console and select **Clusters** under **EMR on EC2** on the left menu. Select the cluster that was just created (with the ID matching the one returned in *step 1*). In the **Steps** tab, there is a Spark application to be executed once the cluster is ready. It should finish in less than a minute and after that, the cluster will shut down. Note that we didn't specify any resources for Spark to use.

3. Once the previous step has been completed, on the cluster summary section of the page, select the **YARN timeline server** link. After a few seconds, it will open a new tab with the YARN application history. If you don't see the SparkPl application listed, there might be a bit of a delay in publishing the logs. Refresh the browser page until it appears.

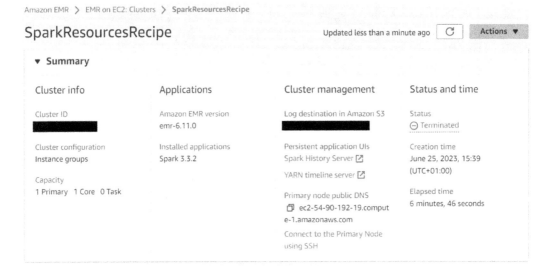

Figure 5.9 – The cluster summary

4. Click on the application ID, which is named in the format: `application_<timestamp>_0001`. This will take you to the table of attempts. There should be only one; click on the attempt ID link. Now, you should see the containers that this app was allocated in. There should be two. The one ending in `000001` is the driver and the other is the executor:

Figure 5.10 – The application container list

5. Click on each container ID. It will take you to the container details. Notice that each one has a different memory allocated, as indicated in the **Resource** line.

6. Now go back to the **Cluster summary page** tab and use the **Spark History Server** link. After a few seconds, a new tab will open wherein the SparkPI application should be listed. Use the link on the app ID to open SparkUI.

7. In SparkUI, navigate to the **Executors** tab. The executor is given six cores and over 7 GB of storage memory, which is half of the JVM heap size by default.

8. You can use the options to control some of the aspects such as the cores, memory, and overhead; the system algorithm will try to adapt, but that often leads to confusing results. Now, you are going to start a cluster where you will disable the automation logic and run it with custom executor resources:

```
aws emr create-cluster --name SparkResourcesRecipe2 \
  --release-label emr-6.15.0 --applications Name=Spark\
  --instance-type=m5.2xlarge --instance-count 2\
  --use-default-roles --ec2-attributes \
  SubnetId=${SUBNET} --log-uri $S3_LOGS_URL \
--auto-terminate --configurations \
'[{"Classification":"spark-defaults","Properties":
{"spark.yarn.heterogeneousExecutors.enabled":"false"}}
]' --steps Type=Spark,Name=SparkPi,ActionOnFailure\
```

```
=TERMINATE_CLUSTER,Args=[--deploy-mode,cluster,\
--executor-cores,4,--executor-memory,6G,--conf,\
'spark.yarn.executor.memoryOverhead=1G',\
--class,org.apache.spark.examples.SparkPi,\
/usr/lib/spark/examples/jars/spark-examples.jar,12]
```

9. Check the Spark resources like in *steps 4* and *5*. Notice the differences in the core node resource usage; now you have three executors on the instance using 21 GB of RAM and 12 Spark cores (YARN only allocates one per executor, this is explained in the next paragraph).

Executor ID	Address	Status	RDD Blocks	Storage Memory	Disk Used	Cores
driver	ip-172-30-0-150.ec2.internal:35337	Active	0	0.0 B / 1008.9 MiB	0.0 B	0
1	ip-172-30-0-150.ec2.internal:37287	Active	0	0.0 B / 3 GiB	0.0 B	4
2	ip-172-30-0-150.ec2.internal:37513	Active	0	0.0 B / 3 GiB	0.0 B	4
3	ip-172-30-0-150.ec2.internal:34007	Active	0	0.0 B / 3 GiB	0.0 B	4

Figure 5.11 – Spark executors' details

10. Clean up the log files if you don't want to keep them:

```
aws s3 rm --recursive $S3_LOGS_URL
```

How it works...

In *step 1*, you created an EMR on EC2 cluster with default resource policies and the minimum number of nodes: a primary node and a core node (the difference between primary and task is that the latter doesn't run the HDFS filesystem). The cluster was created with a step already set up and configured to shut itself down automatically to save on cost. It's also possible to create the cluster, leave it running (remove the `--auto-terminate` flag), and add steps later.

The cluster was created using YARN and Spark configurations fit for the kind of node used. In addition, the `heterogeneousExecutors` feature dynamically adjusts to try to fit a single executor per node, even if nodes have different sizes or are added after the cluster is configured (for instance, if you add a task group of a different type of instance).

When the Spark application was executed, it first allocated the AM container. In this recipe, that container also runs the Spark driver. This is because of `--deploy-mode cluster`. Then it automatically created an executor for the instance trying to use the rest of the available memory.

In YARN, by default, it allocates the memory available but doesn't enforce vcores, which are sharable thanks to CPU threads. Spark cores are really threads, so they don't need to match the cores that YARN has assigned, since it won't enforce vcore usage by default.

If you check the SparkUI **Environment** tab, you will see that EMR set the `spark.executor. memory` property to 4,743M for the cluster, but that was dynamically overridden by the `heterogeneousExecutors` feature to use a JVM of nearly 15 GB (you can know that because by default, half of the heap is used for storage, which you can see in the **Executors** tab).

There is also a discrepancy with the container size allocated on YARN, which is 16,896M. That difference is called the **memory overhead** and is used for memory outside of the heap such as the stack, I/O buffers, unmanaged memory, or Python daemons used for Python UDFs. The driver also has some memory allocated by YARN outside of the JVM. If a process exceeds the memory allocated, YARN will kill it, but it will give you a message suggesting increasing the overhead.

You might have noticed that the two containers on YARN amount to less than 20 GB of RAM, while `m5.2xlarge` has 32 GB. Bear in mind that YARN cannot allocate all the memory on the machine for containers, since there are other processes such as Hadoop itself that need memory. It depends on the kind of node but, in general, about 80% of the memory is used for containers (which is adjustable in the EMR configuration but is risky to lower).

In the end, the automatic allocation used less than 20 GB of the 24 GB available in the worker node and six out of the eight cores. It was very conservative.

In *step 7*, you started a cluster but this time with the `heterogeneousExecutors` feature disabled. In this case, it was disabled at the cluster level. However, it is also possible to disable it in the Spark submit options, like any other Spark configuration property.

Once that feature was disabled, you could control the number of cores, the size of the heap, and the overhead used exactly – 1 G in this example, which is plenty for a job that doesn't require unmanaged memory use. The only thing that was automatic in this case is that by default, EMR is configured to maximize resources (assuming that the cluster is not shared and that you want the best performance for your app), so it allocated as many executors as it could fit. Each executor container was configured with 6 GB heap and 1 GB overhead for a total of 7 GB. 2.5 GB was used by the driver/AM, so that leaves room for three executors.

Also, notice that each executor is configured with four Spark cores, so in total, the executors will use 12 when the machine has eight vcores and there are other processes running. This kind of overcommitting can help maximize CPU usage when the job is low on computing and high on I/O. Otherwise, it might hurt performance a bit due to the CPU context switching. In this case, it was done for demonstration purposes, since the job is purely doing computing so it won't benefit from this core configuration.

There's more...

There is another element of fine-tuning that you can do if you don't require the Spark driver to be failure-tolerant. If your job has a plan that is more complex than the trivial example in this recipe, it is likely that you will require multiple GB of memory, while the primary node memory is underused at the same time (especially as you move to larger instances).

If you run Spark without the `--deploy-mode cluster` argument (which we used in this recipe), the Spark driver will run on the primary node, and the first container is just a small AM using less than 1 GB of memory (you can reduce that further with configuration if the job doesn't need many containers to manage). This can free up worker node resources for executors.

Now that you know how you can fine-tune the resource usage. Instead of trial and error, it's better to get system-level information in addition to SparkUI. For that, you can use an application named **Ganglia**, which you can enable in the cluster application list and then access via a browser once you set up a ssh tunnel to access it (see the documentation at `https://docs.aws.amazon.com/emr/latest/ReleaseGuide/view_Ganglia.html`). With Ganglia, you can get detailed information about system CPU and IO usage both overall and per host.

You might also want to try to run clusters with multiple types of core and task nodes without disabling `spark.yarn.heterogeneousExecutors.enabled` and see how the algorithm reacts and adjusts. This depends on the EMR version, but it has a component of trial and error whereby the algorithm requests containers that fail to allocate or are discarded.

See also

- If you run a job without special requirements and want to simplify resource usage, you can just use EMR serverless or EMR on EKS. If this is the case for you, see the corresponding recipes:

  - *Running jobs using AWS EMR serverless*

  - *Running your AWS EMR cluster on EKS*

Code development on EMR using Workspaces

Developing data processing code on complex distributed frameworks is much more productive when it is done in an interactive way by using representative data and seeing the results of the transformations done on each step. This has led to an increase in the popularity of languages that can be interpreted interactively, such as Python or Scala.

While you can do some interactive development via a shell, as the code becomes larger, it stops being practical. The productive way to do this is via a notebook with cells, where each cell holds and executes a block of code, but the variables are common to the notebook so the work you do in one cell is visible to the others. That way, you can develop and test a small piece of code at a time and see the results.

EMR has traditionally supported this style of development with Apache Zeppelin, which can be installed on the cluster to run multiple types of notebooks including Spark or Bash, with multiple languages such as Python or Scala. Accessing the notebooks requires opening a port to the primary node (or setting up a proxy such as Apache Knox).

The rise in popularity of Python brought its preferred notebook environment, Jupyter, to EMR, which was added as an alternative to Zeppelin.

EMR Workspaces (previously called notebooks) put the Jupyter notebooks at the center of the service, instead of a cluster that runs Jupyter. This service consists of a JupyterLab app, which can be linked to existing or new clusters. In this recipe, you will see how to do this.

How to do it...

1. In a shell, create the role that you will use for EMR Studio:

```
ROLE_NAME=EMR_Notebooks_RecipeRole

echo '{"Version":"2012-10-17","Statement":[{"Effect":
"Allow","Principal":{"Service":
"elasticmapreduce.amazonaws.com"},"Action":
"sts:AssumeRole"}]}' > role-assume.json

aws iam create-role --role-name $ROLE_NAME \
--assume-role-policy-document file://role-assume.json

aws iam attach-role-policy --policy-arn \
arn:aws:iam::aws:policy/service-role/\
AmazonElasticMapReduceRole --role-name $ROLE_NAME

aws iam attach-role-policy --role-name $ROLE_NAME \
--policy-arn arn:aws:iam::aws:policy/service-role/\
AmazonElasticMapReduceEditorsRole

aws iam attach-role-policy --policy-arn \
arn:aws:iam::aws:policy/AmazonS3FullAccess \
--role-name $ROLE_NAME

rm role-assume.json
```

2. Open the AWS console in a browser and navigate to EMR. On the left menu, under **EMR Studio**, select **Studios** and then **Create Studio**.

3. Choose the **Custom** option, name it RecipeStudio and choose **EMR_Notebooks_RecipeRole** (you created this in the first step).You can let it create an S3 bucket or choose an existing one.

4. Enter `RecipeStudio_Workspace` in the **Workspace name** field, extend the **Networking and security** section, and select a VPC and subnets (by default, each region has one VPC with three subnets). Choose **Create Studio**.

5. Select **Workspaces (Notebooks)** on the left, choose the one that was just created, and then choose **Quick Launch**. It will open a new tab with JupyterLab.

6. In JupyterLab, the left bar shows the file explorer with a notebook file named like the workspace by default. Double-click on the name to open it; it will prompt you to select a kernel. Choose **PySpark**.

7. Now we need a cluster that Spark can run in. Open the second tab on the left bar; it will show the **Compute** panel:

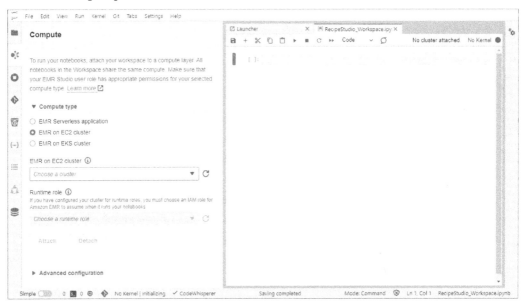

Figure 5.12 – JupyterLab showing the compute configuration

8. Choose the **EMR on EC2 cluster** type and expand the **Advanced configuration** section. Enter the name `RecipeCluster`, pick a release and a bucket to store logs, and two `m5.xlarge` nodes or similar. For this kind of cluster, it is especially important to have an auto termination on idle setup, since it is easy to forget about the cluster and build up charges. Choose **Create cluster** and wait until it is ready. It will give you the cluster ID, which starts with `j-`, and then you can go to EMR and monitor the progress.

9. Once it is running, it should show a message and update the **EMR on EC2 cluster** dropdown. In that dropdown, choose **RecipeCluster | Attach**.

10. Once the action is completed, you will have a cluster powering the PySpark notebook. To prove this, on the first cell, type `spark` and use the **Run** button on the bar (or use the *Shift + Enter* shortcut). The icon on the left of the cell will become an asterisk to indicate that it's in process. Once it finishes, it will show the YARN ID of the Spark application running on the EMR cluster.

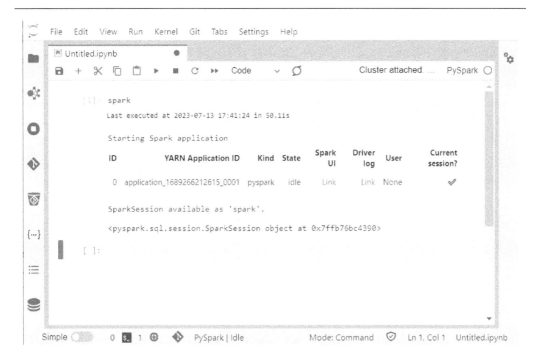

Figure 5.13 – JupyterLab showing the notebook executor results

11. Stop the EMR cluster. At the time of writing this book, you cannot do it from JupyterLab. Go to the EMR console, choose **Clusters** under **EMR on EC2** in the left menu, select the cluster named **RecipeCluster**, and terminate it.

12. Go back to the **Workshop** screen, select the **RecipeWorkspace** row, and choose **Delete** in the **Action** menu (or **Stop** if you want to keep it). You might also want to empty any temporary files on S3, depending on which S3 bucket you chose in *step 3*.

There's more...

In the first step, you created a role to use on EMR Studio. In this example, it has unrestricted access to all the S3 buckets in the account. In a production system, it is better to narrow down the permissions to the bucket and path that will be used by Studio.

Then you created an EMR Studio. This is something you normally only need to do once and can then reuse for all your Workspaces.

Afterward, you created a Workspace. This was previously called a notebook but has since been renamed, as now it provides an environment where you can have multiple notebooks and integrations with other tools such as Git or Presto (if you have them installed on the attached cluster).

Next, you launched the Workspace. In practice, it starts automatically when you create it. When using **Launch** with options instead of the quick launch, you could have specified a cluster to attach to. The cluster must already be running in one of the subnets configured on Studio and have the `JupyterEnterpriseGateway` application installed (so the Workspace can attach to it). Then, if you want to use Spark, you would need Spark, Hadoop, and Livy installed. Normally, it is easier to just create a fit-for-purpose cluster inside the Workspace as you did in the recipe and then terminate it or let it expire. You can create a cluster template so you do not have to configure it every time.

Bear in mind that it is possible to share the Workspace with other users to enable collaboration on the development, although working independently is often desirable.

See also

- If you only use PySpark in your notebooks, you have the option to use a Glue Notebook instead, which provides an integrated Jupyter notebook and cluster with minimum configuration and a cluster created on demand. You can see an example of Glue notebook usage in the *Optimizing your catalog data retrieval using pushdown filters and indexes* recipe in *Chapter 3, Ingesting and Transforming Your Data with AWS Glue*.

- If you are a heavy user of SageMaker, you can link your SageMaker notebooks with EMR or even easier to use a Glue kernel and run a serverless cluster. You can learn more about this at `https://docs.aws.amazon.com/sagemaker/latest/dg/studio-notebooks-emr-spark-glue.html`.

Monitoring your cluster

The convenience of using EMR to create fit-for-purpose, discardable clusters has significantly reduced the maintenance needs for Hadoop clusters compared to long-lived, multitenant on-premises clusters.

However, there is still a need to monitor how the cluster is doing in detail, in cases where you need to optimize the use or troubleshoot an issue. For instance, you might wonder what the limiting factor to performance is: is it the CPU, memory, network, disk, or something else?

In this recipe, you will see how to go deep into the cluster metrics and monitoring tools that it provides out of the box.

Getting ready

To carry out this recipe, you need to set up the SUBNET, S3_LOGS_URL, and KEYNAME shell environment variables (see the *Technical requirements* section at the beginning of this chapter to learn how to set them up).

To complete the recipe, you will need a **SOCKS5** proxy in your browser to access the cluster. Follow the instructions depending on your browser, so that when you access a `**.amazonaws.com` URL, it uses the localhost proxy on port `8157` that you'll create. You can learn more at `https://docs.`

`aws.amazon.com/emr/latest/ManagementGuide/emr-connect-master-node-`
`proxy.html.`

Figure 5.14 – A proxy profile

Figure 5.15 – Auto-switch in SwitchyOmega

How to do it...

1. Create a tiny cluster that will keep running so we can monitor it:

```
CLUSTER_ID=$(aws emr create-cluster --name \
MonitorRecipe --release-label emr-6.15.0 \
--instance-type=m5.xlarge --instance-count 2 \
--use-default-roles --use-default-roles \
--ec2-attributes SubnetId=$SUBNET,KeyName=$KEYNAME \
--applications Name=Spark Name=Ganglia --steps \
Type=Spark,Name=SparkPi,ActionOnFailure=\
TERMINATE_CLUSTER,Args=--deploy-mode,cluster,\
--class,org.apache.spark.examples.SparkPi\
,/usr/lib/spark/examples/jars/spark-examples.jar\
,100000 --log-uri $S3_LOGS_URL \
--auto-termination-policy IdleTimeout=3600 \
| grep ClusterArn | grep -o 'j-.*[^",]')

echo $CLUSTER_ID
```

2. Keep running this command until you get the public DNS name:

```
aws emr describe-cluster --cluster-id $CLUSTER_ID \
  | grep 'Dns'
```

3. Open the AWS console, navigate to EMR, and find the cluster with the same ID as the one printed in *step 1* (on the same region you are using the AWS CLI in). Notice that just by using the console, you can do basic monitoring tasks conveniently. You can open the historical YARN and Spark UIs in the summary, check the list and status of nodes on the **Instances** tab, and view several metrics on the **Monitoring** tab. For instance, in this case, the app that is running uses two YARN containers: a Spark driver and an executor. The app does not move any data but causes significant system load and memory usage.

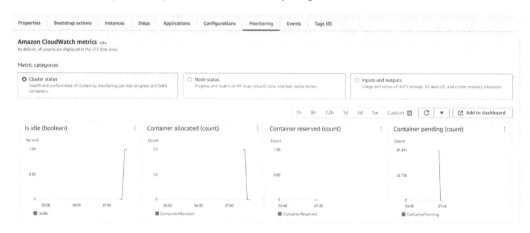

Figure 5.16 – EMR cluster status metrics

4. To dive deeper and explore timely metrics, you'll need to access the cluster. For that, you need to allow SSH access first. In the cluster page, switch to the **Properties** tab, expand the **EC2 Security groups** section, and open the link under **Primary node EMR managed security group**.

5. On the security group page, select **Edit inbound rules** | **Add rule**. In the new role on the dropdowns, select **SSH** and **My IP**. Save the rule changes.

Figure 5.17 – Choosing a new SSH rule

6. Run the following command to enable port forwarding on your machine toward the primary cluster node:

```
aws emr socks --cluster-id $CLUSTER_ID\
  --key-pair-file $KEYNAME.pem
```

If the command results in **Connection timed out**, it means that the previous step wasn't completed correctly or that the IP that was detected by the browser is not the one that the shell is really using to access AWS. In that case, you would need an alternative method to find the IP to allow, for instance, running the following in the shell:

```
curl ifconfig.me
```

7. In your browser, switch the proxy manually if you did not define automated rules in the *Getting ready* section of this chapter. Navigate to the master DNS name retrieved in *step 2* and specify the 8088 port, for instance: `http://ec2-X-X-X-X.compute.amazonaws.com:8088`.

8. This should open the live YARN UI. It looks like the screen that you see on the **YARN timeline server** summary link. The difference is that this one has far more details about the cluster status. Under **Cluster Metrics**, you can see the resources that are available and assigned. In the following screenshot, under **Cluster Node Metrics**, you can see the status of the nodes. There should be just one active; you can click on the number list of each state and see details of the nodes (for instance, why a node is unhealthy).

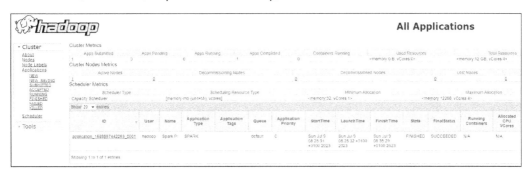

Figure 5.18 – The YARN application list

9. The cluster creation included the Ganglia application. Ganglia can provide you with low-level details of the cluster usage globally and per node, such as I/O, disk, or CPU usage. On the browser with the proxy enabled, enter the master node DNS with the /ganglia location. For instance, you might enter the following: `http://ec2-X-X-X-X.compute-1.amazonaws.com/ganglia/`.

10. Under the time range selector, you should see metrics per node instead of aggregated. Compare the metrics of the two nodes and note the difference between the primary and the core node (where the Spark application ran) CPU usage.

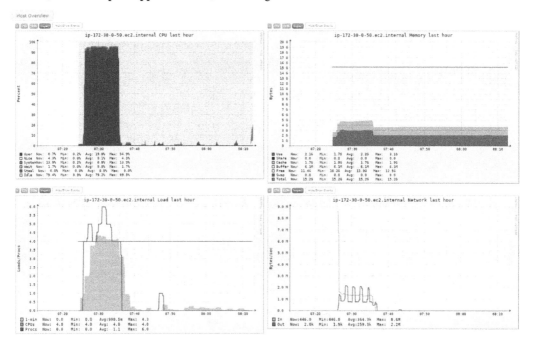

Figure 5.19 – Ganglia cluster metrics

11. Once you have explored the metrics on your own, stop the proxy in the command line using the *Ctrl + C* key shortcut, remove the rule you added in *step 5*, terminate the cluster, and clean up the logs:

```
aws emr terminate-clusters –cluster-id $CLUSTER_ID
aws s3 rm --recursive $S3_LOGS_URL
```

There's more

In this recipe, you created a tiny cluster with a Spark step to generate some workload. The cluster also has an automatic idle auto-shutdown of 30 minutes. This is good practice to limit expenses in case you forget to terminate the cluster.

Then, you created a ssh tunnel using the AWS CLI tool. Notice that the command indicates the openssh command it is running ssh -o StrictHostKeyChecking=no -o ServerAliveInterval=10 -ND 8157 -I yourkey.pem hadoop@yourserver.com.

This command tells ssh to accept the server key without asking for confirmation, to keep the connection alive, and to do dynamic port forwarding from the local `8157` port. If you wanted to use a different local port for some reason, you would need to install and run the ssh command yourself. This port acts as a `socks5` proxy.

Remember that you can always ssh into the primary node, copy and paste the `.pem` file content, and then use that file to ssh to other hosts on the cluster (you can get the internal names of the other nodes using one of the tools that you have seen in this recipe). You have the option of using OS-specific tools such as `iostat` or `vmstat`.

See also

- Instead of monitoring the cluster actively, in most cases, you will want to receive an alert when something is abnormal (for instance, if you have unhealthy YARN nodes). You have the metrics that you see on the console monitor screen on CloudWatch, but the challenge is that they are cluster-specific, so you would need to create these specific alarms on the same script that automates the cluster creation. You can learn more in the documentation: `https://docs.aws.amazon.com/emr/latest/ManagementGuide/UsingEMR_ViewingMetrics.html`

Protecting your cluster from security vulnerabilities

Historically, each version of EMR was provided with an AMI, on which it was tested. In addition, on provisioning the image, it automatically checks the repository for security updates. You can only disable that behavior using `repo-upgrade-on-boot=NONE`. Remember that when using a custom AMI, you must take the responsibility of keeping the image patched. However, patches that affect the kernel require a restart to be installed. Thus, in the past, you had two options:

- Upgrade to a newer version of EMR, which means taking upgrades in all components and services. This could cause your application to need to be retested and would potentially require a migration.

- Indicate EMR to use a newer AMI and risk running the services on an image on which it has never been tested.

That has changed since EMR 5.36 and 6.6, if using EMR 5 or 6, respectively. From those versions onwards, you can let EMR upgrade the AMI automatically with a patched AmazonLinux2 OS without having to do any testing or upgrading yourself.

Getting ready

This recipe assumes that you have defined the `SUBNET` variable in the shell (see the *Technical requirements* section at the beginning of this chapter for more information).

How to do it...

1. Create a cluster that will terminate as soon it runs:

```
CLUSTER_ID=$(aws emr create-cluster --release-label \
emr-6.15.0 --auto-terminate --instance-type=\
m5.xlarge  --instance-count 2 --use-default-roles \
--use-default-roles --ec2-attributes \
SubnetId=$SUBNET | grep ClusterArn \
| grep -o 'j-.*[^",]')
```

2. Check which Amazon Linux version was used:

```
aws emr describe-cluster --cluster-id $CLUSTER_ID\
 | grep OS
```

3. List the EMR version details; at the end, there is the `AvailableOSReleases` section. Notice that the first OS label listed (the most recent one) is the one that was returned in the previous step:

```
aws emr describe-release-label --release-label\
 emr-6.15.0
```

4. Stop the cluster to avoid further charges:

```
aws emr terminate-clusters –cluster-id $CLUSTER_ID
```

How it works...

The cluster in this example had no steps and was set to auto-terminate when it had completed all tasks. So, it terminated as soon as it was started. The point here was to prove that the cluster automatically used the most up-to-date version of AmazonLinux2 for EMR.

The cluster was created with a newer version than the minimum required (6.11 versus 6.6) and you did not specify the `--os-release-label` flag. Thus, EMR was free to automatically select the latest verified release for EMR with all the security patches. At the time of writing this book, the latest version automatically selected for 6.11 is `AmazonLinux 2 2.0.20230628.0`.

This not only provides automatic security but also reduces the EMR provisioning time, since the number of patches that must be applied on start is significantly reduced compared to an old image.

There's more...

You can find the details of the OS release picked by EMR on the AmazonLinux2 release notes at `https://docs.aws.amazon.com/AL2/latest/relnotes/relnotes-al2.html`.

See also

- In this case, you have seen how short-lived clusters get OS security patches. However, that does not cover applications that need patching, since in most cases, patches for applications are not security-related. If that is the case, you will need to consider upgrading to a version of EMR or doing custom patching using bootstrap action; see the *Customizing the cluster nodes easily using bootstrap actions* recipe for reference.

6

Governing Your Platform

Ensuring data integrity, compliance, and security is fundamental in data engineering, providing a reliable foundation for accurate analysis and informed decision-making. Data governance involves establishing clear policies and standards to govern data throughout its life cycle. On the resource management front, governance ensures that best practices and policies are not only defined but also actively applied and enforced. Automation is essential in this process, ensuring consistent adherence to governance policies and facilitating smooth implementation of policy updates.

This chapter explores data governance techniques, such as data masking, implementing quality checks, and classifying data to identify sensitive information. Additionally, it explores resource governance, ensuring effective implementation and enforcement of best practices and policies across the AWS environment.

The following recipes will be covered in this chapter:

- Applying a data quality check on Glue tables
- Automating the discovery and reporting of sensitive data on your S3 buckets
- Establishing a tagging strategy for AWS resources
- Building your distributed data community with Amazon DataZone following data mesh principles
- Handling security-sensitive data (PII and PHI)
- Ensuring S3 compliance with AWS Config

Technical requirements

Several recipes in this chapter require having S3 buckets. Glue tables and Redshift clusters are used in the recipes as well.

You can find the code files for this chapter on GitHub repository using the following link: `https://github.com/PacktPublishing/Data-Engineering-with-AWS-Cookbook/tree/main/Chapter06`.

Applying a data quality check on Glue tables

Glue Data Quality is crucial for maintaining the integrity and reliability of data within the AWS Glue environment. It ensures that data conforms to specified quality standards, allowing organizations to trust and rely on the accuracy of their data-driven insights and decision-making processes. Implementing data quality checks helps identify and rectify issues, such as missing values, inconsistencies, and inaccuracies in datasets, promoting data reliability and reducing the risk of making decisions based on flawed information. AWS Data Quality allows you to enforce quality checks on your data on transit and rest using **Data Quality Definition Language** (DQDL). This allows you to proactively apply data quality rules to your jobs and tables, helping to identify potential issues early. Additionally, you can enforce rules on multiple tables and configure actions or alarms based on detected quality issues, preventing larger problems, such as making decisions based on inaccurate data.

In this recipe, we will learn how to implement data quality checks at rest on a Glue table.

Getting ready

For this recipe, you need to have a Glue table and an IAM role with the following policies (make sure to replace `bucket_name`, `aws_region_id`, and `aws_account_id` for the Glue source with your own values):

- This is the Glue Data Quality rule recommendation policy:

```
{
    "Version": "2012-10-17",
    "Statement": [
        {
            "Sid": "AllowGlueRuleRecommendationRunActions",
            "Effect": "Allow",
            "Action": [
                "glue:GetDataQualityRuleRecommendationRun",
                "glue:PublishDataQuality",
                "glue:CreateDataQualityRuleset",
                "glue:GetDataQualityRulesetEvaluationRun",
                "glue:GetDataQualityRuleset"
            ],
            "Resource": "arn:aws:glue:<aws_region_id>:<aws_
account_id>:dataQualityRuleset/*"
        },
        {
            "Sid":
"AllowS3GetObjectToRunRuleRecommendationTask",
            "Effect": "Allow",
            "Action": [
                "s3:GetObject"
```

```
            ],
            "Resource": [
                "arn:aws:s3:::aws-glue-*"
            ]
        }
    ]
}
```

- This is the Glue Catalog policy:

```
{
    "Version": "2012-10-17",
    "Statement": [
        {
            "Sid": "AllowCatalogPermissions",
            "Effect": "Allow",
            "Action": [
                "glue:GetPartitions",
                "glue:GetTable"
            ],
            "Resource": [
                "*"
            ]
        }
    ]
```

- This is the CloudWatch and S3 policy:

```
{
    "Version": "2012-10-17",
    "Statement": [
        {
            "Sid": "AllowPublishingCloudwatchLogs",
            "Effect": "Allow",
            "Action": [
                "logs:CreateLogStream",
                "logs:CreateLogGroup",
                "logs:PutLogEvents"
            ],
            "Resource": "*"
        },
        {
            "Sid":
    "AllowCloudWatchPutMetricDataToPublishTaskMetrics",
            "Effect": "Allow",
```

```
        "Action": [
            "cloudwatch:PutMetricData"
        ],
        "Resource": "*",
        "Condition": {
            "StringEquals": {
                "cloudwatch:namespace": "Glue Data Quality"
            }
        }
    }
  ]
}
```

- This is the S3 policy:

```
{
    "Version": "2012-10-17",
    "Statement": [
        {
            "Effect": "Allow",
            "Action": [
                "s3:*"
            ],
            "Resource": [
                "arn:aws:s3:::<bucket_name>",
                "arn:aws:s3:::<bucket_name>/*"
            ]
        }
    ]
}
```

How to do it...

1. Log in to the AWS Management Console (https://console.aws.amazon.com/console/home?nc2=h_ct&src=header-signin) and navigate to the Glue service.

2. From the navigation pane on the left, go to **Tables** and open the table you need to set data quality rules for.

3. Go to the **Data quality** tab and click on **Create data quality rules**.

4. Choose the IAM role you have created and, if required, change the values of **Task timeout value** and **Number of workers** (which refers to the number of worker nodes allocated to execute the rule). Optionally, add a filter on the table that the task will run on and click on **Recommended rules**.

5. After a few minutes, the task will finish scanning the table and will generate a set of recommended data quality rules. Click on **Insert rule recommendation**, review the rules, and if you find any rules to be useful for your table, select it and click on **Add selected rules**.

6. In the **Ruleset** editor, you can define your own rule using DDQL.

7. Click on **Save ruleset**. A prompt will appear where you can enter a ruleset name and, optionally, a description and tags for the rule. After that, click again on **Save ruleset** to complete the process.

8. In the **Data quality** section, select your ruleset and click on **Run**. Select the IAM role you have created and select the run frequency for your ruleset. Keep the **Publish run metrics to Amazon CloudWatch** option selected and click on **Run**.

9. Once the run is completed, select the ruleset, and from the **Actions** drop-down list, select **Download results** and review it.

How it works...

Upon the initial establishment of a data quality ruleset, Glue Data Quality scanned our table, formulating recommended rules based on the data within each column of the input table. These rules serve as guidelines for pinpointing potential boundaries where data filtration can be applied to uphold quality standards. We have specified the set of rules to be enforced on our table, initiated a manual run that produced a resultset indicating the status of each rule—whether it has been successfully passed or not, and the **Data quality** snapshot section will show the trend of your data quality score for the last 10 runs.

There's more...

You can set up alerts on data quality issues to take proactive actions, using EventBridge to send notifications to a channel of your choice, such as an SNS topic or Lambda. Follow the outlined steps to achieve this:

1. Navigate to the EventBridge service.

2. Select **Rules** under **Buses** from the navigation pane on the left and click on **Create rule**.

3. Give a name for the rule and optionally a description. Under **Rule type**, select **Rule with an event pattern** and click on **Next**.

4. Under **Event source**, select **AWS events** or **EventBridge partner events**.

5. Choose **Use pattern form** for **Creation method**.

6. Select **AWS service** for **Event source** and then select **Glue Data Quality**. Under **Event type**, select **Data Quality Evaluation Results Available** and then choose **Specific state(s)**, select **Failed**, and click on **Next**.

7. Choose a target for the rule and create it.

See also

- *Data Quality Definition Language (DQDL) reference*: https://docs.aws.amazon.com/glue/latest/dg/dqdl.html#dqdl-syntax-rule-structure

- *Data Quality for ETL jobs in AWS Glue Studio notebooks*: https://docs.aws.amazon.com/glue/latest/dg/data-quality-gs-studio-notebooks.html

Automating the discovery and reporting of sensitive data on your S3 buckets

The identification of sensitive data within Amazon S3 is essential for maintaining data security and compliance in cloud environments. Given that S3 often contains extensive datasets, including PII, pinpointing the locations of sensitive data within S3 buckets is critical for the precise implementation of security measures, the application of relevant access controls, and the effective enforcement of data protection policies. AWS Macie provides an automated solution for the discovery, classification, and protection of sensitive data through the utilization of machine learning algorithms. This proactive approach aids in mitigating the risks associated with data breaches and ensures compliance with regulatory standards. It enables organizations to respond promptly to potential threats, constructing a resilient infrastructure to safeguard the confidentiality and integrity of their data in the cloud. Macie is particularly useful for AWS-based environments with large, unstructured datasets due to its scalability, automated data discovery, and seamless integration with other AWS services. It provides built-in data identifiers, while also allowing the creation of custom ones to meet specific needs.

In this recipe, we will learn how to create AWS Macie's job to identify sensitive data in our S3 bucket, specifically email addresses.

Getting ready

For this recipe, you need to have an S3 dataset with email addresses (dummy data) and other potential PII data (such as credit card information), and you have to enable AWS Macie if you are using it for the first time following the outlined steps:

1. Log in to the AWS Management Console and navigate to the AWS Macie service.
2. Click on **Get started**.
3. Click on **Enable Macie**.

How to do it...

1. Log in to the AWS Management Console and navigate to the AWS Macie service.
2. From the left navigation pane, choose **Jobs** under the **S3 buckets** section.

3. Click on **Create job.** You will get a confirmation message saying that creating a job is not included in the free trial; acknowledge the message by clicking on **Yes.**

4. Under **Choose S3 buckets**, click on **Select specific buckets**, choose your S3 bucket, and click on **Next** as shown:

Figure 6.1 – Selecting an S3 bucket for Macie's job

5. On the **Review S3 buckets** page, your S3 bucket will be listed with its estimated cost. Review the values and click on **Next.**

6. In the **Refine the scope** page, under **Sensitive data discovery options**, you choose either to have a scheduled job where you can set up the update frequency for the job or have a one-time job that will run the job only once. Optionally, you can set up a criterion for the object the job will run on by expanding the **Additional settings** section and choosing **Object criteria.**

Figure 6.2 – Setting up the job run frequency

7. For **Select managed data identifiers**, under **Managed data identifier options**, you can either choose to use all of the managed data identifiers Macie provides by clicking on **Recommended**, you can choose to select the specific identifiers you want to apply, or even choose not to apply any identifiers by clicking on **Custom**. For this recipe, we will go with the recommended option to identify any potential PII data. Click on **Next**.

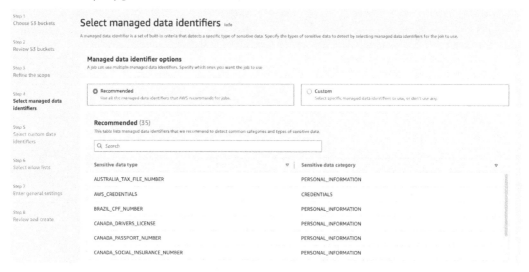

Figure 6.3 – Choosing the managed data identifiers

8. Managed data identifiers will not detect email addresses; so, we will create a custom identifier for that by clicking on **Manage custom identifiers** under **Custom data identifiers** and do the following:

I.   Click on **Create**.

II. Give a name and description to your identifier.

III. Enter (?:[a-z0-9!#$%&'*+/=?^_`{|}~-]+(?:\.[a-z0-9!#$%&'*+/=?^_`{|}~-]+)*|"(?:[\x01-\x08\x0b\x0c\x0e-\x1f\x21\x23-\x5b\x5d-\x7f]|\\[\x01-\x09\x0b\x0c\x0e-\x7f])*")@(?:(?:[a-z0-9](?:[a-z0-9-]*[a-z0-9])?\.)+[a-z0-9](?:[a-z0-9-]*[a-z0-9])?|\[(?:(?:(2(5[0-5]|[0-4][0-9])|1[0-9][0-9]|[1-9]?[0-9]))\.){3}(?:(2(5[0-5]|[0-4][0-9])|1[0-9][0-9]|[1-9]?[0-9])|[a-z0-9-]*[a-z0-9]:(?:[\x01-\x08\x0b\x0c\x0e-\x1f\x21-\x5a\x53-\x7f]|\\[\x01-\x09\x0b\x0c\x0e-\x7f])+)\]) in the **Regular expression** field, which will be used to identify the email addresses.

IV. Optionally, you can choose to add keywords and **ignore words**, which are words that Macie will ignore in the text even if they match the pattern, and the **maximum match distance**, which is the number of characters that can appear between the keyword to ignore and the sensitive data. If the distance between the keyword and the sensitive data is within this specified limit, Macie will ignore the match.

V. For **Severity**, you can keep the default value of using the **Minimum severity** setting for any number of matches or use a custom setting to determine severity.

VI. Optionally, add tags to label your identifier and click on **Submit**.

VII. Go back to the main page and click on the refresh mark to reflect the new identifier, select it, and click on **Next**.

9. Under **Select allow lists**, you can optionally create an allow list to ignore specific text or text patterns by clicking on **Manage allow lists**.

10. Under the **Review and create** page, review all the values you have provided and click on **Submit**.

11. The job will be in running status for some time depending on the data size. Once it moves to completed status, click on **Show results** and then **Show findings**, which will give you a report of the identified sensitive data within your S3 bucket.

How it works...

We kickstarted the job creation process by defining the targeted data scope for Macie to scan, specifying a particular S3 bucket. Moving forward with the job configuration, we chose to leverage all of Macie's managed data identifiers and introduced our custom data identifier since the existing ones lack a pattern for recognizing email addresses. To address this, we defined a regular expression: the first part (before @) matches the local part of email addresses, allowing both alphanumeric and special characters, which can be enclosed in quotes or not. The second part matches the domain, supporting both standard domain names with subdomains or an IP address. After submitting the job, Macie scanned the data within our designated bucket to locate sensitive data in accordance with our chosen data identifiers and its machine learning algorithms. Upon completion, it gave us a comprehensive report detailing the identified sensitive data.

There's more...

You can ensure that sensitive data discovered by Macie is promptly handled by configuring alerts and remediation actions. You can set up an EventBridge rule to be triggered when Macie discovers sensitive data, which in turn triggers a Lambda function that remediates it, such as by encrypting the data or deleting it. For alerts, you can configure SNS topics to notify you of Macie's findings.

See also

- *Processing Macie findings with Amazon EventBridge*: https://docs.aws.amazon.com/macie/latest/user/findings-monitor-events-eventbridge.html

- *Storing and retaining sensitive data discovery results with Amazon Macie*: https://docs.aws.amazon.com/macie/latest/user/discovery-results-repository-s3.html

Establishing a tagging strategy for AWS resources

Establishing a comprehensive resource tagging strategy is pivotal in the efficient organization, management, and optimization of AWS resources. It simplifies the identification and oversight of resources while facilitating precise cost attribution to designated projects, teams, or business units. To implement an effective tagging strategy, it is imperative to assign metadata labels, structured as key-value pairs, to diverse AWS resources based on specific attributes or criteria. This strategic approach is fundamental for achieving operational efficiency, cost-effectiveness, and governance within the cloud environment. Subsequently, employing proactive and reactive measures aligning with the established tagging strategy ensures the systematic tagging of resources.

In this recipe, we will learn how to proactively tag resources (EC2, S3, Lambda) using Lambda function.

Getting ready

This recipe assumes you have AWS CLI installed and configured with the IAM profile, and the IAM role used in Lambda must have the following outlined policy:

```
{
    "Version": "2012-10-17",
    "Statement": [
        {
            "Sid": "AllowLambdaTagging",
            "Effect": "Allow",
            "Action": [
                "lambda:TagResource"
            ],
            "Resource": "*"
        },
        {
            "Sid": "AllowEC2Tagging",
            "Effect": "Allow",
            "Action": [
                "ec2:CreateTags",
                "ec2:DescribeInstances",
                "ec2:DescribeVolumes"
```

```
        ],
        "Resource": "*"
    },
    {
        "Sid": "AllowS3Tagging",
        "Effect": "Allow",
        "Action": [
            "s3:GetBucketTagging",
            "s3:PutBucketTagging"
        ],
        "Resource": "arn:aws:s3:::*"
    },
    {
        "Sid": "AllowCloudTrailLookup",
        "Effect": "Allow",
        "Action": [
            "cloudtrail:LookupEvents"
        ],
        "Resource": "*"
    }
  ]
}
```

Additionally, CloudTrail must be enabled. You can follow this guide to enable it: `https://docs.aws.amazon.com/awscloudtrail/latest/userguide/cloudtrail-create-a-trail-using-the-console-first-time.html`.

How to do it...

1. Create a Lambda function with the help of the following steps:

    I.   Log in to the AWS Management Console (`https://console.aws.amazon.com/console/home?nc2=h_ct&src=header-signin`) and navigate to the Lambda service.

    II.   Click on **Create function**.

    III. Select **Author from scratch** in the **Basic information** section, give a name for the function, set the **Runtime** option to **Python 3.11**, and select **x86_64** as **Architecture**.

    IV.   Click on **Create function**.

V.   In the code editor, replace the existing code with the following code, which only handles tagging Lambda functions (you can refer to the GitHub code for the full code to handle S3 and EC2 tagging):

```python
import json
import boto3
import logging
from botocore.exceptions import ClientError


logger = logging.getLogger()
logger.setLevel(logging.INFO)

def get_user_name(event):
    if 'userIdentity' in event['detail']:
        if event['detail']['userIdentity']['type'] ==
'AssumedRole':
            user_name = str('UserName: ' + event['detail']
['userIdentity']['principalId'].split(':')[1] + ', Role:
' + event['detail']['userIdentity']['sessionContext']
['sessionIssuer']['userName'] + ' (role)')
        elif event['detail']['userIdentity']['type'] ==
'IAMUser':
            user_name = event['detail']['userIdentity']
['userName']
        elif event['detail']['userIdentity']['type'] ==
'Root':
            user_name = 'root'
        else:
            logging.info('Could not determine username
(unknown iam userIdentity) ')
            user_name = ''
    else:
        logging.info('Could not determine username (no
userIdentity data in cloudtrail')
        user_name = ''
    return user_name

def lambda_handler(event, context):
    client = boto3.client('cloudtrail')

    resource_type = event["detail"]["eventSource"]
    user_name=get_user_name(event)

    if resource_type == "lambda.amazonaws.com":
        resource_arn = event["resources"][0]
```

```
            resource_name = event["detail"]["configurationItem"]
    ["configuration"]["functionName"]
        try:
            client = boto3.client('lambda')
            client.tag_resource(
                Resource=resource_arn,
                Tags={'Created_by': user_name}
            )
            logging.info(f"Lambda function {resource_name}
    tagged with username : {user_name}")
        except ClientError as e:
            logging.error(f"Error tagging Lambda function
    {resource_name}: {e}")
```

VI.   Click on **Deploy**.

VII.   Go to the **Configuration** tab and open the IAM role under **Execution Role**.

VIII.   Click on **Add permissions**.

2.   Create EventBridge rule with the next steps:

I.   Navigate to the EventBridge service and click on **Create rule**.

II.   Give a name and optionally a description for the rule, keep **Event bus** as default, select **Rule with an event pattern**, and then click on **Next**.

III.   Under **Creation method**, select **Custom pattern (JSON editor)**, paste the following rule, and click on **Next**:

```
{
    "source": ["aws.s3", "aws.lambda", "aws.ec2"],
    "detail-type": ["AWS API Call via CloudTrail"],
    "detail": {
        "eventSource": ["s3.amazonaws.com", "lambda.amazonaws.
com", "ec2.amazonaws.com"],
        "eventName": ["CreateBucket", "CreateFunction",
"RunInstances"]
    }
}
```

IV.   Under **Target 1**, choose **AWS service**, select **Lambda function**, then select the function you have created, and then click on **Next** and **Create rule**.

If you go back to your Lambda function, an EventBridge rule trigger will be added.

How it works...

We enabled CloudTrail to log all write actions to our services, including the creation of Lambda functions, S3 buckets, and EC2 instances. Then, we set up an EventBridge rule that listens for these CloudTrail events related to the creation of these resources. This rule triggers a Lambda function, which checks the type of event it receives. Based on the event type, the Lambda function retrieves the current tags for the resource and appends a tag with the user identity that created the resource, as obtained from the CloudTrail event.

There's more...

You can have additional tags added to your resources that you can get from CloudTrail events, such as the created time or AWS region, and you can add other taggable resources to the process.

See also

- *Building your tagging strategy*: `https://docs.aws.amazon.com/whitepapers/latest/tagging-best-practices/building-your-tagging-strategy.html`

Building your distributed data community with Amazon DataZone following data mesh principles

Data mesh is a decentralized approach to data architecture, aiming to address the challenges of traditional centralized models by distributing responsibility for data and treating it as a product, which shifts the responsibility from a central team to domain-specific teams, leveraging domain expertise. Key principles in data mesh involve the following:

- Domain-oriented decentralization where each domain will be responsible for creating their data as a product and making it available to others
- Self-serve data infrastructure that empowers each domain to independently create, manage, and utilize their own data pipelines without relying on a central team
- A federated computational governance model where each domain is responsible for the governance of its data, ensuring compliance with global standards while maintaining domain-specific requirements

Amazon DataZone, based on data mesh principles, provides a managed platform for data governance and access control.

In this recipe, we will learn how to use DataZone to set up a domain, publish Glue tables, and make it accessible to others.

Getting ready

To complete this recipe, you need to have a Glue table that is managed by Data Lake (Data Lake permission mode, not hybrid mode), which means the S3 location of the table must be registered on Lake Formation and permission to `IAMAllowedPrincipals` must be revoked (refer to the *Getting ready* section of the *Synchronizing Glue Data Catalog to a different account* recipe in *Chapter 2, Sharing Your Data Across Environments and Accounts*).

How to do it...

This recipe involves different tasks, as discussed under each subsection.

Creating a domain with the next steps

1. Log in to the AWS Management Console (`https://console.aws.amazon.com/console/home?nc2=h_ct&src=header-signin`) and navigate to the DataZone service.
2. Click on **Create domain**.
3. Enter a name and optionally a description for your domain.
4. Under the **Quick setup** section, select **Set-up this account for data consumption and publishing**, which will enable Data Lake and Data Warehouse blueprints. It will create default environment profiles that are open to all users.
5. Click on **Create domain**. Once your domain is created, you can click on **Open data portal** to start building the catalog.

Figure 6.4 – Domain summary

Publishing data to your domain

1. **Create a project**:

   I. Open the data portal of your domain.

   II. Click on **Create project**.

III.   Enter a name and optionally a description for your project.

Figure 6.5 – Creating a project

2. **Create environment**:

    I.   On your project page, click on **CREATE ENVIRONMENT**.

    II.   Enter a name and optionally a description for your environment.

    III. Under **Environment profile**, choose `DataLakeProfile`.

    IV.   Click on **CREATE ENVIRONMENT**.

Figure 6.6 – Creating an environment for the project

3. **Create data source**:

    I.   Go to the **Data** tab in the portal.

    II.   Enter a name and optionally a description for your data source.

    III.   Under **Data Source type**, choose **AWS Glue**.

    IV.   Select the environment you created in the previous step in the **Select an environment** section.

    V.   Under **Data Selection**, write down the name of the Glue database you need to add, and under **Table selection criteria**, choose **Include *** to include all the tables in your database and click on **Next**.

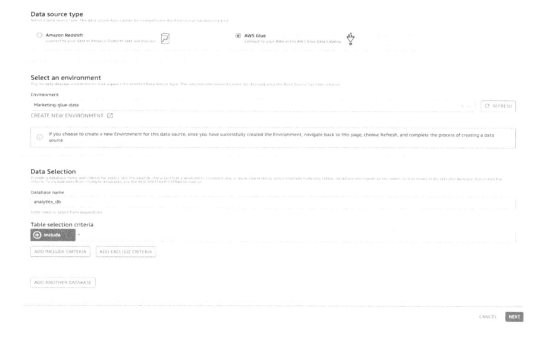

Figure 6.7 – Selecting a data source type

VI.   Under the **Publishing setting** section, select **No** for **Publish assets to the catalog**, which will allow us to review and edit the assets before publishing them. Select **automated business name generation** and click on **Next**.

Figure 6.8 – Publishing setting

VII.   Under **Run preference**, you can choose to run the data source on demand or to run it based on a schedule, and then click on **Next**.

VIII.   Review all the details you have provided and click on **Create**.

IX.   Click on **RUN** to trigger the data source. Once the run is complete, your Glue database tables will be added as assets.

X.   Open each asset created on your data source and review the business name that was generated for you and the business attribute names in the **Schema** tab. Accept or edit them if required. You can also provide metadata to your assets such as a README file and glossary terms. Once done with the changes, click on **PUBLISH ASSET**.

XI.   To allow DataZone to grant access to consumers on your behalf, grant the DataZone access role, `<AmazonDataZoneGlueAccess-<region>-<domainId>`, the following permissions through Lake Formation, as shown in *Figure 6.9*:

- Database permissions
- Table permissions

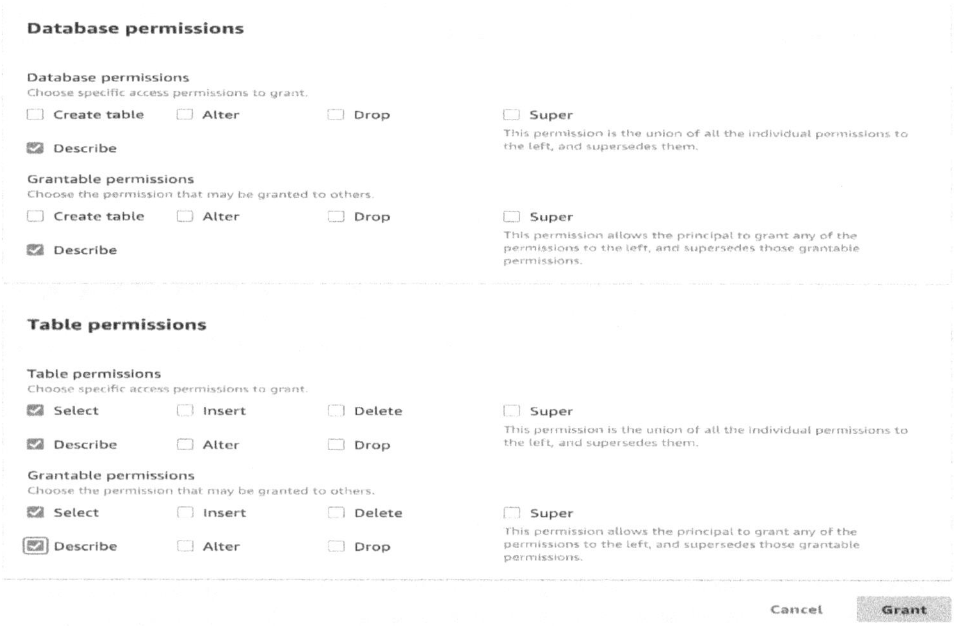

Figure 6.9 – DataZone Lake Formation permissions

Consuming data from the domain

1. Log in to the AWS Management Console using the consumer user or role and navigate to the DataZone service.

2. Create a new project and environment for the consumer following the same steps done for the producer.

3. In the search bar, search for the table/asset you have added as a producer, add a justification for subscribing to the data under the **Reason for request** field, and click on **SUBSCRIBE**.

Approving consumer access request

1. Navigate to the DataZone portal and choose the project with the published data asset.

2. Go to the **Data** tab and click on **Incoming requests** from the navigation pane on the left. You should be able to see the request under **REQUESTED** – click on **View request** and approve it.

How it works...

We initiated a domain through the quick setup, allowing DataZone to automate the creation of IAM roles and facilitate domain access to AWS resources such as Glue, Athena, and Redshift. As publishers, we configured a collaborative project for managing asset access and ownership. Within the project, we created an environment using a predefined template profile to host our resources. DataZone background processes generated IAM roles, S3 suffixes, Athena workgroups, and AWS Glue databases, streamlining Data Lake operations. We proceeded to create a data source that defined the data to bring, adding assets visible only to project members until published to the catalog for wider discovery. As a subscriber, we established our own project and environment, subscribing to assets that notify publishers for approval. Once approved by publishers, assets were then accessible for querying in Athena within the environment.

There's more...

After publishing an asset, you can still edit its business or technical values, which will create a new version of your asset that you will have to publish to make it discoverable.

See also

- *Security in DataZone*: https://docs.aws.amazon.com/datazone/latest/userguide/security.html

- *Monitoring Amazon DataZone*: https://docs.aws.amazon.com/datazone/latest/userguide/monitoring-overview.html

Handling security-sensitive data (PII and PHI)

Redshift data tokenization is a crucial aspect of data security and privacy within AWS Redshift to protect the confidentiality of critical data. Tokenization involves substituting sensitive information with unique identifiers or tokens, preserving the format and length of the original data without revealing the actual sensitive information. Tokenization is necessary to mitigate the risk of data breaches, comply with regulatory requirements, and maintain customer trust by ensuring their personal information is secure. Leveraging AWS Lambda **user-defined functions** (**UDFs**) for data tokenization in Redshift provides a scalable and efficient solution for protecting the data.

In this recipe, we will learn how to create a Lambda UDF to use in Redshift for tokenization.

Getting ready

To complete this recipe, you must have a Redshift cluster with a table in which you need to tokenize its data. You need to have a Cloud9 environment as well, please create one following the outlined steps:

1. Log in to the AWS Management Console and navigate to the Cloud9 service.

2. Click on **Create environment**.

3. Give a name and a description for the environment.

4. Select **New EC2 instance**, then select the t2.micro instance and **Amazon Linux 2** for the **Platform** setting.

5. Keep everything else set to default and click on **Create**.

How to do it...

1. **Create a secret key in Secrets Manager**:

    I.   Log in to the AWS Management Console (https://console.aws.amazon.com/console/home?nc2=h_ct&src=header-signin) and navigate to **Secret manager**.

    II. Click on **Store new secret**.

    III. Choose **Other type of secret** for **Secret type**. Under **Key/value pairs**, type secret_key and the value you would like to use for the key. It should be a long, complex value, at least 32 characters of mixed uppercase and lowercase characters, digits, and special characters. Then, click on **Next**.

    IV. Give a name for the secret and then click on **Next**. Review the details and click on **Store**.

2. **Create a Lambda layer**:

    I.   Navigate to Cloud9 service.

    II. Open a Linux environment and run the following commands. Replace aws_region with the region you will use to create the Lambda function:

```
sudo amazon-linux-extras install python3.8
curl -O https://bootstrap.pypa.io/get-pip.py
python3.8 get-pip.py --user
mkdir python
python3.8 -m pip install pyffx -t python/
zip -r layer.zip python
aws lambda publish-layer-version --layer-name pyffx-layer
--zip-file fileb://layer.zip --compatible-runtimes python3.8
--region <aws_region>
```

3. **Create a Lambda function**:

    I.   Navigate to the Lambda service.

    II.   Click on **Create function**.

    III.   Select **Author from scratch**.

    IV.   In the **Basic information** section, give a name for the function, set the **Runtime** setting as **Python 3.11**, and select **x86_64** as **Architecture**.

    V.   Click on **Create function**.

    VI.   Under **Layers** in your Lambda function's **Code** tab, select **Add a layer**.

    VII.   Select **Custom layers**, choose `pyffx-layer` created in the previous step, and select the latest version you have created.

    VIII.   In the code editor, replace the existing code with the following code:

```
import boto3
import json
import numbers
import pyffx
from botocore.exceptions import ClientError, BotoCoreError

def get_secret():
    try:
        client = boto3.client('secretsmanager')
        response = client.get_secret_
value(SecretId='RedshiftTokenizationSecretKey')
        secret = json.loads(response['SecretString'])
        return secret['secret_key']
    except (ClientError, BotoCoreError) as e:
        return json.dumps({'success': False, 'message':
f'Failed to retrieve secret key: {e}'})


def encrypt_data(data, secret_key):
    alphabet = '0123456789abcdefghijklmnopqrstuvwxyz'
    try:
        if isinstance(data, numbers.Number):
            e = pyffx.Integer(secret_key.encode(),
length=len(str(data)))
        else:
            e = pyffx.String(secret_key.encode(),
alphabet=alphabet, length=len(data))

        encrypted_text = e.encrypt(data)
```

```
        return encrypted_text
    except Exception as e:
        return json.dumps({'success': False, 'message':
f'Failed to encrypt text: {e}'})


def lambda_handler(event, context):
    secret_key = get_secret()

    return_value = dict()
    response = []
    for argument in event['arguments']:
        msg = argument[0]
        encrypted_text = encrypt_data(msg, secret_key)
        response.append(json.dumps(encrypted_text))

    return_value['success'] = True
    return_value['results'] = response

    return json.dumps(return_value)
)
```

IX.   Click on **Deploy**.

4. **Create an IAM role for Redshift**:

I.   Navigate to the IAM service.

II.   Select **Policies** under **Access management** from the navigation pane on the left, and click on **Create Policy**.

III. Choose the **JSON** tab and in the **Policy** editor, replace the code with the following code (make sure to add your Lambda function ARN, which you can get from the function overview), and click on **Next**:

```
{
    "Version": "2012-10-17",
    "Statement": [
        {
            "Sid": "Invoke",
            "Effect": "Allow",
            "Action": [
                "lambda:InvokeFunction"
            ],
            "Resource": "<lambda_function_arn>"
        }
    ]
}
```

IV.   Give a policy name and optionally a description, and then click on **Create policy**.

V.   Select **Roles** under **Access management** from the navigation pane on the left, and click on **Create Role**.

VI.   Under **Trusted entity type**, choose **AWS service**.

VII. Under **Use case**, choose **Redshift - Customizable** and click on **Next**.

VIII. Select the policy you have created and click on **Next**.

IX.   Enter a role name and optionally a description, and click on **Create role**.

5. **Associate the IAM role with the Redshift cluster**:

I.   Navigate to the Redshift service.

II.   Open your Redshift cluster and go to the **Properties** tab.

III. Under the **Cluster permissions**, select **Associate IAM roles** from the **Manage IAM roles** list. Then, select the IAM role you have created in the previous step and click on **Associate IAM roles**.

6. **Create a UDF in Redshift**:

I.   Connect to your Redshift cluster.

II.   Use the following commands to create the UDF:

```
CREATE OR REPLACE EXTERNAL FUNCTION PII_tokenize_str (value
varchar)
RETURNS varchar STABLE
LAMBDA 'your_lambda_function_name'
IAM_ROLE 'your_redshift_role_arn';

CREATE OR REPLACE EXTERNAL FUNCTION PII_tokenize_int (value
int)
RETURNS varchar STABLE
LAMBDA 'your_lambda_function_name'
IAM_ROLE 'your_redshift_role_arn';
```

How it works...

We developed a Lambda function capable of encrypting text or integers using the pyffx library, which implements **format-preserving encryption** (**FPE**). This ensures that the encrypted data retains the same structure, type, and length as the original input. We used Secrets Manager to securely store our secret key, which is used by pyffx internally to generate a pseudo-random permutation that is applied to the input text. We provided the alphabet for encrypting strings, which is used by pyffx to define which characters should appear in the encrypted text. For numeric data identified using the Numbers library, no alphabet is needed, as pyffx will keep the numeric representation of the input. Since Lambda doesn't include the pyffx library by default, we addressed this by installing and packaging it within a Lambda layer, which has been added to the code. To enable Redshift to invoke this Lambda function, we established an IAM role providing the necessary permissions, which has been associated with the Redshift cluster. Within the Redshift cluster, we created a function that references the Lambda function, utilizing the IAM role we established. As a result, Redshift is now equipped to call the function and encrypt the data.

There's more...

If you want to be able to reverse the tokenization, you can create a decrypt function by following the same steps from *How to do it...* but in the Lambda function; replace the encrypt_data function with a decrypt_data function and update the Lambda handler using the following code (refer to DataDecryptionLambda.py file in GitHub for the full Lambda function code):

```python
def decrypt_data(token, is_number, secret_key):
    alphabet = '0123456789abcdefghijklmnopqrstuvwxyz'
    try:
        if is_number:
            d = pyffx.Integer(secret_key.encode(),
length=len(str(token)))
        else:
            d = pyffx.String(secret_key.encode(), alphabet=alphabet,
length=len(token))

        decrypted_text = d.decrypt(token)
        return decrypted_text
    except Exception as e:
        return {
            'success': False,
            'error': str(e)
        }


def lambda_handler(event, context):
    secret_key = get_secret()

    return_value = dict()
    response = []
    for argument in event['arguments']:
        token = argument[0]
        is_number = argument[1]
        try:
            result = decrypt_data(token, is_number, secret_key)
            response.append(json.dumps(result))
```

```
        except Exception as e:
            return {
            'success': False,
            'error': str(e)
        }
    return_value['success'] = True
    return_value['results'] = response

    return json.dumps(return_value)
```

See also

- *The pyffix code repository*: https://github.com/emulbreh/pyffx

Ensuring S3 compliance with AWS Config

Ensuring that your AWS resources are configured according to your specifications and best practices is crucial. AWS Config facilitates the governance of your resources by continuously evaluating your resources against your predefined rules. It provides a comprehensive view of your resources, enabling you to monitor and take corrective actions if any resource deviates from these rules and becomes non-compliant.

In *Chapter 1*, *Managing Data Lake Storage*, we discussed the importance of encryption, life cycle policies, and access control for S3 buckets. In this recipe, we will learn how to implement AWS Config rules to verify the compliance of S3 buckets with these standards.

Getting ready

For this recipe, you'll need S3 buckets that you will monitor for compliance, and you must enable AWS CloudTrail, as AWS Config relies on CloudTrail logs to track and record resource configurations.

How to do it...

1. **Set up the rules**:

    I.   Log in to the AWS Management Console (https://console.aws.amazon.com/console/home?nc2=h_ct&src=header-signin) and navigate to the AWS Config service.

    II. Click on **Get started**.

III.   Under **Recording strategy**, select **Specific resource types** and choose **AWS S3 Bucket** for **Resource type**. For the **Frequency** setting, select **Daily**.

IV.   Under **Data governance**, choose **Create AWS Config service-linked role**.

V.   Under **Delivery method**, either select **Choose a bucket from your account** if you have an existing bucket for storing config data or select **Create a bucket** to create a new one. Specify the bucket name and an optional prefix for storing the data, which is helpful if you are using an existing bucket.

VI.   Click on **Next**. On the **Rule** page, under **AWS Managed Rules**, select `s3-bucket-level-public-access-prohibited` and `s3-bucket-server-side-encryption-enabled`.

VII.   Select **Next**. Review the setting and click on **Confirm**.

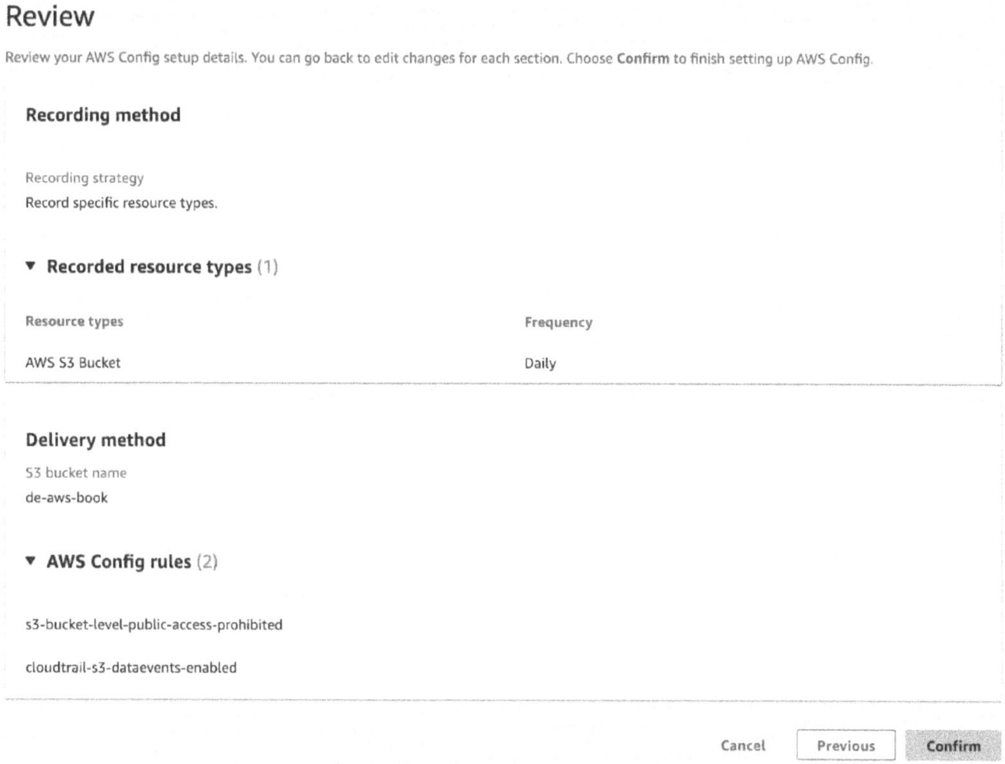

Review

Review your AWS Config setup details. You can go back to edit changes for each section. Choose **Confirm** to finish setting up AWS Config.

Recording method

Recording strategy
Record specific resource types.

▼ **Recorded resource types** (1)

Resource types	Frequency
AWS S3 Bucket	Daily

Delivery method

S3 bucket name
de-aws-book

▼ **AWS Config rules** (2)

s3-bucket-level-public-access-prohibited

cloudtrail-s3-dataevents-enabled

Cancel   Previous   **Confirm**

Figure 6.10 – Reviewing config settings

2. **Enable remediation**:

    I.   From the navigation pane on the left of your AWS Config dashboard, select **Rules**.

    II.   Open one rule at a time, and from the **Actions** drop-down list, select **Manage remediation**.

    III. Choose either **Automatic remediation** for AWS Config to take action automatically on your behalf or **Manual remediation** if you want to review the non-compliant bucket before remediating it.

    IV. Under **Remediation action details**, choose the following settings:

- For `s3-bucket-level-public-access-prohibited`, choose `AWSConfigRemediation-ConfigureS3BucketPublicAccess Block`

- For `s3-bucket-server-side-encryption-enabled` choose `aws-enables3bucketencryption` and **Bucketname** under **Resource ID parameter**

    V.   Click on **Save changes**.

How it works...

We enabled AWS Config specifically to track S3 buckets and selected two managed rules to check if the buckets are publicly accessible and if server-side encryption is enabled. AWS Config generated a dashboard displaying the resources in the account, focusing on S3 buckets in this recipe. It showed the compliance status of your S3 buckets based on the rules set, scanning them daily and flagging compliant buckets accordingly while highlighting non-compliant ones for remediation. Integration with AWS Systems Manager Automation documents allowed us to define remediation actions for each rule to rectify non-compliant buckets. If you identify a non-compliant bucket within a rule, you can easily remediate it by selecting it and clicking on the **Remediate** button.

Figure 6.11 – S3 bucket remediation

Additionally, AWS Config maintains a historical record of configuration changes for your S3 buckets, providing valuable audit trails. You can review these changes by accessing the S3 bucket via **Resource Inventory** in the dashboard and examining **Resource Timeline**, which details CloudTrail events and configuration changes.

Events All times are in Europe/Luxembourg (UTC+02:00)	Start date 2024/07/08	Event type All event types

July 7, 2024

23:35:30   ⊞ CloudTrail Event

July 6, 2024

July 5, 2024

22:39:06   ⊞ Configuration change   0 field change(s)

Figure 6.12 – S3 bucket historical events

There's more...

If you require a custom rule that isn't available among AWS-managed rules, you can create it using AWS Lambda. Let's create a rule to monitor whether our S3 buckets have life cycle policies configured to either delete objects after a specified time or archive them:

1. **Create a Python Lambda function**:

    I.   Create a Python function from scratch with a new IAM role.

    II. Add the following policies for the function IAM role:

      - `AWSConfigRulesExecutionRole` AWS-managed policy

      - Run the following command in AWS CLI or AWS CloudShell to create a resource-based policy that will allow Config to invoke our function:

```
aws lambda add-permission --function-name your_function_
arn \
--principal config.amazonaws.com \
--statement-id AllowConfigInvoke \
--action lambda:InvokeFunction \
--source-account your_aws_account_id
```

- Create an inline policy with the following permission:

```
{
"Version": "2012-10-17",
"Statement": [
        {
            "Effect": "Allow",
            "Action": [
                "s3:GetLifecycleConfiguration"
            ],
            "Resource": [
                "*"
            ]
        }
    ]
}
```

III.   Add the following code, which AWS Config will invoke for each bucket:

```
import boto3
import json
from botocore.exceptions import ClientError

def lambda_handler(event, context):
invoking_event = json.loads(event['invokingEvent'])
configuration_item = invoking_event['configurationItem']
# Extract the bucket name from the configurationItem
bucket_name = configuration_item['configuration']['name']

s3_client = boto3.client('s3')
compliance_type = 'NON_COMPLIANT'

try:
        lifecycle_policy = s3_client.get_bucket_lifecycle_
configuration(Bucket=bucket_name)
        for rule in lifecycle_policy.get('Rules', []):
            if ('Transitions' in rule and any('StorageClass'
in transition and transition.get('StorageClass', '') in
['GLACIER', 'GLACIER_DEEP_ARCHIVE'] for transition in
rule['Transitions'])) or \
                ('NoncurrentVersionTransition'
in rule and any('StorageClass' in transition and
transition.get('StorageClass', '') in ['GLACIER',
'GLACIER_DEEP_ARCHIVE'] for transition in rule.
get('NoncurrentVersionTransition', []))) or \
                ('Expiration' in rule and 'Days' in rule.
```

```
get('Expiration', {})) or \
                    ('NoncurrentVersionExpiration' in rule and
'NoncurrentDays' in rule.get('NoncurrentVersionExpiration',
{})):
                    compliance_type = 'COMPLIANT'
                    message = 'The bucket has a lifecycle
policy with expiration after specified days.'
                    break
            else:
                message = 'The bucket lifecycle policy does
not archive or delete objects.'

except ClientError as e:
        message = f"Error getting lifecycle configuration for
bucket {bucket_name}: {e}"
evaluation_result = {
'ComplianceResourceType': 'AWS::S3::Bucket',
'ComplianceResourceId': bucket_name,
'ComplianceType': compliance_type,
'OrderingTimestamp': configuration_
item['configurationItemCaptureTime'],
'Annotation': message
}

config_client = boto3.client('config')
response = config_client.put_evaluations(
        Evaluations=[evaluation_result],
        ResultToken=event['resultToken']
)

return response
```

2. **Set up a Config custom rule**:

I.   From the left navigation pane of AWS config, select **Rules**.

II. Select **Add rule**, choose **Create custom Lambda rule** for the **Rule type**, and click on **Next**.

III. Provide a name for the rule and enter the ARN of the Lambda function under **Lambda function ARN**.

IV.   Enable **Turn on detective evaluation** under **Evaluation mode**, select **When configuration changes** for **Trigger type**, and choose **Resources** for **Scope of change**. Select **AWS resource** under **Resource category** and specify **AWS S3 bucket** for **Resource type**.

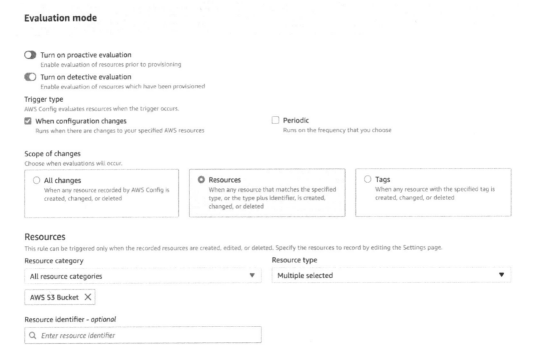

Figure 6.13 – Custom rule evaluation mode

V.   Select **Next**, review the details, and click on **Save**.

You will be able to track this rule along with the managed config rules.

See also

- *Conformance Packs*: https://docs.aws.amazon.com/config/latest/developerguide/conformance-packs.html

7

Data Quality Management

Unreliable data can lead to incorrect insights, misguided business decisions, and a significant loss of resources. As organizations treat data as a product and rely more on data freshness, data engineers and analysts must implement robust data quality control mechanisms to ensure the data's accuracy, completeness, consistency, and reliability to maintain high data quality standards.

In this chapter, we will explore various methods and tools available on AWS for maintaining data quality. We'll provide step-by-step recipes to help you implement these tools effectively in your data engineering workflows. The recipes will guide you through practical examples, starting with data quality control using AWS DataBrew, Deequ, and Glue. Before diving into the chapter, it is important to work with your stakeholders to build a data quality control framework and an SLA for your data quality. When you lead a data quality project, besides identifying the data owners, you need to work with the data owners to create a process to manage and ensure the data quality. For this chapter, we won't explore the theoretical details of building the data quality scorecards or defining the metrics and process for a data quality project. It would be beneficial for you to explore the book *DMBoK – Data Management Body of Knowledge* for supplementary knowledge on how to work with stakeholders on running a data quality project.

We will cover the following recipes in this chapter:

- Creating data quality for ETL jobs in AWS Glue Studio notebooks
- Unit testing your data quality using Deequ
- Schema management for ETL pipelines
- Building unit test functions for ETL pipelines
- Building data cleaning and profiling jobs with DataBrew

Technical requirements

You can find the code files for this chapter on GitHub at `https://github.com/PacktPublishing/Data-Engineering-with-AWS-Cookbook/tree/main/Chapter07`.

The dataset for this chapter is available at `https://github.com/tidyverse/dplyr/blob/main/data-raw/starwars.csv`.

Creating data quality for ETL jobs in AWS Glue Studio notebooks

From the *Applying a data quality check on Glue tables* recipe in *Chapter 6, Governing Your Platform*, we learned how to set a ruleset for the Glue pipeline. In this recipe, we will dive deeper into how to use Glue Studio notebooks to build a Data Quality template. Using Glue Studio is useful because you can see the output along with the dataset that you are testing. We will also introduce how to use caching and produce row-level and rule-level outputs. The row-level output would be suitable for using data quality rule violations for each of the records.

Getting ready

Before proceeding with this recipe, go through the *Applying a data quality check on Glue tables* recipe in *Chapter 6, Governing Your Platform*, and ensure that you have basic knowledge of how Glue works as covered in *Chapter 3, Ingesting and Transforming Your Data with AWS Glue*. In this recipe, we will provide the code to run the quality check scenarios, so you do not need to follow the steps to create ruleset steps from *Chapter 6*.

In this recipe, we will assume that you already have the relevant IAM for Glue and a dataset against which to run Data Quality. The public dataset we will use as an example is `starwars.csv` that is renamed as `star_wars_characters.csv`.

How to do it...

1. Head to the Glue console and click on **Author using an interactive code notebook**:

Figure 7.1 – Clicking on the script editor to edit the code

2. Instead of using the console to create a ruleset as we did in *Chapter 6*, in this recipe, we will examine how to use the Glue Studio notebook. The first step is to import the sample data quality notebook from `Recipe1` in this chapter's GitHub folder and select the relevant IAM role that you created in *Chapter 3* or *Chapter 6*.

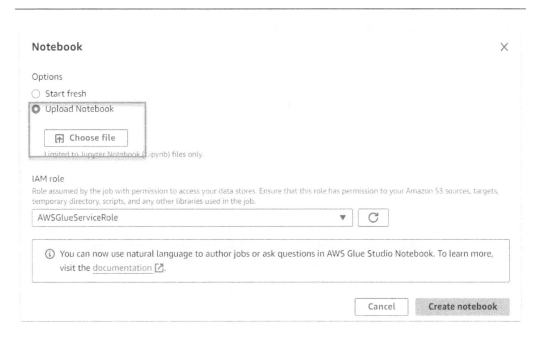

Figure 7.2 – Uploading the sample notebook from GitHub

3. The next step is to set up an interactive session and import relevant libraries. For this recipe, the code is edited.

```
Run this cell to set up and start your interactive session.

In [1]:  %idle_timeout 500
         %glue_version 4.0
         %worker_type G.1X
         %number_of_workers 2

         import sys
         from awsglue.transforms import *
         from awsglue.utils import getResolvedOptions
         from pyspark.context import SparkContext
         from awsglue.context import GlueContext
         from awsglue.job import Job
         from awsglue.dynamicframe import DynamicFrame

         from awsgluedq.transforms import *

         sc = SparkContext.getOrCreate()
         glueContext = GlueContext(sc)

Welcome to the Glue Interactive Sessions Kernel
For more information on available magic commands, please type %help in any new cell.

Please view our Getting Started page to access the most up-to-date information on the Interactive Sessions ker
nel: https://docs.aws.amazon.com/glue/latest/dg/interactive-sessions.html
Installed kernel version: 1.0.5
Current idle_timeout is None minutes.
idle_timeout has been set to 500 minutes.
```

Figure 7.3 – Setting up a Glue session

4. For this recipe, we will create rules to compare two DataFrames with each other as shown. We will call these two DataFrames `main` and `reference`.

```
# Script generated  main dataframe
df_main = glueContext.create_dynamic_frame.from_options(
    connection_type="s3",
    format_options={"quoteChar": "\"",
                    "withHeader": True,
                    "separator": ","},
    connection_options={
        "paths": ["s3://sample-test-wf09/star_wars_characters.csv"],
        "recurse": True,

    },
    format="csv",
    transformation_ctx="main",
)


# Script generated for  reference dataframe
df_reference = glueContext.create_dynamic_frame.from_options(
    connection_type="s3",
    format_options={"quoteChar": "\"",
                    "withHeader": True,
                    "separator": ","},
    connection_options={
        "paths": ["s3://sample-test-wf09/star_wars_characters.csv"],
    },
    format="csv",
    transformation_ctx="reference",
)
```

Figure 7.4 – Creating two DataFrames for evaluating

5. The following code in this recipe will go through various data quality check scenarios by creating a ruleset in the **Data Quality Definition Language** (**DQDL**) format. We will first go through the rules that will be used for comparing multiple DataFrames. Depending on your organization, you need to set your own rules. The ruleset defines the following rules:

- `RowCount = 42`: Checks that the number of rows in the dataset is `42`

- `ColumnCount = 14`: Checks that the number of columns in the dataset is `42`

- `ReferentialIntegrity "homeworld" "reference.homeworld" = 1`: Checks the data integrity with the `homeworld` column in the `reference` dataset

- `ReferentialIntegrity "species" "reference.species" = 1`: Checks the data integrity with the `species` column in the `reference` dataset

```
# Script generated for Evaluate Data Quality

EvaluateDataQualityRule = """
    Rules = [RowCount = 42,
             ColumnCount = 14,
             ReferentialIntegrity "homeworld" "reference.homeworld" = 1,
             ReferentialIntegrity "species" "reference.species" = 1
             ]
"""
```

Figure 7.5 – Setting up data quality rules

6. For the previous rules to work, you need to add the data frame that is required for the data quality evaluation. In this case, the reference dataset will be used as an additional source:

```
EvaluateDataQuality_additional_sources = {
    "reference": df_reference
    
}
```

7. Next, we will validate the datasets against the ruleset. We will add the following options:

- Change the option enableDataQualityResultsPublishing': False. This option should not be True while using Glue Studio Notebook because you do not have a Glue job to run within a notebook. Thus, if this option is set to True, it will produce an error.

- Set observations.scope": "ALL". Specify the scope of the observations that will be collected during the execution of the AWS Glue job. The ALL value indicates that all observations will be collected.

- Use cache input to improve the performance of the evaluation. This is useful if the input data is large or the assessment needs to be run multiple times by enabling the CACHE_INPUT option.

```
EvaluateDataQuality_output = EvaluateDataQuality().process_rows(
    frame= df_main,
    additional_data_sources=EvaluateDataQuality_additional_sources,
    ruleset= EvaluateDataQualityRule,
    publishing_options={
        'enableDataQualityResultsPublishing': False,
        "dataQualityEvaluationContext": "EvaluateDataQuality",
    },
    additional_options={"observations.scope": "ALL","performanceTuning.caching": "CACHE_INPUT"})
```

Figure 7.6 – Using the EvaluateDataQuality function to evaluate two DataFrames

8. Instead of producing only a single dataset, we will also use `SelectFromCollection` to produce `rowLevelOutcomes` and `ruleOutcomes`, as shown from cell 21 and onward in the notebook. You can use the notebook to reference how these two methods' outputs would be different from each other.

```
ruleOutcomes = SelectFromCollection.apply(
    dfc=EvaluateDataQuality_output,
    key="ruleOutcomes",
    transformation_ctx="ruleOutcomes",
).toDF()



ruleOutcomes.show()

+------------------+-------+-----------------+-----------------+-----------------+
|              Rule|Outcome|    FailureReason| EvaluatedMetrics|    EvaluatedRule|
+------------------+-------+-----------------+-----------------+-----------------+
|     RowCount = 42| Failed|Value: 87 does no...|{Dataset.*.RowCou...|     RowCount = 42|
|   ColumnCount = 14| Passed|             null|{Dataset.*.Column...|   ColumnCount = 14|
|ReferentialIntegr...| Passed|             null|{Column.reference...|ReferentialIntegr...|
|ReferentialIntegr...| Passed|             null|{Column.reference...|ReferentialIntegr...|
+------------------+-------+-----------------+-----------------+-----------------+
```

Figure 7.7 – A sample of result of ruleOutcomes

For more information on the output of Glue Data Quality, please check the *Applying a data quality check on Glue tables* recipe in *Chapter 6, Governing Your Platform*.

How it works...

When the `CACHE_INPUT` option is enabled, the input DynamicFrame is cached in memory during the first execution of the data quality evaluation. Subsequent executions of the data quality evaluation on the same input DynamicFrame will use the cached data, which can significantly reduce the processing time.

This caching mechanism can be particularly useful in the following scenarios:

- When running the data quality evaluation multiple times on the same input data, such as during iterative development or testing

- When the input data is large and loading it from the source every time can be time-consuming

- When you want to optimize the performance of the data quality evaluation process and reduce the overall processing time

The key difference between `rowLevelOutcomes` and `ruleOutcomes` is that `rowLevelOutcomes` contains the original input data with additional columns that indicate the data quality evaluation results at the row level.

This output is useful when you want to identify specific rows that failed the data quality checks and understand the reasons for the failures.

Remember, `ruleOutcomes` contains the overall data quality evaluation results at the rule level. For each rule in the data quality ruleset, it provides the following information:

- `Rule`: The name of the data quality rule
- `Outcome`: The pass/fail status of the rule
- `FailureReason`: The reason why the rule failed (if applicable)
- `EvaluatedMetrics`: Any metrics or statistics calculated as part of the rule evaluation

This output is useful when you want to understand the overall performance of the data quality rules and identify which rules are failing, along with the reasons for the failures.

There's more...

`EvaluateDataQuality.DATA_QUALITY_RULE_OUTCOMES_KEY` is used to access the rule-level outcomes from the output of the `EvaluateDataQuality` transform in AWS Glue. It provides access to the overall data quality evaluation results at the rule level. This includes information such as the following:

- The name of the data quality rule
- The pass/fail status of the rule
- The reason for rule failure (if applicable)
- Any metrics or statistics calculated as part of the rule evaluation

The following code snippet will use the assert statement to evaluate the check if there are any failed data quality rules and then raise an exception error. This code is useful when integrated into an **Extract, Transform, Load** (ETL) pipeline to ensure the pipeline passes the data quality check. To turn the code from Glue Studio into an ETL job, click on the **Script** tab. You will see the `job.commit()` line added:

```
assert EvaluateDataQuality_output[EvaluateDataQuality.DATA_QUALITY_
RULE_OUTCOMES_KEY].filter(lambda x: x["Outcome"] == "Failed").count()
== 0, "The job failed due to failing DQ rules"
```

In the Glue Studio notebook , you can save the output to S3 with the following code:

```
glueContext.write_dynamic_frame.from_options(
        frame = rowLevelOutcomes_data,
        connection_type = "s3",
        connection_options = {"path": "s3://sample-target/dq_outcome"},
        format = "csv")
```

See also

- *Measuring the performance of AWS Glue Data Quality for ETL pipelines*: `https://aws.amazon.com/blogs/big-data/measure-performance-of-aws-glue-data-quality-for-etl-pipelines/`

- *Evaluating data quality for ETL jobs in AWS Glue Studio*: `https://docs.aws.amazon.com/glue/latest/dg/tutorial-data-quality.html`

Unit testing your data quality using Deequ

Amazon Deequ is an open source data quality library developed internally at Amazon. The purpose of Deequ is to **unit test** data before feeding it to analytics use cases. Several analytics products such as DataBrew and Glue Data Quality were built upon the Deequ library to help serve the needs of data engineers and data scientists. See the Deequ GitHub page (`https://github.com/awslabs/deequ`) for more information.

In the previous recipe, we learned about Glue Data Quality. There are several key considerations when choosing between AWS Glue Data Quality and Deequ:

- **Managed service versus open source library**: AWS Glue Data Quality is a fully managed service built on top of the open source Deequ framework. Deequ is an open source library that you can use to implement data quality checks in your applications. Also, since Deequ is an open source library, there are metrics that might be available on Deequ but are not (yet) available on AWS Glue Data Quality, such as **RatioOfSums** (at the time of writing this book).

- **Deployment and integration**: AWS Glue Data Quality mainly focuses on the Glue pipeline. In contrast, Deequ is an open source library that can be used in any environment that supports the JVM, which allows users to have more flexibility in deployment scenarios such as if you need to check for data quality in an on-prem/hybrid environment. If you're looking for a fully managed and scalable data quality solution that integrates well with other AWS services, AWS Glue Data Quality would be the better option.

Getting ready

Deequ depends on Java 8. Deequ version 2.x only runs with Spark 3.1. Deequ metrics are computed using Apache Spark running on Glue or EMR. Scala or Python can be used over Spark for this purpose. As Deequ is an open source library, the update of new metrics might not be the same across the supported language. Thus, there might be metrics that are available in Scala prior to being available in Python. In this recipe, we will use Scala to demonstrate examples of how Deequ and Scala work together in a Sagemaker endpoint.

How to do it...

1. Click on **Author code with a script editor** in the console:

Figure 7.8 – Clicking on the script editor in the console

2. Choose **Spark engine** and create a script.

3. In the next screen, type `DQ-test-scala` under **Name** and choose the relevant IAM role, **Glue 3.0** as **Glue version**, and **Scala** as **Language**, as shown:

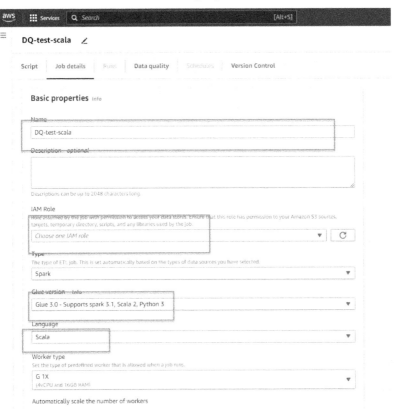

Figure 7.9 – Setting up the Glue pipeline with Scala option

4. Scroll to the **Advance** option. Under the **Libraries** section, add the Deequ package to the JARs path. The Deequ package is available on the MVN repository at `https://mvnrepository.com/artifact/com.amazon.deequ/deequ`. Make sure that you select the version that is compatible with your Spark version.

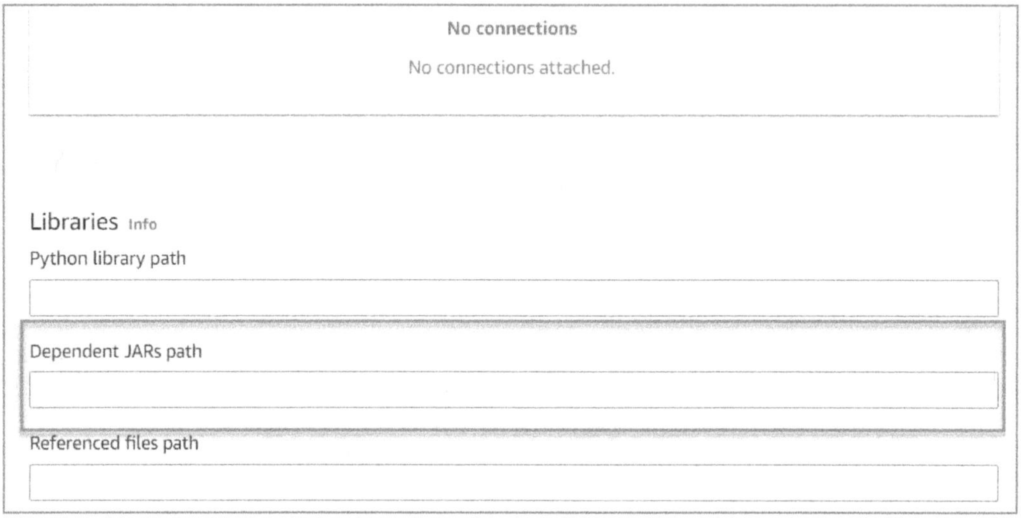

Figure 7.10 – Importing the S3 bucket to the dependent JARs path

5. In the **Script** section, paste the following script:

```
import com.amazonaws.services.glue.GlueContext
import com.amazonaws.services.glue.util.GlueArgParser
import com.amazonaws.services.glue.util.Job
import org.apache.spark.SparkContext
import org.apache.spark.SparkConf
import scala.collection.JavaConverters._
import com.amazonaws.services.glue.log.GlueLogger

import scala.util.matching.Regex
import com.amazon.deequ.{VerificationSuite, VerificationResult}
import com.amazon.deequ.VerificationResult.
checkResultsAsDataFrame
import com.amazon.deequ.checks.{Check, CheckLevel}
import com.amazon.deequ.constraints.ConstrainableDataTypes

object GlueApp {
  def main(sysArgs: Array[String]) {
    val spark: SparkContext = new SparkContext()
```

```
    val glueContext: GlueContext = new GlueContext(spark)

    val args = GlueArgParser.getResolvedOptions(sysArgs,
Seq("JOB_NAME").toArray)
    Job.init(args("JOB_NAME"), glueContext, args.asJava)
    val datasource0 = glueContext.getCatalogSource(database
= "dq-test", tableName = "test", redshiftTmpDir = "",
transformationContext = "datasource0").getDynamicFrame()

//Deequ start
    val dataset: DataFrame = dframe.toDF()
    val verificationResult : VerificationResult = {
VerificationSuite()
    .onData(df)
    .addCheck(
        Check(CheckLevel.Error, "Customer Code Check")
            .isComplete("CUSTOMER_CODE")
            .isUnique("CUSTOMER_CODE"))
 .addCheck(
        Check(CheckLevel.Error, "Mobile Check")
            .isComplete("PHONE_1")
            .hasPattern("PHONE_1","""^[0][\d]{9}$""".r, _>=0.9)
            .isUnique("PHONE_1"))
 .addCheck(
        Check(CheckLevel.Error, "Tax code Check")
            .isComplete("TAX_ID")
            .isUnique("TAX_ID")
            .satisfies("length(`TAX_ID`) = 10 or length(`TAX_
ID`) = 14", "is 10 or 14 digits", Check.IsOne, None))
    .run()
// retrieve successfully computed metrics as a Spark data frame
val resultDataFrame = checkResultsAsDataFrame(spark,
verificationResult)

glueContext.getSinkWithFormat(
  connectionType = "s3",
  options = JsonOptions(Map("path" -> "s3://dq-outputs/
results/")),
  format = "csv"
).writeDynamicFrame(resultDataFrame)
Job.commit()
```

How it works...

When you build a Glue pipeline, you can define which stage of the pipeline Deequ will be used to capture data quality metrics.

Deequ has three main components:

- **Metrics computation**: Deequ computes data quality metrics, that is, statistics such as completeness, maximum, or correlation.

- **Constraint verification**: As a user, you focus on defining a set of data quality constraints to be verified. Deequ takes care of deriving the required set of metrics and verifying the constraints.

- **Constraint suggestion**: You can choose to define your custom data quality constraints, or use the automated constraint suggestion methods that profile the data to infer useful constraints with `ConstraintSuggestionRunner`.

The code example in *step 5* showcases how to use constraint verification.

There's more...

The following table shows some examples of Deequ rules that you can use to check your data:

Column	Description	DQ Rule
First Name	First Name is not null	.isComplete("First_Name")
Emp_ID	Emp_ID has no negative values	.isNonNegative("Emp_ID")
Emp_ID	EMP_ID is integer	.hasDataType("EMP_ID")
EMAIL	Email is unique	.isUnique("EMAIL")
EMAIL	Email follows a pattern	.hasPattern("EMAIL","""[\S]+[@][\S]+[.][\S]{1,10}$""".r, Check.IsOne)
LEGAL_ID	Legal ID length either 12 or 9	.satisfies("length(`LEGAL_ID`) = 12 or length(`LEGAL_ID`) = 9", "is 9 or 12 digits", Check.IsOne, None))
work_position	the column work_position should not contain null	isComplete("work_position")
individual income	the individual income has a completeness of at least 30%	.hasCompleteness("individual income", _>=0.3)

Figure 7.11 – Example of some common data quality checks

See also

- *Deequ GitHub*: https://github.com/awslabs/deequ/tree/master

- *Programming AWS Glue ETL scripts in Scala*: https://docs.aws.amazon.com/glue/latest/dg/aws-glue-programming-scala.html

- *Using Scala to program AWS Glue ETL scripts*: https://docs.aws.amazon.com/glue/latest/dg/glue-etl-scala-using.html

- *Building a serverless data quality and analysis framework with Deequ and AWS Glue*: https://aws.amazon.com/blogs/big-data/building-a-serverless-data-quality-and-analysis-framework-with-deequ-and-aws-glue/

Schema management for ETL pipelines

In this recipe, we will learn how to perform schema validation using Apache Spark and AWS Glue. Schema validation is essential to ensure the consistency and integrity of your data as it moves through various stages of the data pipeline. By validating schemas, you can prevent data quality issues and ensure that downstream applications receive data in the expected format.

Without schema validation, once the data reaches Redshift or Athena, it will cause data corruption errors from, for example, duplicate columns or using a wrong datatype. Schema-on-read is a feature of the modern data lake, which contrasts with the schema-on-write that is traditionally used in on-prem data warehouses. In the data lake environment, when the data moves through layers of the data lake, you typically need to define the schema and store the defined scheme either in a JSON config file or in a database that the ETL pipeline could later use to verify the schema of a file. Using Spark's Infer schema is not always a good option, especially when dealing with columns representing currency. Thus, this recipe will introduce to you how to manage your schema and what is the benefit of using Glue Schema Registry.

Getting ready

This recipe assumes that you have a Glue pipeline along with a Glue catalog set up from previous recipes.

How to do it...

1. Head to the Glue console and click on **Schemas** under **Data Catalog**:

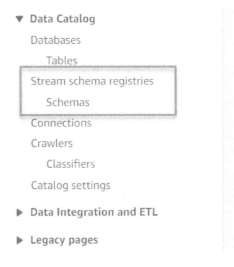

Figure 7.12 – Selecting schemas under stream schema registries

2. Click on **Create schema**, fill in **Schema name** and **Registry**, and select **JSON** as **Data format**, as shown:

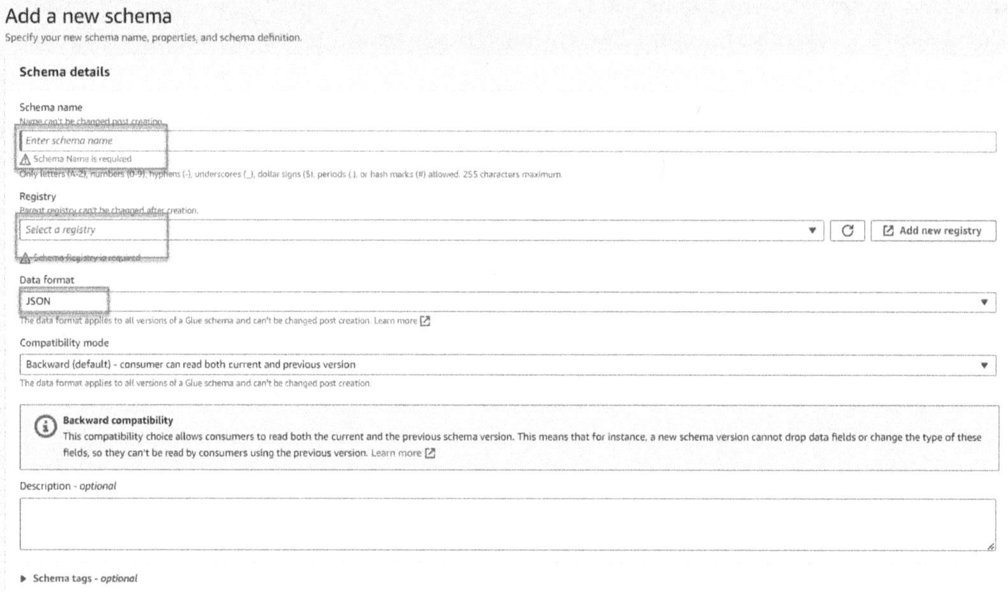

Figure 7.13 – Adding the new schema name, registry, and data format

3. In the first schema version, add in your schema. Make sure you read the *Create Schema* article (`https://docs.aws.amazon.com/glue/latest/webapi/API_CreateSchema.html`) so that you adhere to the syntax. Here is one example of a schema in JSON format:

```
{
  "$schema": "http://json-schema.org/draft-07/schema#",
  "type": "object",
  "properties": {
    "firstname": {
      "type": "string"
    },
    "ID": {
      "type": "integer"
    }
  }
}
```

4. Click on **Create schema** and choose **Version 1**.

5. Once you have created a schema, you can use `SchemaReference` in the request syntax when calling the `CreateTable` or `UpdateTable` APIs, or through the Glue Data Catalog console under the relevant database, as shown:

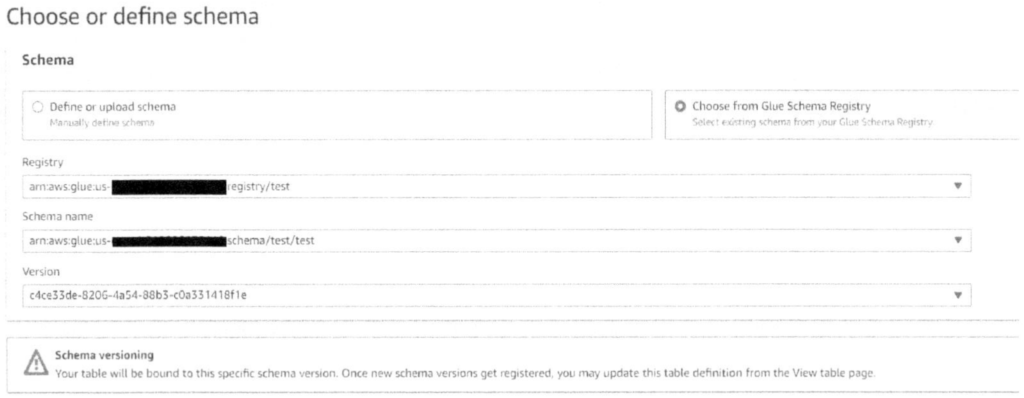

Choose or define schema

Schema

| ○ Define or upload schema | ● Choose from Glue Schema Registry |
| Manually define schema | Select existing schema from your Glue Schema Registry |

Registry

arn:aws:glue:us-███████████registry/test   ▾

Schema name

arn:aws:glue:us-██████████schema/test/test   ▾

Version

c4ce33de-8206-4a54-88b3-c0a331418f1e   ▾

⚠ **Schema versioning**
Your table will be bound to this specific schema version. Once new schema versions get registered, you may update this table definition from the View table page.

Figure 7.14 – Choosing a schema from Glue

How it works...

The AWS Glue Schema Registry allows you to centrally discover, control, and evolve data schemas. It helps enforce schema compatibility checks when registering new schema versions.

Glue Schema Registry also helps to manage the schema evolution. When a new version of a schema is registered, Schema Registry will check the compatibility based on the configured mode. If the new version is compatible, it will be registered as a new version. You can also specify which version of the schema you want a table to have. This allows you to evolve your schemas over time without breaking downstream applications. Schema evolution comes with the following modes:

- **NONE**: Any change is allowed
- **BACKWARD**: New fields can be added, but existing fields cannot be removed or renamed
- **FORWARD**: Existing fields can be removed or renamed, but new fields cannot be added
- **FULL**: Changes must be backward- and forward-compatible

There's more...

Schema Registry is a specialized product for schema management. It can integrate with streaming data applications such as Apache Kafka, Amazon MSK, Amazon Kinesis Data Streams, Apache Flink, and AWS Lambda, providing schema management capabilities for these platforms. Data Catalog can also reference schemas stored in Schema Registry when creating or updating AWS Glue tables or partitions.

To expand the architecture further, investigate an architecture of ETL framework where you integrate a control table to keep track of Glue job run status and a technical and business metadata dictionary.

See also

- *AWS Glue Schema Registry – AWS Glue*: https://docs.aws.amazon.com/glue/latest/dg/schema-registry.html

- *Integrating with AWS Glue Schema Registry*: https://docs.aws.amazon.com/glue/latest/dg/schema-registry-integrations.html

Building unit test functions for ETL pipelines

In this recipe, we will learn how to build some unit test functions for your ETL pipeline to help identify and fix issues at an early stage of your pipeline development. By incorporating unit tests, you can catch errors early in the development process, leading to more robust and reliable data workflows. This recipe is particularly useful for data engineers who need to validate the functionality of their ETL jobs and ensure data integrity before deploying them to production.

The goal of this recipe is to introduce some code snippets of functions to test Glue Jobs in a unit testing context that you can use to integrate into your Glue pipeline or your company's internal libraries.

How to do it...

1. You should create a file name, such as `unit_test.py`, that is separate from your ETL code. This file will contain various functions for unit testing.

2. Import the relevant libraries. These are the libraries that we will use for this recipe:

```
import logging
import os
import subprocess
import sys
from functools import reduce
from itertools import zip_longest
from pyspark.sql import DataFrame, Row, SparkSession
from pandas.testing import assert_frame_equal
```

```
from pyspark import SparkConf, SparkContext
from pyspark.sql.types import ArrayType, MapType
from pyspark.testing import assertDataFrameEqual,
assertSchemaEqual
```

3. In this code, we will raise an exception error as follows:

```
class DataFrameNotEqualError(Exception):
    """The created dataframes are not equal"""
    pass
```

4. We will create a `GlueUnitTest` class, as shown, to compare schemas and row data:

```
class GlueUnitTest:
    @staticmethod
    def assert_df_equality(df1, df2, ignore_col_order=True,
ignore_row_order=True, ignore_nullable=True):
        """
        Check two PySpark dataframes, and raise an exception if
they are not equal
        Handles list/dict style columns, and prints detailed
error messages on failure.

        Do not use this to compare a PySpark dataframe to a
Python dict in a test. Instead, collect() the dataframe rows and
do a comparison in Python.

        :param df1: The 1st dataframe to check
        :param df2: The 2nd dataframe to check
        :param ignore_col_order: If true, the columns will be
sorted prior to checking the schema
        :param ignore_row_order: If true, the rows will be
sorted prior to checking the data
        :param ignore_nullable: If true, differences in whether
or not the schema allows None/null will be
            ignored.
        """
        transforms = []
        if ignore_col_order:
            transforms.append(lambda df: df.select(sorted(df.
columns)))
        if ignore_row_order:
            transforms.append(lambda df: df.sort(df.columns))
        df1 = reduce(lambda acc, fn: fn(acc), transforms, df1)
        df2 = reduce(lambda acc, fn: fn(acc), transforms, df2)
        GlueUnitTest.assert_df_schema_equality(df1, df2)
```

```
        GlueUnitTest.assert_df_data_equality(df1, df2)

    @staticmethod
    def assert_df_data_equality(df1, df2):
        """
        compare two PySpark dataframes and return if rows are
different and the percentage of difference
        :param df1: the first dataframe
        :param df2: the second dataframe
        """
        try:
            assertDataFrameEqual(df1, df2)
        except AssertionError as e:
            raise AssertionError(f"Dataframes are not equal:
{e}") from e



    @staticmethod
    def assert_df_schema_equality(df1, df2):
        """
        Assert that two pyspark dataframe schemas are equal.
this function only compares two schemas , column names,
datatypes and nullable property
        :param df1: the first dataframe
        :param df2: the second dataframe
        """
        try:
            assertSchemaEqual(actual=df1.schema, expected=df2.
schema)
        except AssertionError as e:
            raise AssertionError(f"Schema are not equal: {e}")
from e
```

5. In your Glue Studio notebook, you can add an additional `unit_test.py` file with the following code:

```
%%configure
{
  "--extra-files": "s3://path/to/additional/unit_test.py/"
}
```

How it works...

There are three types of testing that you can implement for your pipelines to ensure their accuracy and reliability:

- **Unit test that tests individual unit code**: The standard unit test libraries are `Pytest`, `uniitest`, or `scalatest` if you write your code in Scala. Unit testing your pipeline is different from unit testing your data. This is why we have two different recipes for unit testing in this chapter.

- **Integration test that tests for components of your pipeline**: It is not recommended to build a pipeline with too many components. The best practice is to standardize your pipeline into a single and modular action and use an orchestration tool such as Glue Workflow or Step Functions to orchestrate the steps of these actions.

- **End-to-end testing whereby you test the entire pipeline from start to finish**: You will need to adapt the knowledge from *Chapter 8, DevOps – Defining IaC and Building CI/CD Pipelines*, where you set up a code deployment pipeline for CI/CD workflow.

In this recipe's code snippets, we went through two unit test scenarios that helped you evaluate your DataFrame before passing it to the next layer of the data lake as follows:

- Comparing whether the information of two data frames is identical
- Comparing whether the schema of two data frames is identical

Each organization will have different test scenarios to meet your data quality needs. You can build a comprehensive testing framework with Deequ and Spark's UDF. You can then add it to your pipeline as extra files.

There's more...

To take this a step further, you should write unit test cases and automate the running of these test cases with AWS CodePipeline. Please see *Chapter 8, DevOps – Defining IaC and Building CI/CD Pipelines*, for more details.

As an example of writing a unit test, you would need to build mock data and pass it into a test case, as shown here:

```
import pytest
tableName    = "marketing_data"
dbName       = "marketing"
# Does the table exist?
def test_tableExists():
  assert tableExists(tableName, dbName) is True
```

See also

- *Using Python libraries with AWS Glue*: `https://docs.aws.amazon.com/glue/latest/dg/aws-glue-programming-python-libraries.html`

- *PySpark Testing*: `https://spark.apache.org/docs/latest/api/python/reference/pyspark.testing.html`

- *Configuring AWS Glue interactive sessions for Jupyter and AWS Glue Studio notebooks*: `https://docs.aws.amazon.com/glue/latest/dg/interactive-sessions-magics.html`

Building data cleaning and profiling jobs with DataBrew

AWS Glue DataBrew is a no-code data preparation tool that simplifies data profiling, cleansing, and validation, making it an excellent choice for data engineers looking to automate data quality checks. In this recipe, we'll use DataBrew to perform data profiling and PII detection.

Getting ready

This recipe assumes that you have a dataset in S3 for testing out DataBrew.

How to do it...

1. Navigate to the AWS DataBrew console, click on **Projects | Create project**, and finally click on **Provide project name**.

Figure 7.15 – Clicking on Create Project

2. Select the appropriate dataset that you would like to use for building the DataBrew project:

Figure 7.16 – Selecting a relevant dataset

3. In the **Permission** section, select the appropriate name under **Role name**, then click on **Create project**. If you have not created an IAM role for DataBrew, you can click on **Create new IAM role** and then click on **Create project**.

Permissions Info

DataBrew needs permission to connect to data on your behalf. Use an IAM role with the **required policy** ⬀ attached.

Role name

Choose the role that has access to connect to your data. Refresh to see the latest updates.

Create new IAM role

New IAM role suffix

Your role will be prefixed with "**AWSGlueDataBrewServiceRole-**"

AWSGlueDataBrewServiceRole-pii-test

By clicking "Create project" you are authorizing creation of this role.

Figure 7.17 – Selecting an IAM role

4. You will need to wait a few minutes for the console to start. Once the console finishes loading, click on **RECIPE**, then click on **Add step**:

Figure 7.18 – Adding steps to your recipe

5. Next, we will see how to build some steps in the recipe. Assume that we have a PII column of 10 number and we only want to keep the first number and the last 5 numbers for analytics purposes. First, search for **REDACT VALUES**:

Figure 7.19 – Choosing the REDACT VALUES option

6. Next, choose the value for **Source columns**, **Value to redact**, **Starting position**, and **Redact symbol**.

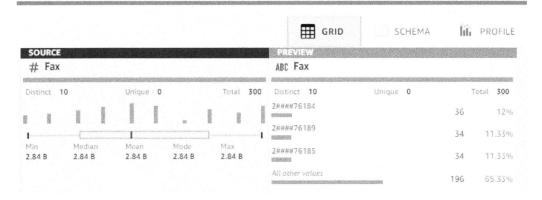

Figure 7.20 – Selecting rules for Redact Values

7. Then click on **Preview changes** to preview the effect. Once you are happy with the result, click on **Apply**.

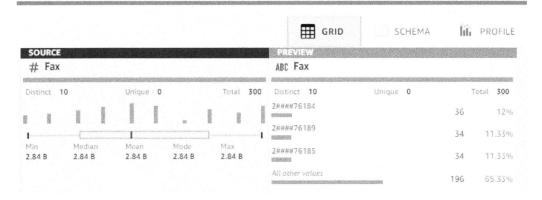

Figure 7.21 – Previewing output

8. You can also use **Substitute values** to create new values from your existing record. In the following example, you can use shuffle value to shuffle customers' first or last names. The **Name of columns to group by with** field could be used in a similar way to the GROUP_BY command in SQL when you want the data to shuffle only within the Group by value.

Figure 7.22 – Creating rules for substitute values

9. The following screenshot shows an example of a recipe with multiple steps, including re-arranging columns to rename, filtering value, and masking data. Depending on your needs, you can also add hashing and encryption for your dataset. Once you finish with the recipe steps, click on **Publish**.

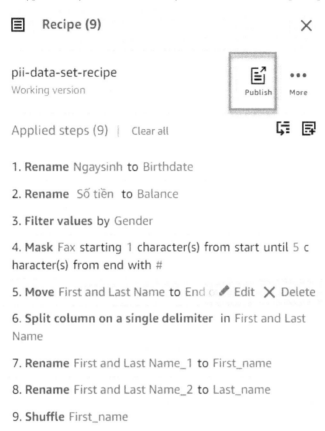

Figure 7.23 – Reviewing the steps and publishing

10. Once you publish a recipe, you can head to the **Job** tab and fill in the job name, dataset, and previously created recipe, as shown. Once you fill in the necessary details, click on either **Create job** or **Create and run job**.

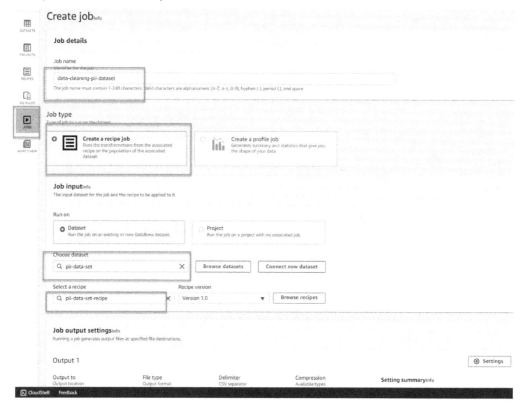

Figure 7.24 – Creating a recipe job

How it works...

AWS DataBrew simplifies data quality control by providing a visual interface for data profiling and transformation. The steps outlined in this recipe show how to leverage AWS Glue DataBrew to automate data quality checks by creating a recipe. A recipe could be adjusted to be reused in multiple datasets, such as creating a recipe to mask PII data for columns with ID.

There's more...

Profiling helps you understand the state of your data by generating statistics and identifying potential issues. It is advised to run a profile job first to understand your dataset before applying a ruleset. Click on **Create job** on the **Profile jobs** tab to start your profiling job for the dataset.

Figure 7.25 – Clicking on Create job to automate the DataBrew recipe

As discussed in the *Unit testing your data quality using Deequ* recipe, Data Quality is all built upon the Deequ library. When choosing between AWS Glue DataBrew and Deequ for your data workloads, there are a few key factors to consider:

- **Data transformation capabilities**: AWS Glue DataBrew is a visual data preparation tool that allows you to easily transform and clean your data, while Deequ is more of a data quality checking tool.

- **Ease of use**: AWS Glue DataBrew provides a user-friendly, no-code interface that makes it easy for non-technical users to prepare and transform data. On the other hand, Deequ requires more technical expertise, as it is a library that needs to be integrated into your data processing pipelines.

- **Integration with other AWS services**: Deequ can be used with a variety of data processing frameworks, including Apache Spark. It can also be used within your Glue data pipeline with DataBrew. However, it would be a separate step and would need an orchestration tool such as Airflow or Step function to trigger the Glue pipeline and DataBrew together.

See also

- *GROUP_BY*: `https://docs.aws.amazon.com/databrew/latest/dg/recipe-actions.GROUP_BY.html`

- *AWS Glue DataBrew adds binning, skewness, binarization, and transpose transformations for pre-processing data for machine learning and analytics*: `https://aws.amazon.com/about-aws/whats-new/2021/03/aws-glue-databrew-adds-binning-skewness-binarization-transformations-pre-processing-data/`

8

DevOps – Defining IaC and Building CI/CD Pipelines

The idea of combining software development with production operations has been around for decades, but it's since the rise of cloud computing that it has been established as a best practice, aiming to shorten the software development cycles from development to deployment while improving their reliability and lowering operational costs.

This methodology involves many aspects, including team organization, culture, and processes. In this chapter, we will focus on implementing two of the key technical components for successfully implementing DevOps on AWS:

- **Infrastructure as Code** (**IaC**): IaC fulfills the key role of eliminating or reducing manual operations by coding the deployment and maintenance of infrastructure and services. This reduces human errors, improves scalability, and reduces the maintenance burden in general. In recent years, AWS **Cloud Development Kit** (**CDK**) has become the reference IaC for AWS. It allows you to code the AWS infrastructure on one of the multiple languages it supports, after which it's converted into an AWS CloudFormation template. We will also cover the main alternative, Terraform, which aims to be a cloud-agnostic tool for IaC.

- **Continuous integration/continuous deployment** (**CI/CD**): This is the set of techniques that takes care of reliably deploying code to production systems. The aim is that once the developer code has been merged into the code repository, a pipeline takes care of all the deployment automatically. This includes not only deploying the code alongside any infrastructure required but also following a deployment process that normally includes tests, stages, and potentially manual approvals (full CI/CD is when there are none).

  The traditional tool to implement CI/CD pipelines is Jenkins. However, here, we will use AWS CodePipeline since it is easier to set up and is well-supported by CDK.

The following recipes will be covered in this chapter:

- Setting up a code deployment pipeline using CDK and AWS CodePipeline
- Setting up a CDK pipeline to deploy on multiple accounts and regions
- Running code in a CloudFormation deployment
- Protecting resources from accidental deletion
- Deploying a data pipeline using Terraform
- Reverse-engineering IaC
- Integrating AWS Glue and Git version control

Technical requirements

The recipes in this chapter assume you have a bash shell or equivalent available with the AWS CLI installed (refer to the instructions at `https://docs.aws.amazon.com/cli/latest/userguide/getting-started-install.html`) with access to AWS. If you're using Microsoft Windows, you can enable WSL (`https://learn.microsoft.com/en-us/windows/wsl/install`) or install Git (`https://git-scm.com/downloads`) to use the bash shell it brings.

Configure the default region and user credentials, ensuring you have enough permissions to use the different services. You can use `aws configure` or an AWS CLI profile. Alternatively, you can use environment variables to provide the credentials: `AWS_ACCESS_KEY_ID`, `AWS_SECRET_ACCESS_KEY`, and optionally `AWS_SESSION_TOKEN`.

For the recipes that need CDK, you need to install NPM first by downloading it from `https://nodejs.org/en/download` (in Linux, it's highly dependent on the GNU Libc version that your OS uses, so you might need to use an older NPM version or compile it yourself) and then, in the command line, run `npm install -g aws-cdk` and then `cdk --version` to verify it's working. More information and details can be found in the AWS documentation: `https://docs.aws.amazon.com/cdk/v2/guide/cli.html`.

You can find the code files for this chapter on GitHub: `https://github.com/PacktPublishing/Data-Engineering-with-AWS-Cookbook/tree/main/Chapter08`.

Setting up a code deployment pipeline using CDK and AWS CodePipeline

In this recipe, you will create a CDK project that defines a deployment pipeline and a service infrastructure stack with a simple Glue Shell job as an example. This recipe uses Python for both the Glue Shell script and the CDK script.

The pipeline will monitor the CodeCommit repository to automatically deploy both changes to the pipeline itself and the solution (the Glue job in this case).

Getting ready

For this recipe, you'll require the AWS CLI, CDK, Python 3.8+, and Git to be installed in your system, as well as a command-line interface to invoke them. Check the *Technical requirements* section for indications and guidance on installing the CLI and CDK. You can install Python and Git from their respective websites (www.python.org and git-scm.com). On Windows, make sure that when you install Python, you enable the option to add it to PATH so that it can be easily run from the command line. You can verify the Python version by running python --version.

How to do it...

1. Bootstrap the CDK infrastructure on the previously configured account and region. This is required so that you can deploy the CDK project later:

   ```
   cdk bootstrap
   ```

2. Prepare the CDK Python project by creating a directory for the project and preparing the Python virtual environment:

   ```
   mkdir cdk-deployment-recipe
   cd cdk-deployment-recipe
   cdk init --language=python
   source .venv/bin/activate
   echo git-remote-codecommit >> requirements.txt
   pip install -r requirements.txt
   ```

3. Create the Git CodeDeploy repo where the project will be stored:

   ```
   aws codecommit create-repository --repository-name \
     cdk-deployment-recipe --repository-description \
     "CDK deployment pipeline recipe"
   git remote add codecommit codecommit::$(aws \
     configure get region | \
     sed 's/\s//g')://cdk-deployment-recipe
   ```

4. Create a script for the Glue job script:

   ```
   mkdir glue
   echo "print('Hello from GlueShell')" >\
     ./glue/GlueShellScript.py
   ```

5.   Override the default CDK script and add the imports:

```
CDK_FILE=\
cdk_deployment_recipe/cdk_deployment_recipe_stack.py
cat > $CDK_FILE << EOF
from aws_cdk import (
    Stack,
    Stage,
    aws_codecommit as codecommit,
    aws_glue as glue,
    aws_iam as iam,
    aws_s3 as s3,
    aws_s3_deployment as s3_deploy
)
from aws_cdk.pipelines import (
    CodePipeline,
    CodePipelineSource,
    ShellStep
)
from constructs import Construct
EOF
```

6.   Add the deployment stack to the script. Ensure that you respect the indentation:

```
cat >> $CDK_FILE << EOF
class CdkDeploymentRecipeStack(Stack):
    def __init__(self, scope: Construct,
                 construct_id: str, **kwargs):
        super().__init__(scope, construct_id, **kwargs)

        repo = codecommit.Repository.\
            from_repository_name(self,"DeployRecipeRepo",
                repository_name="cdk-deployment-recipe")
        git_source = \
        CodePipelineSource.code_commit(repo, "master")
        pipeline = CodePipeline(self, "Pipeline",
            pipeline_name="RecipePipeline",
                synth=ShellStep("Synth",
                        input=git_source,
                        commands=[
                          "npm install -g aws-cdk",
                "python -m pip install -r requirements.txt",
                          "cdk synth"]
                        )
```

```
            )
        pipeline.add_stage(GlueStage(self, "prod"))

    class GlueStage(Stage):
      def __init__(self, scope: Construct,
                    construct_id: str, **kwargs):
        super().__init__(scope, construct_id, **kwargs)
        GlueAppStack(self, "GlueAppStack",
                    stage=self.stage_name)

    EOF
```

7. Add the first part of the Glue stack to the script:

```
    cat >> $CDK_FILE << EOF
    class GlueAppStack(Stack):
      def __init__(self, scope: Construct,
                construct_id: str, stage: str, **kwargs):
        super().__init__(scope, construct_id, **kwargs)
        bucket_name = f"deployment-recipe-{self.account}-{stage}"
        bucket = s3.Bucket(self, id="GlueBucket",
                        bucket_name=bucket_name)
        deployment = s3_deploy.BucketDeployment(self,
            "DeployCode", destination_bucket=bucket,
            sources=[s3_deploy.Source.asset("./glue")])
        role_name = \
            f"AWSGlueServiceRole-CdkRecipe-{stage}"
        job_role = iam.Role(self, id=role_name,
                role_name=role_name, managed_policies=[
          iam.ManagedPolicy.from_managed_policy_arn(self,
          "glue-service", "arn:aws:iam::aws:policy\
    /service-role/AWSGlueServiceRole")
                        ],
                        assumed_by=iam.ServicePrincipal(
                                "glue.amazonaws.com")
                    )
        job_role.add_to_policy(iam.PolicyStatement(
                    effect=iam.Effect.ALLOW,
                    resources=[
                        f'arn:aws:s3:::{bucket_name}',
                        f'arn:aws:s3:::{bucket_name}/*'
                    ],
                    actions=[
                        's3:ListBucket',
```

```
                              's3:GetObject',
                              's3:PutObject'
                    ]
                )
            )
        EOF
```

8. Complete the stack definition:

```
cat >> $CDK_FILE << EOF
    job = glue.CfnJob(
            self,
            "glue_CDK_job",
            command = glue.CfnJob.JobCommandProperty(
                    name = "pythonshell",
                    python_version= '3.9',
                    script_location = \
f's3://{bucket_name}/GlueShellScript.py'
                ),
                role= job_role.role_arn,
                name= "deployment_recipe_glueshell",
                glue_version="3.0"
    )
EOF
```

9. Test the CDK script to ensure it's valid. If it isn't, review the previous steps while paying attention to the Python indentation, especially on new statements:

```
cdk synth
```

10. Commit the changes and push them to the project via the Git repo:

```
git add *
git commit -m "Added cdk and Glue code"
git push --set-upstream codecommit master
```

11. Deploy the pipeline and the stack:

```
cdk deploy
```

12. Open the AWS console, navigate to CodePipeline in the same region you have the AWS CLI configured to use, and locate the RecipePipeline pipeline. If it hasn't completed yet, wait until all steps up to the prod stage are green.

Note that you can retry a failed action, but in most cases, you will need to make a correction to the CDK code and push to the repository to make corrections:

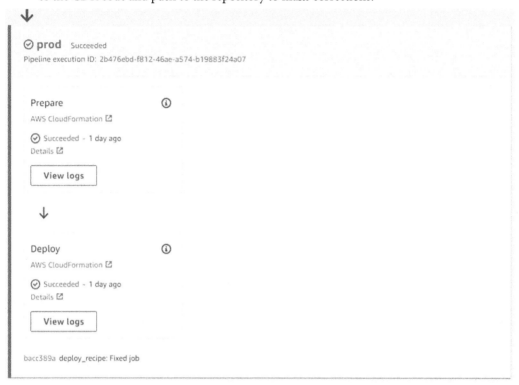

Figure 8.1 – Pipeline prod stage

13. Now, you can navigate to Glue on the AWS console, find the **deployment_recipe_glueshell** job, select it, and run it. Once it completes successfully, in the **Run details** tab, using the **Output** link, you can view the message that the script printed.

14. If you don't want to keep it, remove the stack, which will delete the resources:

```
cdk destroy
```

How it works...

In this recipe, you created a CDK project and the sample Glue script. Because the language of choice was Python, it needed some CDK dependencies. You downloaded these via pip based on the requirement.txt file, loaded the Python virtual environment created by CDK, and activated them by running source .venv/bin/activate. This allowed Python to install the specific dependencies just for this project and not globally so that other CDK projects could use different versions. Other languages have their own way of using CDK and managing dependencies.

The `cdk bootstrap` command deployed the **CDKToolkit** CloudFormation stack on the configured region and account. This is a one-time resource setup and is required to be able to deploy CDK stacks onto that region and account; you only need to bootstrap again if you're upgrading the project's CDK version (the tool will detect that and remind you if needed).

Then, you created a Git repository on AWS CodeCommit to store the project. This is not required by CDK since you could run the CDK project locally, but this is best practice to ensure the infrastructure gets treated like any other code – that is, it gets peer-reviewed, stored in a version management system, and automatically deployed.

After updating the code, you pushed the changes to the Git remote server and deployed the CDK project. This was needed because the pipeline doesn't exist yet. Once the pipeline is in place, it will monitor the repository project to trigger the pipeline on any commit that's done. The pipeline has a stage to self-update, so if you have manually deployed changes that have not been pushed to the Git server, they will be undone. Remember to only invoke `cdk deploy` the very first time you create the pipeline.

To do the deployment (either when you do it manually or with the pipeline), CDK generates a CloudFormation stack or changeset, as needed.

Once the pipeline completes the deployment, the Glue job is ready to use. All the work to create a bucket, upload the script, and create a role was taken care of by the CDK stack. In the same way, the stack could have deployed other AWS components or services, such as an AWS Lambda, an RDS database, or an EMR cluster.

There's more...

This recipe combines using CDK and AWS CodePipeline. If you wanted to use the traditional tool for pipelines instead, such as Jenkins, you would need a Jenkins task that also monitors the Git repository and then runs CDK in a project, where the main stack would be GlueAppStack instead of CdkDeploymentRecipeStack, which is needed for CodePipeline.

If something fails in the pipeline, you can retry the failed stages – for instance, if the issue is that it couldn't deploy because there was a conflict with some existing resource (such as the S3 bucket) and you were able to solve it manually without code changes.

At the bottom of each stage in the pipeline, the Git commit is listed with a message stating that it has been deployed or is currently being deployed. This means that in most cases, you can roll back the changes by rolling back the changes in Git.

See also

- For a more advanced use case of this recipe involving multiple stages and regions, please refer to the *Setting up a CDK pipeline to deploy on multiple accounts and regions recipe*.

- If you're going to need to use multiple cloud vendors, consider using Terraform for your Glue stack on the pipeline instead of CDK. See the *Deploying a data pipeline using Terraform* recipe for an example of how to use that cloud-vendor-agnostic tool.

Setting up a CDK pipeline to deploy on multiple accounts and regions

Once your pipeline reaches a certain complexity, it's good practice to deploy it to different environments using different accounts or at least different regions; this separates the environments (especially when using different accounts), which reduces the blast radius if something goes wrong and helps with security separation. For instance, you can allow developers to log in to pre-production accounts for troubleshooting or investigation, but only administrators can access production accounts. Another reason to do this is to avoid having different environments competing for service quotas, which in most cases are per account and region.

Deploying on multiple regions is often used to serve customers in that area or if your service needs to survive a regional disaster.

Getting ready

This recipe is an extension of the *Setting up a code deployment pipeline using CDK and AWS CodePipeline* recipe; therefore, you need to follow the *Getting ready* instructions there to set up the tools required for this recipe.

In addition, when using a multi-account deployment, you need the CDK infrastructure stack to trust the role that's used by the pipeline. To do so for each account involved (even the main one for consistency), you need to do an upgraded CDK bootstrap to add this trust to the pipeline. If that account and region have already been CDK bootstrapped, the command will just make the updates needed to add that trust.

As indicated in the *Technical requirements* section, you are expected to have AWS credentials configured by default, either using the configuration files or environment variables. When you need to access the alternative account(s), if you use the configuration option, you can update the `~/.aws/config` and `~/.aws/credentials` file to add a profile for each account so that you can specify different credentials and configurations for each profile. At this point, you can reference them in the commands. Please check the AWS documentation on CLI profiles for more details.

If you're using environment variables, just override the variables while you run the command and don't specify any profile (omit the `--profile` parameter).

Decide on the accounts and regions you will deploy as part of this recipe. Ideally, you should use two accounts and regions to fully explore this recipe. For each pair of accounts and regions, run the following bootstrap command, replacing my values with your own:

```
cdk bootstrap 1234567890/us-west-1 --profile myprofile \
  --trust 11111111111 --cloudformation-execution-policies \
  arn:aws:iam::aws:policy/AdministratorAccess
```

Here, `1234567890` and `us-west-1` are the account and region where the stack will be deployed, and `myprofile` is the AWS CLI profile with credentials for the `1234567890` account. In this case, it assumes the account is dedicated to the pipeline deployments and it gives full-trust administrator access. If the account is shared for other purposes, the best practice is to grant the minimum required permissions.

How to do it...

1. Complete *steps 2 to 7* of the *Setting up a code deployment pipeline using CDK and AWS CodePipeline* recipe. If you have the resulting project after completing that recipe, it's okay if you run the last step to destroy the stack. Move into the project directory:

    ```
    cd cdk-deployment-recipe
    ```

2. Edit the application file to specify an account and region for the main stack – that is, the one that hosts the pipeline. Use a text editor to open the file (for example, `vi app.py`) and uncomment one of the lines that assigns the env variable either using the AWS CLI defaults or by specifying your own account and region, like so:

    ```
    env=cdk.Environment(
        account=os.getenv('CDK_DEFAULT_ACCOUNT'),
        region=os.getenv('CDK_DEFAULT_REGION')),
    ```

3. Edit the main CDK file, `cdk_deployment_recipe/cdk_deployment_recipe_stack.py`, in the CodePipeline constructor editor and add a new parameter called `cross_account_keys=True` so that it looks like this:

    ```
    pipeline = CodePipeline(self, "Pipeline",
                            pipeline_name="RecipePipeline",
                            cross_account_keys=True,
                            synth=ShellStep("Synth",
                            . . . . . . . . . . .
    ```

4. In the file, just after pipeline creation, replace the line containing addStage with one line for each of the accounts and regions you want to deploy, specifying an env argument for each one. For instance, if you want to deploy the GlueStack to 1234567890 with the us-east-1 region set to "integration" and 0987654321 with the same region but set to "production", you must do the following (make sure you respect the Python indentation so that both lines are aligned with the pipeline variable assignment):

```
pipeline.add_stage(GlueStage(self, "integration",
    env={"region":"us-east-1","account":"1234567890"}))
pipeline.add_stage(GlueStage(self, "prod2",
    env={"region":"us-east-1","account":"0987654321"}))
```

5. Test whether the CDK script is valid. If not, review the previous steps while paying attention to the Python indentation being aligned:

```
cdk synth
```

6. Commit the changes and push them to Git:

```
git add *
git commit -m "Made the pipeline multiregion"
git push --force --set-upstream codecommit master
```

7. Deploy the pipeline and the stack:

```
cdk deploy --all
```

    If you get an "invalid principals" error, revise that the accounts and regions you specified in *step 4* match the bootstrap you did in the *Getting ready* section.

8. Open the AWS console, navigate to CodePipeline in the same region where you have AWS CLI configured by default, and locate the RecipePipeline pipeline. If it hasn't been deployed yet, wait until all steps up to the prod stage are green.

9. Note that you can retry a failed action, but in most cases, you will need to make a correction to the CDK code and push it to the repository to make corrections.

10. Now, navigate to Glue in the console and go to the accounts in one of the regions you chose. Review and optionally run the **deployment_recipe_glueshell** Glue shell job.

11. If you no longer need it, remove the stack (which deletes the pipeline as well):

```
cdk destroy --all
```

How it works...

This recipe extends the previous recipe, *Setting up a code deployment pipeline using CDK and AWS CodePipeline*, so the concepts and explanations there apply here as well. We'll focus on the differences when deploying on multiple accounts and regions.

When CodePipeline detects that multiple accounts are involved, the pipeline is considered by CDK as multi-account. This requires an explicit environment configuration, even if you use environment variables to take it from the AWS CLI defaults. Also, the pipeline will work while assuming roles, even for the current account, thus the need for the special bootstrap you did in the *Getting ready* section for each of the accounts and regions involved. This trusts the pipeline to do the deployment.

In the CodePipeline code, you changed the `cross_account_keys` flag to `True`. This is required so that you can deploy artifacts cross-account using trust policies. It creates a KMS key that can be shared between the accounts to encrypt the artifact bucket. In older versions, this was always done by default, but then it was disabled to avoid the cost of maintaining the key (typically $1/month). This isn't needed when the pipeline deploys just in the same account, so it doesn't need to impersonate roles.

Then, in the main script, you added additional deployments of the Glue stack for multiple accounts and regions. Note how simple this is compared to manually propagating changes across environments, even if you're using CloudFormation changesets.

Finally, you deployed the project like any other CDK stack.

There's more...

CDK does most of the security setup using defaults. You can examine the changes and the CloudFormation templates regarding roles and permissions. It is possible to configure the pipeline so that it uses specific roles that already exist. Alternatively, you can create roles in the pipeline with more fine-grained control or have a role name predefined.

Check the CDK documentation for more options. Each new version provides both more automation and more customization if you need it.

In this example, we used the **s3_deployer** utility, which is an easy way to put files from the repository into S3. So, in the step where it builds the artifacts, it also deploys them to S3. At the time of writing, it doesn't optimize the build for reuse between the different environments and causes more builds than needed.

See also

- If you want to learn more about Glue jobs and the different usages for building data pipelines, you can check the recipes dedicated exclusively to AWS Glue in *Chapter 3, Ingesting and Transforming Your Data with AWS Glue*.

Running code in a CloudFormation deployment

Before the availability of more sophisticated code-based solutions such as CDK or Terraform, AWS CloudFormation was the standard to automate AWS deployments for more than a decade. It started as a purely template solution but then introduced the ability to call functions with custom resource deployment.

While CloudFormation templates are not a fully fledged IaC solution, since it was the only AWS deployment solution for years, they are still present on many projects. Often, the effort of migrating to a tool such as CDK doesn't justify the benefit, so instead, it makes sense to use CloudFormation's advanced features and automation infrastructure, despite its limitations.

In this recipe, you'll learn how to use CloudFormation custom resources to run your own code on deployment, which overcomes the limitations of templates. In this case, the code will be used to set up a file on S3, but the same concept can be used for any of the features of the AWS SDKs.

Getting ready

For this recipe, you just need the AWS CLI installed and configured, a text editor, and access to the AWS console to view the results. See the *Technical requirements* section for guidance on setting up the command line.

How to do it...

1. Using a text editor, create a file named `recipe3_template.json` with the following JSON content. There are line breaks to fit the code on the page. When you enter it in the file, make sure you keep it as valid JSON by not introducing line breaks within the string double quotes, `" "`:

```
{
  "Resources": {
    "S3Bucket": {
      "Type": "AWS::S3::Bucket",
      "Properties": {
        "BucketName": {
          "Fn::Sub": "recipe-deploy-action-${AWS::AccountId}"
        }
      }
    },
    "DeployLambdaRole": {
      "Type": "AWS::IAM::Role",
      "Properties": {
        "AssumeRolePolicyDocument": {
          "Statement": [
```

```json
          {
            "Effect": "Allow",
            "Principal": {
              "Service": "lambda.amazonaws.com"
            },
            "Action": "sts:AssumeRole"
          }
        ]
      },
      "ManagedPolicyArns": [
        "arn:aws:iam::aws:policy/service-role/
AWSLambdaBasicExecutionRole"
      ]
    }
  },
  "DeployerPolicy": {
    "Type": "AWS::IAM::ManagedPolicy",
    "Properties": {
      "Path": "/",
      "PolicyDocument": {
        "Version": "2012-10-17",
        "Statement": [
          {
            "Effect": "Allow",
            "Action": "s3:PutObject",
            "Resource": {
              "Fn::Sub": "arn:aws:s3:::${S3Bucket}/*"
            }
          },
          {
            "Effect": "Allow",
            "Action": "s3:DeleteObject",
            "Resource": {
              "Fn::Sub": "arn:aws:s3:::${S3Bucket}/*"
            }
          }
        ]
      },
      "Roles": [
        {
          "Ref": "DeployLambdaRole"
        }
      ]
    }
```

```json
        },
        "DeployLambda": {
          "DependsOn": [
            "S3Bucket",
            "DeployLambdaRole",
            "DeployerPolicy"
          ],
          "Type": "AWS::Lambda::Function",
          "Properties": {
            "FunctionName": "DeployActionLambda",
            "Description": "Intended to run during deploment",
            "Code": {
              "ZipFile": {
                "Fn::Sub": "import boto3\
nimport cfnresponse\n\ndef handler(event,
context):\n  print(f'Event:')\n  print(event)\n  bucket_
name = '${S3Bucket}'\n  file_path = 'config/setup.txt'\
n\n  if event['RequestType'] == \"Create\":\n    print('The
bucket ${S3Bucket}')\n    print(f'Using bucket {bucket_
name}')\n    s3 = boto3.resource('s3')\n    config =
('''property1=value1\n                property2=value2\
n''')\n    s3.Object(\n        bucket_name=bucket_name,
\n        key=file_path\n    ).put(Body=config)\n  elif
event['RequestType'] == \"Delete\":\n    print(f'Deleting
file: s3://'\n        '{bucket_name}/{file_path}')\n    boto3.
client('s3').delete_object(\n        Bucket=bucket_
name,\n        Key=file_path\n    )\n  cfnresponse.send(event,
context, cfnresponse.SUCCESS, {})\n"
              }
            },
            "Handler": "index.handler",
            "Role": {
              "Fn::GetAtt": [
                "DeployLambdaRole",
                "Arn"
              ]
            },
            "Runtime": "python3.9",
            "Timeout": 5
          }
        },
        "DeployRun": {
          "Type": "AWS::CloudFormation::CustomResource",
          "DependsOn": "DeployLambda",
          "Version": "1.0",
          "Properties": {
```

```
      "ServiceToken": {
        "Fn::GetAtt": [
          "DeployLambda",
          "Arn"
        ]
      }
    }
  }
},
"Outputs": {
  "RecipeBucketName": {
    "Description": "Name of the bucket created by the recipe",
    "Value": {
      "Ref": "S3Bucket"
    }
  }
}
}
```

2. Validate the template by running the following command:

```
aws cloudformation validate-template \
  --template-body file://recipe3_template.json
```

3. Check for any errors and address the issue in the template. If everything is okay, it will return a JSON advising the capabilities required:

```
{
    "Parameters": [],
    "Capabilities": [
        "CAPABILITY_IAM"
    ],
    "CapabilitiesReason": "The following
resource(s)                    require capabilities: [AWS::IAM::Role]"
}
```

4. Deploy the stack and wait until it completes successfully:

```
aws cloudformation deploy --template-file\
  recipe3_template.json --stack-name RecipeCF-Action\
  --capabilities CAPABILITY_IAM
```

5. Check the bucket that was created based on the configured account and that the text file was created and updated by the stack deployment:

```
BUCKET_NAME="recipe-deploy-action-$(aws sts get-\
caller-identity --query 'Account' --output text)"
aws s3 ls --recursive $BUCKET_NAME
```

6. Optionally, navigate to the AWS console to check the **RecipeCF-Action** stack. Explore the **Resources** and **Events** tabs.

7. Delete the stack. This will remove all resources that were created by the stack, including the S3 object and bucket:

```
aws cloudformation delete-stack \
  --stack-name RecipeCF-Action
```

8. The `delete-stack` command triggers the deletion process but doesn't return anything immediately. You can check the deletion's progress by checking the last events until the command errors because the stack no longer exists. The deletion can sometimes take a couple of minutes:

```
aws cloudformation describe-stack-events \
--stack-name RecipeCF-Action --max-items 3
```

How it works...

When you deployed the CloudFormation template, it applied the `Fn::Sub` and `Fn::GetAtt` functions, resolving the actual values at deployment time. Then, it proceeded to create the resources, considering the dependencies defined. The last to be created was the `DeployRun` custom resource, which causes `DeployLambda` to run on stack creation and deletion.

This Lambda defines the Python code inline for convenience, within a call to `Fn::Sub` so that it can dynamically insert the name of the bucket using `${S3Bucket}`. In addition, the code is defined to be deployed as a ZIP file via the `ZipFile` property, so it automatically includes module dependencies, which are detected by CloudFormation using the top `import` lines.

The code creates a text file on S3 with some configuration for an application.

To create the template via the command line, you had to specify the `--capabilities` argument with a value of `CAPABILITY_IAM`, which confirms that accepting the template might create resources and affect permissions. This is the equivalent of the confirmation check box when you create the stack via the console.

When the stack is deleted, it deletes the resources in the inverse order of deployment. To delete the `DeployRun` custom resource, it invokes the Lambda, indicating in the parameters that it is a deletion run. The code then deletes the file it created on deployment (so that the bucket can be deleted by CloudFormation) and executes the callback provided in the parameters to specify that the deletion is complete. If this call isn't made, the deletion will be stuck in progress until it eventually times out.

There's more...

CloudFormation also supports using the YAML format, which is more compact but also stricter with indentation and spaces. You can convert a template from one format into the other using the template editor in the CloudFormation console.

Custom resources also allow you to specify an SNS ARN instead of a Lambda, which can be used to notify listeners that the stack has been deployed and can take asynchronous actions.

> **Tip**
>
> While you develop deployment actions, you will likely make mistakes and need to make several attempts since deleting and redeploying each time isn't practical.
>
> Instead, you can use the AWS Lambda console to open the specific Lambda and do test invocations while passing a similar JSON to what CloudFormation uses until you're happy with the code. Then, you can do a full redeployment to validate.

See also

- CloudFormation templates have limitations because you need to follow the template structure instead of doing your own custom code, and thus are not ideal for complex scenarios. When starting a new project, it's preferable to use CDK, even if you don't need complex logic yet. See the recipes in this chapter on using CDK – that is, *Setting up a code deployment pipeline using CDK and AWS CodePipeline* and *Setting up a CDK pipeline to deploy on multiple accounts and regions*.

Protecting resources from accidental deletion

In general, automation is the best way to avoid mistakes that can result in loss of service, data, or both. However, there are cases where human intervention is needed to address exceptional situations. In a situation of urgency, the operator might need to build an ad hoc script quickly or take manual action under pressure.

Some AWS resources allow resource protection to prevent costly mistakes that can lead to serious or even irreversible damage. In this recipe, you'll learn how to protect RDS databases, DynamoDB tables, and CloudFormation stacks from accidental deletion.

Getting ready

To complete this recipe, you need a bash command line with the AWS CLI, as indicated in the *Technical requirements* section at the beginning of this chapter.

How to do it...

1. Create a simple RDS database. This will return the full configuration:

```
aws rds create-db-instance --db-instance-class \
  db.t3.micro --db-instance-identifier \
  recipe-db-protected --deletion-protection --engine \
  postgres --no-publicly-accessible \
  --allocated-storage 20 --master-username postgres \
  --master-user-password Password1
```

2. Try to delete the instance – it should refuse to carry out the action:

```
aws rds delete-db-instance --db-instance-identifier \
  recipe-db-protected --delete-automated-backups \
  --skip-final-snapshot
```

3. Disable the deletion protection:

```
aws rds modify-db-instance --db-instance-identifier \
  recipe-db-protected --no-deletion-protection
```

4. Repeat *step 2*. It should delete the database instance.

5. Create a DynamoDB table:

```
aws dynamodb create-table --table-name\
  recipe-protected --deletion-protection-enabled\
  --key-schema AttributeName=id,KeyType=HASH\
  --attribute-definitions\
  AttributeName=id,AttributeType=S\
  --billing-mode PAY_PER_REQUEST
```

6. Try to delete the table – it should refuse:

```
aws dynamodb delete-table --table-name \
  recipe-protected
```

7. Disable the deletion protection on the table:

```
aws dynamodb update-table --table-name \
  recipe-protected --no-deletion-protection-enabled
```

8. Repeat *step 6*. Now, it should delete the table. You can verify that it's been deleted:

```
aws dynamodb list-tables | grep recipe-protected
```

9.   Define your variables and create the CloudFormation template:

```
CFN_TEMPLATE_FILE=cfn_sample.yaml
STACK_NAME=CfnProtectRecipe

cat > $CFN_TEMPLATE_FILE << EOF
AWSTemplateFormatVersion: '2010-09-09'
Resources:
  CfnProtectionSample:
    Type: 'AWS::DynamoDB::Table'
    Properties:
      BillingMode: PAY_PER_REQUEST
      AttributeDefinitions:
        - AttributeName: id
          AttributeType: S
      KeySchema:
        - AttributeName: id
          KeyType: HASH
EOF
```

10. Deploy the template just created:

```
aws cloudformation deploy --template-file \
  $CFN_TEMPLATE_FILE --stack-name $STACK_NAME
```

Protect the stack from deletion:

```
aws cloudformation update-termination-protection \
  --stack-name $STACK_NAME \
  --enable-termination-protection
```

11. Try to delete the protected stack:

```
aws cloudformation delete-stack\
  --stack-name $STACK_NAME
```

12. Disable deletion protection on the stack:

```
aws cloudformation update-termination-protection\
  --stack-name $STACK_NAME\
  --no-enable-termination-protection
```

13. Repeat *step 11*. It should confirm the stack was deleted correctly.

How it works...

In this recipe, you created examples for three kinds of AWS data-related resources. For each of the three cases – RDS, DynamoDB, and CloudFormation – you verified that when protected, they cannot be deleted by issuing the `delete` command, even if your user has permission; deletion protection had to be explicitly disabled first.

For RDS and DynamoDB, the resources were created with the protection already enabled, while in the case of CloudFormation, it required a separate call.

There's more...

This recipe covered RDS, DynamoDB, and CloudFormation, the services that support this kind of protection at the time of writing. In the case of S3, the bucket cannot be deleted if there are still objects. Note that by enabling versioning or replication, you can mitigate the damage of accidental S3 deletion.

This still leaves out important systems such as Redshift databases or Kinesis streams. In those cases (and in general), the best practice is applying the principle of least privilege. This means that both for automation and manual access (both via the console and the command line), the user role should be as restricted as possible to accomplish its purpose, which also serves to help reduce mistakes.

For instance, the operations team could have access to read-only roles, but to get an admin role, they might require approval, review, or a two-person process.

See also

- Ideally, you want to have environments where deployments and changes are only made by pipelines and not manually. With that automation in place and highly restricted manual intervention on production, the accidental protections seen in this recipe could be considered unnecessary. To learn more about deployment automation, see the *Setting up a code deployment pipeline using CDK and AWS CodePipeline* and *Deploying a data pipeline using Terraform* recipe.

Deploying a data pipeline using Terraform

In this recipe, you will see the Terraform cloud-agnostic IaC automation tool. Agnostic doesn't mean you can abstract yourself from the underlying vendor but that the tool has support for multiple cloud providers, so you can apply the knowledge of the tool to other providers as well as build multi-cloud projects. But you still need to know about the different services on each cloud provider and how to use them. See the official vendor documentation for further details: `https://www.terraform.io/`.

The trade-off of its multi-cloud support is that because it's not provided directly by each vendor, it may take some time until Terraform supports new features.

Getting ready

To complete this recipe, you need a bash command line with the AWS CLI set up, as indicated in the *Technical requirements* section at the beginning of this chapter.

You also need the Terraform executable in your system. You must put it in the system's PATH so that it can be executed easily. Check the HashiCorp website for instructions on how to download the executable for your OS and architecture: https://developer.hashicorp.com/terraform/install.

You also need a user who can log in to the AWS console and use AWS Glue.

How to do it...

1. Create a directory for the Terraform project and change to it:

   ```
   mkdir terraform_recipe && cd terraform_recipe
   ```

2. Create a file with configuration variables (note that the first \ is to break the line and the second one is to escape the $ sign so that bash doesn't try to replace it):

   ```
   cat > var.tf << EOF
   variable "region" {
     description = "AWS region"
     type = string
     default = "us-east-1"
   }

   data "aws_caller_identity" "current" {}

   locals {
     bucket_name = "terraform-recipe-\
   \${data.aws_caller_identity.current.account_id}"
   }

   variable "script_file" {
     type    = string
     default = "ShellScriptRecipe.py"
   }
   EOF
   ```

3. Create a placeholder script for the Glue Shell job:

   ```
   echo 'print("Running Glue Shell job")' > \
     ShellScriptRecipe.py
   ```

4. Create the main Terraform project file so that you can deploy the script to S3 (here, EOF has quotes around it, so it is written literally and doesn't have to escape via $):

```
cat > main.tf << 'EOF'
provider "aws" {
  region = var.region
}

resource "aws_s3_bucket" "bucket" {
  bucket = local.bucket_name
}

resource "aws_s3_object" "script" {
  bucket = local.bucket_name
  key    = "scripts/${var.script_file}"
  source = var.script_file
  depends_on = [
    aws_s3_bucket.bucket
  ]
}
EOF
```

5. Append a Glue Shell job and a new role with minimum permissions it can use to run to the main file (notice >> to append instead of overwriting):

```
cat >> main.tf << EOF
resource "aws_iam_role" "glue" {
  name = "AWSGlueServiceRoleTerraformRecipe"
  managed_policy_arns = [
"arn:aws:iam::aws:policy/AmazonS3ReadOnlyAccess",
"arn:aws:iam::aws:policy/CloudWatchAgentServerPolicy"
  ]
  assume_role_policy = jsonencode(
      {
          "Version": "2012-10-17",
          "Statement": [
            {
                "Action": "sts:AssumeRole",
                "Principal": {
                  "Service": "glue.amazonaws.com"
                },
                "Effect": "Allow",
                "Sid": ""
            }
```

```
            ]
          }
        )
    }

    resource "aws_glue_job" "shell_job" {
      name     = "TerraformRecipeShellJob"
      role_arn = aws_iam_role.glue.arn

      command {
        name = "pythonshell"
        python_version = "3.9"
        script_location = "s3://\${local.bucket_name}\
    /scripts/\${var.script_file}"
        }
    }
    EOF
```

6. Initialize the Terraform project. This can take a minute or so since it needs to download the AWS provider:

   ```
   terraform init
   ```

7. Validate that the files that were generated in the previous steps are valid, as well as the AWS CLI credentials and configuration:

   ```
   terraform plan
   ```

   If all is correct, the plan summary should look like this:

   ```
   Plan: 4 to add, 0 to change, 0 to destroy.
   ```

8. Deploy the stack with the apply command. Using the region variable, set the name of your region – for example, us-east-2:

   ```
   terraform apply -var region=us-east-2
   ```

9. In the AWS console, navigate to AWS Glue on the chosen region. Use the **ETL Jobs** menu and then select **TerraformRecipeShellJobs** from the job table. Use the **Run** button and then open the **Runs** tab. Once it completes successfully, you can use the **Output** link to view the message the job printed:

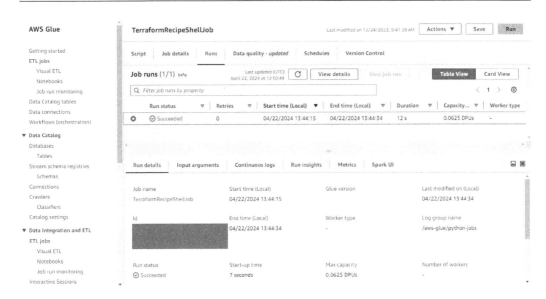

Figure 8.2 – AWS Glue job deployed by Terraform

10. Back on the command line, clean up if you don't want to keep the deployment.

    You need to specify the same region you used on deployment, like so:

    ```
    terraform destroy -var region=us-east-2
    ```

How it works...

In addition to the trivial Python script, you created two files with the `.tf` (Terraform Format) extension. These are text files that resemble JSON but it's not (notice that there are no commas at the end of the lines and the object definition is also different). In addition, this format allows you to easily reference other objects from the same or different files as you don't need to import other files to reference them.

The names of the files are arbitrary. Terraform will pick up any files with the extension and load them together; as shown in this recipe, it's common practice to have a file that acts as a variable/configuration reference so that it's easy to find what can be configured.

Inside the main file, the variables can be used either with a direct reference using the `var/local` prefix or inside a string with a syntax like bash: `${variable-name}`. The bucket name is defined in `locals` instead of `variable` so that it can dynamically retrieve the account ID and make the bucket unique for your account. Variables have default values that can be overridden at deployment time, as you did by specifying the region when applying and destroying the stack.

If you peruse the Terraform file, you will see that it's intuitive to understand how it defines the components. An important element is the `provider "aws"` section, which tells Terraform that the project needs the libraries for AWS resources.

When you ran `terraform init`, it downloaded the AWS provider and put it under the project's hidden `.terraform` folder.

By running the AWS Glue job successfully, you proved that the job was deployed with all the required configurations: it could find and load the script you created locally and was able to impersonate the IAM role created for it.

There's more...

In this recipe, you used `terraform plan` to validate the deployment on the actual AWS account. If you want to check if the config is valid but don't have access to an account (for instance, because you are offline), you can run `terraform validate` instead to make sure the syntax is correct. The `terraform fmt` command can be used to format the `.tf` files to conform to the Terraform conventions and recommendations.

In addition to AWS, Terraform has support for many providers for the popular cloud vendors, as well as generic ones such as the one for Kubernetes. There is a registry where they are clearly labeled, indicating whether they are provided by HashiCorp, a partner, or the community. You can use multiple providers on the same project.

To keep track of deployments, Terraform creates a file named `terraform.tfstat` (as well as a backup when it needs to update it). The file is synchronized with the cloud before deployment. On a production project, it is bad practice to have it locally since it can be lost or get out of sync if different people have their own local copy. You can configure Terraform to save that file in a shared place (for example, S3) and use locks with a tool such as DynamoDB so that the status file is free from conflicts or loss.

See also

- Instead of writing your own Terraform code, you could generate it from existing resources. See the *Reverse-engineering IaC* recipe.

- The main alternative to Terraform is the frameworks and tools provided by each specific cloud vendor. For AWS, that's CloudFormation and CDK. To learn more about them, take a look at the *Setting up a code deployment pipeline using CDK and AWS CodePipeline*, *Setting up a CDK pipeline to deploy on multiple accounts and regions*, and *Running code in a CloudFormation deployment* recipes in this chapter.

- To learn more about data pipelines using AWS Glue, refer to *Chapter 3, Ingesting and Transforming Your Data with AWS Glue*.

Reverse-engineering IaC

Often, we aren't creating a brand-new data infrastructure but extending existing resources that we want to automate. So, future changes are automatically tracked and deployed using IaC. You may also need some help building your IaC and would rather create the resources using the AWS console than generate the corresponding code. You can then use this as the basis instead of writing the code from scratch.

In this recipe, you will learn how Terraformer can reverse engineer existing resources to generate Terraform code. This tool is relatively new and still under development, so it's likely to help you but not do the full work for you.

If you wish to use other IaC tools instead of Terraform, there are reverse-engineering alternatives that you can check in the *See also* section at the end of this recipe.

Getting ready

To complete this recipe, you need a bash command line with the AWS CLI set up, as indicated in the *Technical requirements* section at the beginning of this chapter.

You also need the Terraform executable in our system in the system's PATH so that it can be executed just using its name. Check the HashiCorp indications and download the executable matching the OS and architecture from their website: `https://developer.hashicorp.com/terraform/install`. In addition, you'll need the Terraformer executable, which you can download as a binary or install with a package manager. Check the project on GitHub for instructions: `https://github.com/GoogleCloudPlatform/terraformer`.

> **Disclaimer**
> Despite the project being hosted in the Google Cloud community repository, it is Apache licensed open source and it's not affiliated nor provided by Alphabet Inc.

How to do it...

1. Create the Glue job that you'll reverse engineer later to Terraform code:

    ```
    aws glue create-job --name recipe-shell-reveng\
      --role arn:aws:iam::$(aws sts get-caller-identity\
      --query 'Account' --output text)\
    :role/SomeRoleForGlue --command \
     '{"Name":  "pythonshell", "PythonVersion":"3.9",
       "ScriptLocation": "s3://somebucket/yourscript.py"}'
    ```

2. View the job you just created. You'll see that many properties have been filled with defaults:

```
aws glue get-job --job-name recipe-shell-reveng
```

3. Initialize an empty Terraform project for AWS:

```
mkdir recipe_reveng && cd recipe_reveng
echo "provider "aws" {}" > provider.tf
terraform init
```

4. Import the Glue job you created in *step 1* into the Terraform project:

```
terraformer import aws --path-pattern=.\
 --compact=true --resources=glue --filter\
 "Name=name;Value=recipe-shell-reveng"
```

5. Edit the resources file using a text editor – for instance, vi resources.tf. Make the following changes and save them:

- Remove number_of_workers = "0"

- Change the number of retries to 1 with max_retries = "1"

6. Test the Terraform project with the changes and address any errors:

```
terraform plan
```

At the time of writing, the latest version of Terraformer is v0.8.24 and still references the old Terraform AWS provider. Check for the following error:

```
Error: Failed to load plugin schemas
Error while loading schemas for plugin components: Failed to
obtain provider schema:   Could not load the schema for provider
registry.terraform.io/-/aws: failed to instantiate provider
"registry.terraform.io/-/aws" to obtain schema: unavailable
provider "registry.terraform.io/-/aws"..
```

If you can see this error, you need to address it by upgrading the provider:

```
terraform state replace-provider \
 registry.terraform.io/-/aws \
 registry.terraform.io/hashicorp/aws
```

If you're running the code on Microsoft Windows, you might get the following error:

```
Error: Failed to read state file
The state file could not be read: read terraform.tfstate: The
process cannot access the file because another process has
locked a portion of the file.
```

This is a known issue regarding handling file locks locally. You can work around this by adding `-lock=false` to the failed command, like so:

```
terraform plan -lock=false
```

7. Deploy the changes you made in *step 5* and confirm that the retries is now set to `1`:

```
terraform apply
aws glue get-job --job-name recipe-shell-reveng
```

8. Delete the Glue Shell job using Terraform:

```
terraform destroy
```

How it works...

First, you created a very simple Glue job to reverse engineer. Note that this job can't run since it references a non-existing AWS role and S3 path – it is just for demonstration purposes. To learn more about Glue, check out *Chapter 3, Ingesting and Transforming Your Data with AWS Glue*.

Then you initialized a minimal Terraform project to download the AWS provider under the `.terraform` directory, which Terraformer requires to access the account. It is possible to put the provider on a shared path, so it's shared by projects.

Using Terraformer, you imported the job, and underneath it used the AWS CLI to access the account using the default credentials and configuration, such as `region`. That generated multiple files in the current directory. This is due to the `--path-pattern=` parameter, which places the import files in the current directory instead of multiple subdirectories. Note that `--compact=true` indicates all the Terraform resources are imported into the `resources.tf` file.

In addition, the `import` statement overwrote the `provider.tf` file to pin the AWS provider version, while Terraform 1.0 or later has a combability promise. It is good practice to specify the version for better stability. The `~>` operator indicates that the build version (the last of the three numbers) can change but not the others. This allows you to make fixes to the version with a very low risk of introducing incompatibilities.

It's also noteworthy that it generated a `variables.tf` file, indicating that the configuration is stored locally on the `terraform.tfstate` file. For a production project, this should be replaced with a shared durable repository such as S3.

If you had to replace the AWS provider, it upgraded the `terraform.tfstate` file as needed to bring it up to the Terraform version you're using. The state file is a JSON file, so you can open it and explore its contents.

You used Terraform to deploy a change to the Glue job and finally delete it by destroying the stack. Both commands demonstrated that once you completed the import, you can continue working as a regular Terraform project. Running Terraformer again would override any changes you made.

There's more...

As you've seen, reverse engineering is not trivial and you'll likely need to make adjustments, such as removing the number of workers on a Glue Shell job, since it doesn't have that concept. You should use the generated project as a template and revise all the code that's produced for potential mistakes and inconsistencies. In addition, many resource types are still not available, so you would have to add that part of the code manually.

When you use the Terraformer `import` command, you can specify multiple resources so that they are generated on the same project. However, the filters for resources must be specific to the resource type. Alternatively, you can import all without filters and then manually delete what you don't need.

See also

- To learn how to create a Terraform project from scratch, see the *Deploying a data pipeline using Terraform* recipe.

- If you'd rather use CDK, CloudFormation, or AWS CLI commands instead of Terraform, there are alternative tools to help you generate code automatically:

  - **Former 2** (`www.former2.com`): This connects to your AWS account, detects resources, and generates code for multiple languages, including CDK and CloudFormation. It requires you to provide AWS credentials, so it's best practice to provide a temporary session with read-only limited privileges. The tool is under development and doesn't work for all kinds of resources.

> **Disclaimer**
> The Former 2 tool is provided "as-is" without any guarantees, by a reputable developer (AWS ambassador since 2018).

  - **Console Recorder for AWS**: From the creator of Former 2, it's no longer under development, so it is missing new services but it's very easy to use. It provides a plugin for the main web browsers, which listens to the actions you do on the AWS console and then can provide the equivalent code for many languages, including CDK, CloudFormation, boto3, and the AWS CLI.

Integrating AWS Glue and Git version control

AWS Glue is a serverless data integration service that offers different engines and tools for different personas involved in data engineering. Git is the industry standard source code version control system. Integrating both enables version handling on Glue jobs and improves DevOps in general.

In this recipe, you'll learn how to save and retrieve the status of a Glue job on a Git repository provided by AWS CodeCommit. This feature is supported for other kinds of jobs, including notebooks, and is also available directly on the AWS console. See the *There's more...* section for further details.

Getting ready

To complete this recipe, you need a command line bash with the AWS CLI set up, as indicated in the *Technical requirements* section at the beginning of this chapter. The AWS user needs permission to use AWS CodeCommit.

How to do it...

1. Set up a placeholder Python script on S3 for the Glue job:

    ```
    BUCKET_NAME="recipe-glue-git-\
    $(aws sts get-caller-identity --query 'Account' \
      --output text)"
    AWS_REGION="$(aws configure get region)"
    aws s3api create-bucket --bucket $BUCKET_NAME \
      --create-bucket-configuration \
      LocationConstraint=$AWS_REGION
    echo "print('Running Shell job')" > ShellScript.py
    aws s3 cp ShellScript.py s3://$BUCKET_NAME
    rm ShellScript.py
    ```

2. Create a Glue Shell job, referencing the script you uploaded in *Step 1*:

    ```
    aws glue create-job --name recipe-glue-git-job\
      --role arn:aws:iam::$(aws sts get-caller-identity\
      --query 'Account' --output text)}role/RoleForGlue\
      --command '{"Name" :  "pythonshell",'\
    '"PythonVersion": "3.9", "ScriptLocation":'\
    '"s3://'$BUCKET_NAME'/ShellScript.py"}'
    ```

3. Create a CodeCommit Git repository and push the job onto the `main` branch:

    ```
    aws codecommit create-repository \
      --repository-name RecipeGlueGit
    aws glue update-source-control-from-job --job-name \
     recipe-glue-git-job --provider AWS_CODE_COMMIT \
     --repository-name RecipeGlueGit --branch-name main
    ```

4. Delete the job script and verify it's no longer in the bucket:

    ```
    aws s3 rm s3://$BUCKET_NAME/ShellScript.py
    aws s3 ls s3://$BUCKET_NAME/
    ```

5. Recover the script from Git to S3 so that the job can use it:

    ```
    aws glue update-job-from-source-control --job-name \
      recipe-glue-git-job --provider AWS_CODE_COMMIT \
      --repository-name RecipeGlueGit --branch-name main
    aws s3 ls s3://$BUCKET_NAME/
    ```

6. Download the restored script and verify that it contains the simple `print` statement you created it with in *step 1*:

    ```
    aws s3 cp s3://$BUCKET_NAME/ShellScript.py .
    cat ShellScript.py
    ```

7. Finally, run the cleanup commands:

    ```
    rm ShellScript.py
    aws glue delete-job --job-name recipe-glue-git-job
    aws s3 rm s3://$BUCKET_NAME/ShellScript.py
    aws s3api delete-bucket --bucket $BUCKET_NAME
    ```

How it works...

In the first two steps, you created a sample Glue Shell job with a minimal script. Please note that the role that we used wasn't created in this recipe, so the job can't be run. However, this isn't a requirement for this recipe.

Then, you created a CodeCommit Git repository using the default `main` branch, where you asked Glue to save the job. This created a commit on the branch with two files placed under a directory matching the name of the job: a JSON file with the job definition and the Python script. It is possible to configure the folder on Git under which these files are saved.

Notice how simple this was compared to having to configure a Git client, credentials, and connectivity; it's all integrated using your IAM user permissions.

To demonstrate versioning, you simulated a mistake where the script on S3 was deleted and then recovered from the latest version of the Git repository.

There's more...

This recipe used the command line to do all the actions to demonstrate it can be automated, as well as to make things easier for users not familiar with AWS Glue.

However, normally, this Git integration is used via the AWS console while the user is developing a data solution. This is especially the case for AWS Studio to develop visual jobs, where the user defines the ETL visually instead of writing a script file.

The Glue console provides an action to push and pull changes from Git that can be used in all kinds of Glue jobs, including managed notebooks. When you use this action in the console, the repository's configuration is stored on the **Version Control** tab:

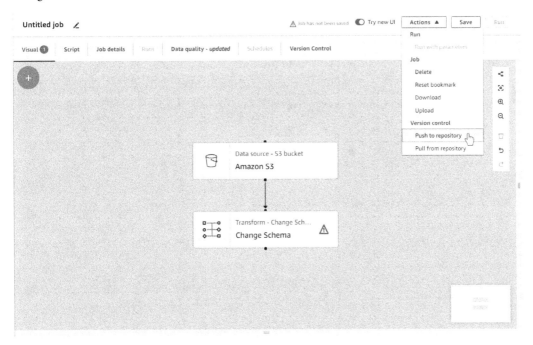

Figure 8.3 – Glue Studio version control actions

If you wish to use one of the other Git repositories supported instead of CodeCommit, such as GitHub or GitLab, you must specify further details and authentication mechanisms, depending on each vendor's requirements.

In this recipe, you used the default `main` branch for simplicity; in practice, development would typically be done on a different branch. Then, once a version is deemed suitable for production, it would be merged onto the branch that reflects production, possibly with an automated process that monitors that branch for commits and triggers a pipeline to get those changes tested and deployed.

At the time of writing, branches must already exist for Glue to be able to use them. You can create one easily if you have the Git client on a checkout project by running `git branch <new branch name>` or by using the AWS CLI on CodeCommit, branching from another branch, like so:

```
aws codecommit get-branch --repository-name <your repo> \
  --branch-name <branch to branch from>
aws codecommit create-branch --repository-name <your repo>\
  --branch-name dev --commit-id <from the first command>
```

See also

- When the Glue job is pushed into the repository, you can have a deployment pipeline automatically trigger and run a CI/CD process. The *Setting up a code deployment pipeline using CDK and AWS CodePipeline* recipe shows how to do this.

- To learn more about Glue and its capabilities, see the recipes dedicated to this AWS data integration service in *Chapter 3, Ingesting and Transforming Your Data with AWS Glue*.

9

Monitoring Data Lake Cloud Infrastructure

In this chapter, we will discuss the essential aspects of tracking and monitoring your data lake infrastructure. A data lake, often a repository for vast amounts of structured and unstructured data, is a critical component of any data-driven organization. However, without effective monitoring, the data lake can quickly become a data swamp, leading to inefficiencies, increased costs, and potential compliance risks. The recipes covered in this chapter are designed to address common challenges and ensure your data lake remains an asset rather than a liability.

This chapter includes the following recipes:

- Automatically setting CloudWatch log group retention to reduce cost
- Creating custom dashboards to monitor Data Lake services
- Setting up System Manager to remediate non-compliance with AWS Config rules
- Using AWS config to automate non-compliance S3 server access logging policy
- Tracking AWS Data Lake cost per analytics workload

By the end of this chapter, you will have acquired the knowledge and skills necessary to monitor your data platform effectively. This will help you maintain a robust, efficient, cost-effective data lake operation.

Technical requirements

The code files for this chapter are available on GitHub at `https://github.com/PacktPublishing/Data-Engineering-with-AWS-Cookbook/tree/main/Chapter09`.

Additional information

Before you start this chapter, you need to understand the following concepts:

- **Logs**: Logs are vital for diagnosing issues, auditing activities, and monitoring the health of your data lake. You can subscribe to specific events within your log files, such as errors and warnings, to stay informed about critical occurrences. This proactive approach enables you to address issues before they escalate, ensuring the smooth operation of your data lake services.

- **Alarms**: Alarms are used to monitor specific metrics and trigger actions based on predefined thresholds. For example, you can set an alarm for CPU utilization to notify the development team when usage exceeds a certain limit. Additionally, alarms can automate responses, such as launching new instances to handle increased load, thereby maintaining performance and avoiding service disruptions.

- **EventBridge**: It contains events, rules, and targets. Amazon EventBridge allows you to manage and respond to events across your AWS environment. It comprises three main components:

  - **Events**: Any significant occurrence in your system, such as an API call, console sign-in, auto-scaling state change, EC2 instance state change, or EBS volume creation, is included here.

  - **Rules**: They define the conditions under which events should trigger actions. For example, you might create a rule that triggers when an EC2 instance changes state.

  - **Targets**: These are the actions that are executed when an event matches a rule. Common targets include AWS Lambda functions, Amazon **Simple Notification Service** (**SNS**), and Amazon **Simple Queue Service** (**SQS**).

  By configuring EventBridge with appropriate rules and targets, you can automate responses to various events, enhancing the resilience and efficiency of your data lake infrastructure.

Automatically setting CloudWatch log group retention to reduce cost

Amazon CloudWatch collects metrics, logs, and events from your resources by default. These logs could then be used to build dashboards, alarms, and alerts. By default, Amazon CloudWatch Logs stores your log data indefinitely, which can add up the cost, particularly when you use detailed monitoring instead of basic monitoring. Using Lambda to automatically check log groups within the regions of your services and data lake can help you save on storage costs.

Getting ready

Before reducing logging costs, you need to have a strategy. Good logging leads to good monitoring. A sizable number enables humans and machines to analyze information. It would be best to have a logging strategy that can answer questions such as "who did what and when?" without including sensitive information such as passwords or secrets before trying to reduce the number of logs.

How to do it...

1. In the **Home** console, click on **IAM**:

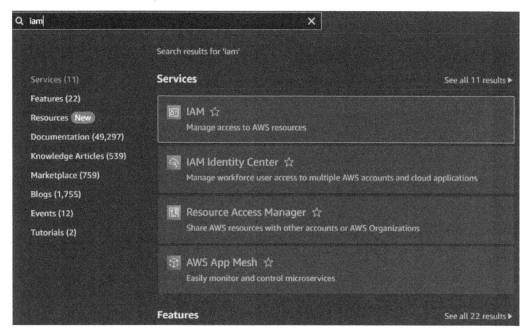

Figure 9.1 – Selecting IAM in the portal

2. Click on **Policies** and then **Create policy**, as shown:

Figure 9.2 – Creating a policy for the Lambda service

3.   Select **Lambda** from the **Select a service** drop-down menu:

Figure 9.3 – Selecting Lambda

4.   On the next page, select **JSON** formatting:

Figure 9.4 – Creating a policy as JSON

5. Add the following Lambda IAM policy in JSON formatting:

```
"Version": "2012-10-17",
"Statement": [
    {
        "Sid": "VisualEditor0",
        "Effect": "Allow",
        "Action": [
            "logs:CreateLogGroup",
            "logs:CreateLogStream",
            "logs:PutLogEvents",
            "logs:DescribeLogGroups",
            "logs:PutRetentionPolicy"
        ],
        "Resource": "*"
    }
]
}
```

6. Name the policy, for example, `lambda-logging`. Then click on **Create policy**:

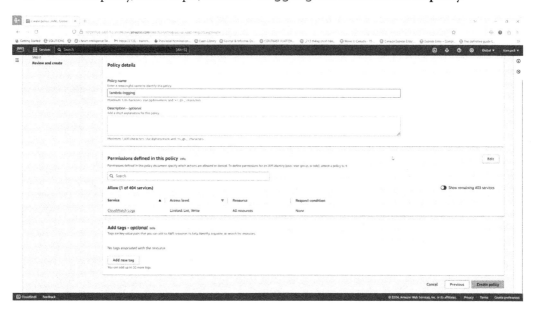

Figure 9.5 – Naming the Lambda policy

7. Next, click on **Create role** from the IAM console:

Figure 9.6 – Selecting roles in the IAM portal

8. Select **AWS Service** and then **Lambda** under the **Service or use case** dropdown, and then click on **Next**:

Figure 9.7 – Selecting the Lambda service for roles

9. Select the `lambda-logging` policy that we created in *step 6* and click on **Next**:

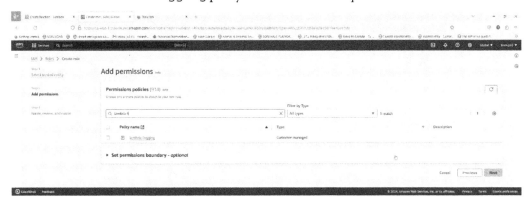

Figure 9.8 – Adding the lambda-logging policy to roles

10. Fill in **Role details** and click on **Create role**. For best supervision, click on **Add tag** and add the purpose of creating the policy:

Figure 9.9 – Clicking on Create role

11. Select **Services | Lambda** as shown:

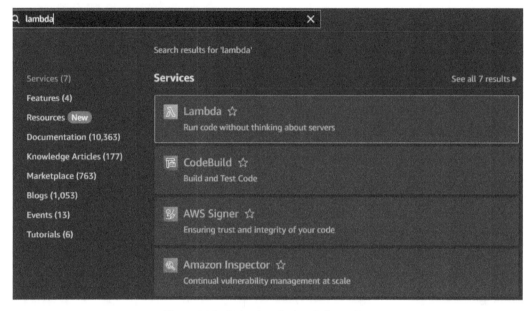

Figure 9.10 – Selecting the Lambda service

12. Then click on the **Create a function** button:

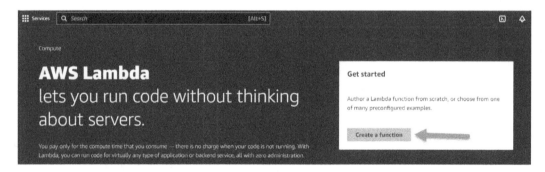

Figure 9.11 – Selecting the Create a function button

13. Fill in basic information, as shown in *Figure 9.12*, and choose the IAM role that was created in *step 10*. Then click on **Next**:

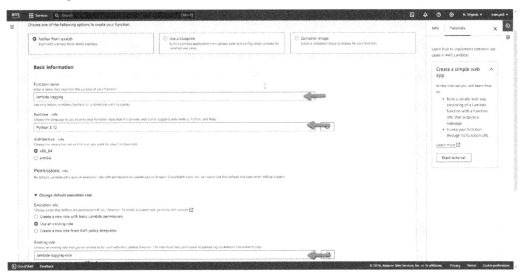

Figure 9.12 – Filling in the function name, runtime, and execution role

14. Go to the code editor and paste the following Python code:

```
import boto3
from botocore.config import Config
import logging
import os


# Set the number of retention days
retention_days = 30

# logging
LOGGER = logging.getLogger()
LOGGER.setLevel(logging.INFO)
logging.getLogger('boto3').setLevel(logging.CRITICAL)
logging.getLogger('botocore').setLevel(logging.CRITICAL)

regions = [item.strip() for item in os.environ['AVAILABLE_
```

```python
    REGION'].split(",") if item]


def lambda_handler(event, context):
    LOGGER.info(f"start checking= {regions}")
    if not regions:
        return {'statusCode': 200, 'body': 'No regions found '}

    for region in regions:

        client = boto3.client('logs', region_name= region)
        response = client.describe_log_groups()
        nextToken = response.get('nextToken', None)
        log_groups = response['logGroups']

        # Continue to fetch log groups if nextToken is present
        while nextToken is not None:
            response = client.describe_log_
groups(nextToken=nextToken)
            nextToken = response.get('nextToken', None)
            log_groups += response['logGroups']

        for group in log_groups:
            if 'retentionInDays' in group.keys():
                print(group['logGroupName'],
group['retentionInDays'], region)
            else:
                print("Retention Not found for ",
group['logGroupName'], region)
                set_retention = client.put_retention_policy(
                    logGroupName=group['logGroupName'],
                    retentionInDays=retention_days
                )
                LOGGER.info(f"PutRetention result {set_
retention}")
    return {'statusCode': 200, 'body': 'completed.'}
```

15. Go to the CloudWatch console and navigate to the **Rules** section:

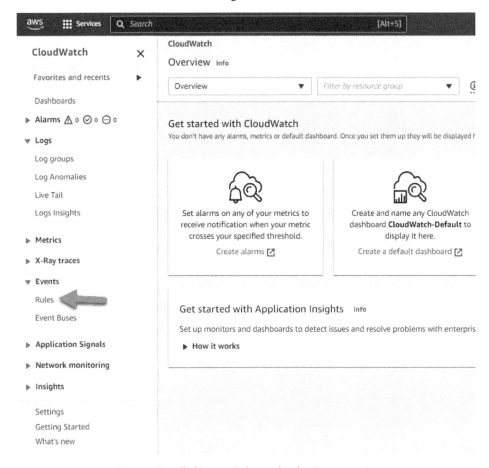

Figure 9.13 – Clicking on Rules under the Events section

16. Click on **Create rule**:

Figure 9.14 – Clicking on Create rule

17. Fill in the **Name** and **Description** fields, as shown, and click on **Next**:

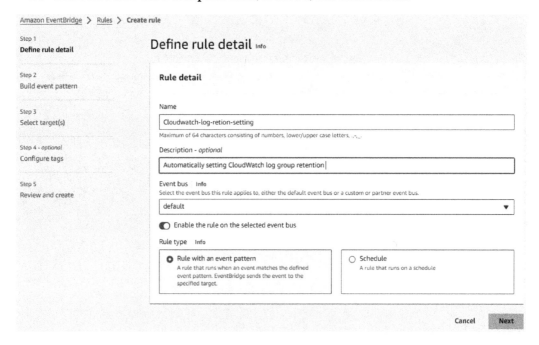

Figure 9.15 – Filling in the rule details

18. When you click on **Next**, you will be led to the page where you can select the event pattern. Select **CloudWatch Logs** under **AWS Service** and **AWS API Call via CloudTrail** under **Event Type** as shown here:

Figure 9.16 – Selecting the AWS service and Event type options

19. In the **Specific operations** field, add `CreateLogGroup`:

events from the following STS
actions: GetFederationToken and
GetSessionToken. Data events (for
example, for Amazon S3 object level
events, DynamoDB, and AWS
Lambda) must have trails configured
to receive those events. Learn more.

Event Type Specification 1

○ Any operation

● Specific operation(s)

Specific operation(s)

CreateLogGroup

Remove

Figure 9.17 – Filling in the Specific operation(s) box

20. On the **Create rule** page, select the Lambda function that you created to link the CloudWatch Events rule with the Lambda function:

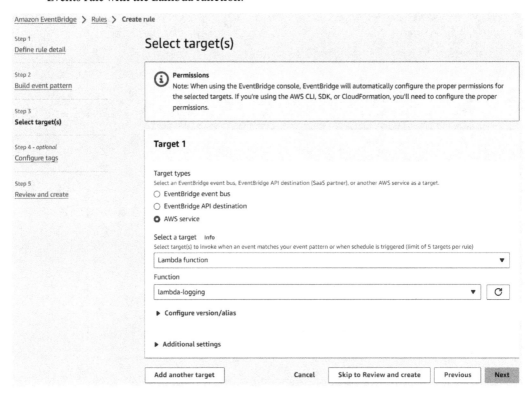

Figure 9.18 – Selecting Lambda as the target service

21. Configure tags if needed. Then, under **Review and create**, review all the previous steps and then click on **Create rule**:

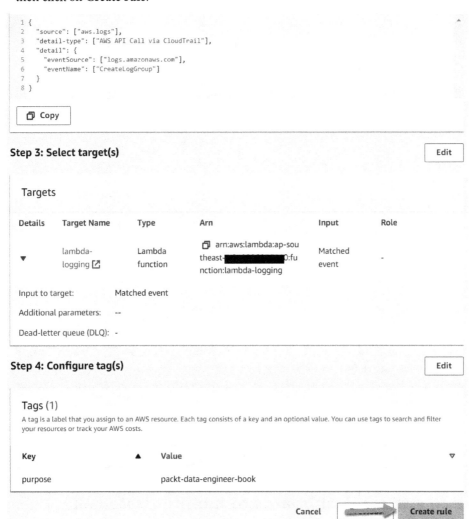

Figure 9.19 – Creating the CloudWatch rule

How it works...

The Lambda function will scan all log groups within the region and apply the 30-day retention rule. CloudWatch will trigger the Lambda function to run every five days. There are two required parameters, logGroupName and retentionInDays, which define the name of the log group you want to target and the number of days you are trying to retain a log.

In the `Event` pattern, once you are used to the manual select, you can re-use the JSON and alter it per use case. The JSON event pattern that was created through *steps 8-10* is as follows:

```
{
  "source": ["aws.logs"],
  "detail-type": ["AWS API Call via CloudTrail"],
  "detail": {
    "eventSource": ["logs.amazonaws.com"],
    "eventName": ["CreateLogGroup"]
  }
}
```

There's more...

Another architecture option is to combine Event Bridge with Lambda. This architecture creates an event-based Lambda trigger to apply the retention rules right away after the log groups are created, such as in the reference link. Alternatively, you can also set the S3 life cycle policy archive log data to S3.

See also

- put_retention_policy: https://boto3.amazonaws.com/v1/documentation/api/latest/reference/services/logs/client/put_retention_policy.html#

- Reduce log-storage costs by automating retention settings in Amazon CloudWatch: https://aws.amazon.com/blogs/infrastructure-and-automation/reduce-log-storage-costs-by-automating-retention-settings-in-amazon-cloudwatch/

- Filter pattern syntax for metric filters, subscription filters, and filter log events: https://docs.aws.amazon.com/AmazonCloudWatch/latest/logs/FilterAndPatternSyntax.html

Creating custom dashboards to monitor Data Lake services

CloudWatch dashboards allow you to create interactive visualizations of your data, giving you a consolidated view of the health and status of your AWS assets. CloudWatch provides several pre-built dashboards for various services, but you can also create custom dashboards to meet your specific needs.

In the daily operation, you should have a high-level dashboard built from CloudWatch log groups metrics to help you understand the general performance and perform drill-down on services if needed, such as creating a high-level dashboard on ETL performance and then drilling down to Glue job run monitoring if required for further investigation.

Getting ready

To proceed with this recipe, you need a CloudWatch log group for the services for which you want to build a dashboard.

How to do it...

1. Open the AWS console and select the **CloudWatch** service, which will present you with a screen that is similar to the following:

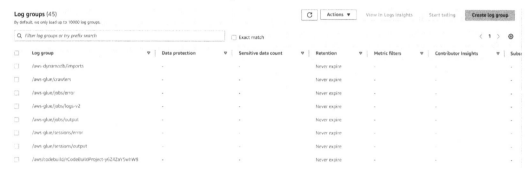

Figure 9.20 – Selecting the CloudWatch log group that you want to monitor

2. Click on **Dashboards**. Here, you will see two options: **Custom dashboards** and **Automatic dashboards**. For this recipe, we will click on **Create dashboard** next to **Custom Dashboards**.

Figure 9.21 – Selecting Create dashboard

3. On the next page, select the widget that is relevant to your use case. The popular ones are as follows:

    •   Line chart, which helps you to see the trend over time

    •   Alarm status, which helps you see a set of alarms

    •   Bar chart, which helps you compare categories of data

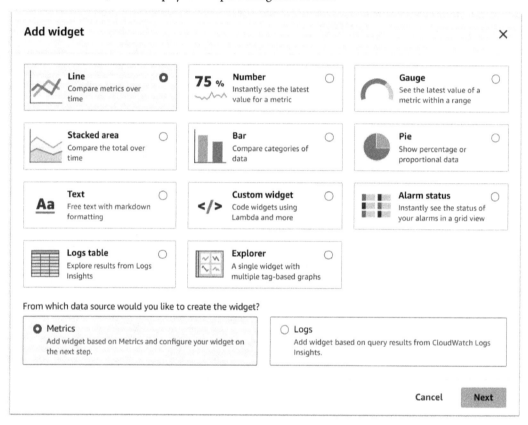

Figure 9.22 – Selecting the type of chart that is appropriate for your dashboard

4. Select whether you want to monitor metrics or log insights, which allow you to query log groups. You can click on **Choose a sample query** to see a sample query that you can run in the CloudWatch console.

Figure 9.23 – Creating a query for your dashboard

How it works...

Amazon CloudWatch dashboards rely on the underlying structure of log groups and streams to organize and present data. A log group acts as a container for log streams that share the same properties, often grouped by application, system component, or similar criteria. These log groups can be associated with specific AWS resources.

When creating a dashboard, users can select specific log groups and streams and then apply metric filters to transform the raw log data into numerical metrics. These metrics can be visualized in various forms, such as graphs, charts, or tables within the dashboard. This process allows users to distill large volumes of log data into meaningful insights.

There's more...

After you create a dashboard, you can migrate it across multiple environments or accounts by simply copying the source of the widget to a new environment using the **Copy source** option, as shown. Remember to check that the properties part aligns with the properties of the new environment.

Figure 9.24 – Copying the source of a widget to migrate the dashboard to another environment

For S3 monitoring, S3 Storage Lens provides metrics and dashboard ability to allow you to understand your storage usage. Some use cases include observing bucket costs across organizations, life cycle rules, and incomplete multiple upload parts that are over seven days old. There are free and advanced metrics for building the dashboard.

Depending on your use case, there would be a lot of metrics to choose from, such as ActiveJobCount and ActiveWorkflowCount, which help you understand Glue resource utilization or YARN utilization metrics for EMR. Before choosing which logs to monitor, your team must decide on relevant use case logs with the goals of monitoring and investigation in mind. Selecting the relevant metrics should help give you an overview of your service and system health with an option to drill down further if needed. There are five key areas to monitor:

- **Performance** such as bottlenecks and hotspots
- **Security** such as assessments and violations
- **Configuration** such as change history and violations
- **Cost** such as drive and control business spending
- **Fault tolerance** such as reliability and availability

Cloudwatch comes with standard metrics such as CPU and network utilization for instances, disk read/write operations for Amazon EBS volumes, and memory and disk activity for managed database (Amazon RDS) instances. You can also create custom metrics that are specific to the function of your instance and have them registered in CloudWatch. For example, if you are running an HTTP server on the instance, you could publish a statistic on service memory usage.

See also

- Monitoring and optimizing the Data Lake environment: `https://docs.aws.amazon.com/whitepapers/latest/building-data-lakes/monitoring-optimizing-data-lake-environment.html`
- Maximizing the value of your cloud-enabled enterprise Data Lake: `https://aws.amazon.com/blogs/apn/maximizing-the-value-of-your-cloud-enabled-enterprise-data-lake-by-tracking-critical-metrics/`
- Cost and usage analysis – AWS Well-Architected Labs: `https://wellarchitected-labs.com/`
- Monitoring workload resources: `https://docs.aws.amazon.com/wellarchitected/latest/reliability-pillar/monitor-workload-resources.html`

Setting up System Manager to remediate non-compliance with AWS Config rules

In this recipe, we will learn how to set up AWS **System Manager** (**SSM**) so that we can use this setting to automate non-compliance remediation in the next chapter. AWS SSM automation simplifies maintenance and deployment tasks for AWS services such as S3. For a data engineer, one of the common challenges is ensuring that all resources adhere to organizational policies and regulatory requirements. Due to the length of the recipe, it will be broken into two parts: setting up SSM in this recipe and using this setup in the *Using AWS config to automate non-compliance S3 server access logging policy* recipe later.

AWS SSM is a comprehensive suite of tools designed to help automate and streamline various management tasks across your AWS infrastructure. It provides a secure, remote management solution for configuring your managed instances, ensuring that they remain compliant with your organization's standards and policies. This automation not only enhances efficiency but also reduces the risk of manual errors, making it an invaluable tool for maintaining the health and security of your AWS environment.

Getting ready

Ensure that you enable AWS Config and AWS SSM in your account.

How to do it...

1. In the AWS IAM console, navigate to **Roles** and click on **Create role**.
2. Under **Select type of trusted entity**, choose **AWS service and System Manager**.
3. Under **Role name**, search for the `SSMServiceRoleForAutomation` policy on the **Attached permissions policy** page.

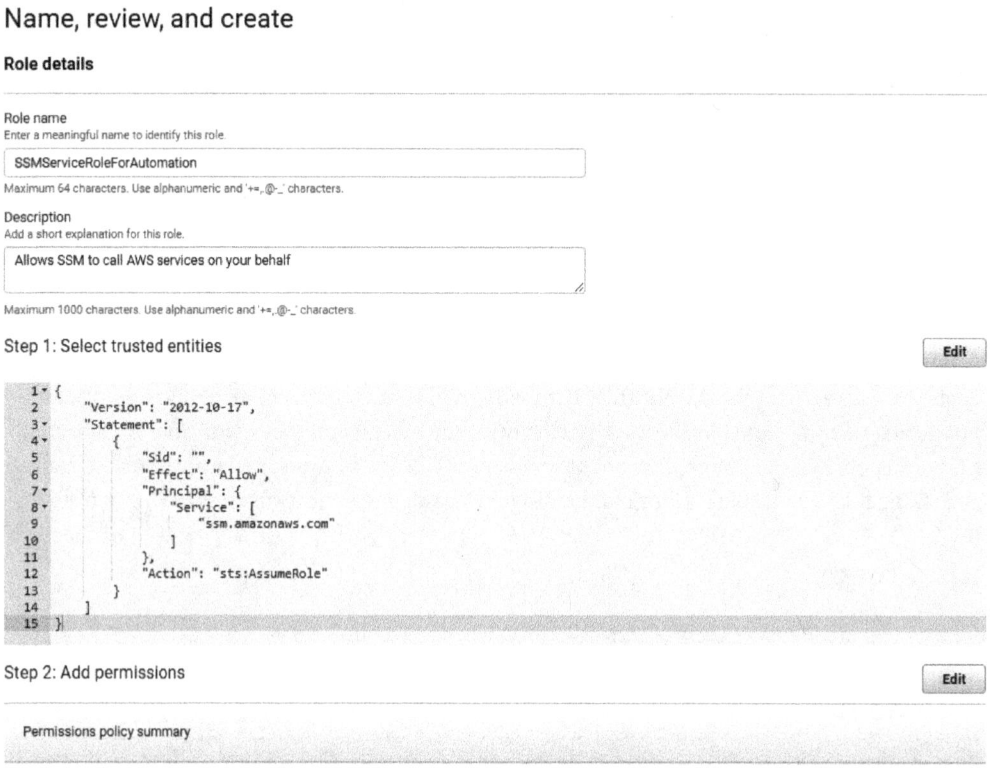

Name, review, and create

Role details

Role name
Enter a meaningful name to identify this role.

```
SSMServiceRoleForAutomation
```

Maximum 64 characters. Use alphanumeric and '+=,.@-_' characters.

Description
Add a short explanation for this role.

```
Allows SSM to call AWS services on your behalf
```

Maximum 1000 characters. Use alphanumeric and '+=,.@-_' characters.

Step 1: Select trusted entities   [Edit]

```
 1 ▾ {
 2        "Version": "2012-10-17",
 3 ▾      "Statement": [
 4 ▾          {
 5 ▾              "Sid": "",
 6                "Effect": "Allow",
 7 ▾              "Principal": {
 8 ▾                  "Service": [
 9                        "ssm.amazonaws.com"
10                    ]
11                },
12                "Action": "sts:AssumeRole"
13            }
14        ]
15   }
```

Step 2: Add permissions   [Edit]

Permissions policy summary

Figure 9.25 – Selecting the SSM role

4. You need to add S3 permissions that SSM automation would use, such as the following policy. You need to tailor it to your organization's security requirements:

```
{
    "Version": "2012-10-17",
    "Statement": [
        {
            "Sid": "VisualEditor0",
            "Effect": "Allow",
            "Action": [
                "s3:PutEncryptionConfiguration",
                "s3:PutBucketLogging",
                "s3:GetBucketLogging"
            ],
            "Resource": "arn:aws:s3:::*"
        }
    ]
}
```

5. Go to **Trust Relationships** and edit the trust entities as shown:

```
{
    "Version": "2012-10-17",
    "Statement": [
        {
            "Effect": "Allow",
            "Principal": {
                "Service": "ssm.amazonaws.com"
            },
            "Action": "sts:AssumeRole",
            "Condition": {
                "StringEquals": {
                    "aws:SourceAccount": "YOUR-ACCOUNT-ID"
                },
                "ArnLike": {
                    "aws:SourceArn": "arn:aws:ssm:*:YOUR-
ACCOUNT-ID:automation-execution/*"
                }
            }
        }
    ]
}
```

Save and proceed to the next recipe for monitoring non-compliant AWS Config rules.

There's more...

AWS SSM Explorer is an operations dashboard presenting information about your resources. This dashboard consolidates operations data views from various AWS accounts and across different AWS regions. By doing so, Explorer offers insights into the distribution of operational issues, their trends over a period, and their differentiation by various categories. It aids in understanding the overall functional health and potential areas that may require attention.

See also

- Use IAM to configure roles for automation: https://docs.aws.amazon.com/systems-manager/latest/userguide/automation-setup-iam.html

Using AWS config to automate non-compliance S3 server access logging policy

In the *Creating custom dashboards to monitor Data Lake services* recipe, we learned that the S3 Storage Lens provides a general dashboard to observe your S3 activities. One of the best practices for more comprehensive monitoring and auditing of your bucket is enabling S3 server access logging. This feature gives you detailed records of the requests made to the buckets, which is helpful in scenarios wherein you need to detect potential security weaknesses and incidents. This recipe will teach you to use AWS Config and AWS SSM to enforce this feature.

You can use the idea in this recipe to create more enforcement not only for S3 but also for other resources in your Data Lake.

Getting ready

Ensure that you enable AWS Config and AWS SSM in your account. You also need to finish setting up the AWS SSM role as covered in the *Setting up System Manager to remediate non-compliant AWS Config rules* recipe.

Besides that, you need to identify the scope of your AWS Config in the general settings as shown:

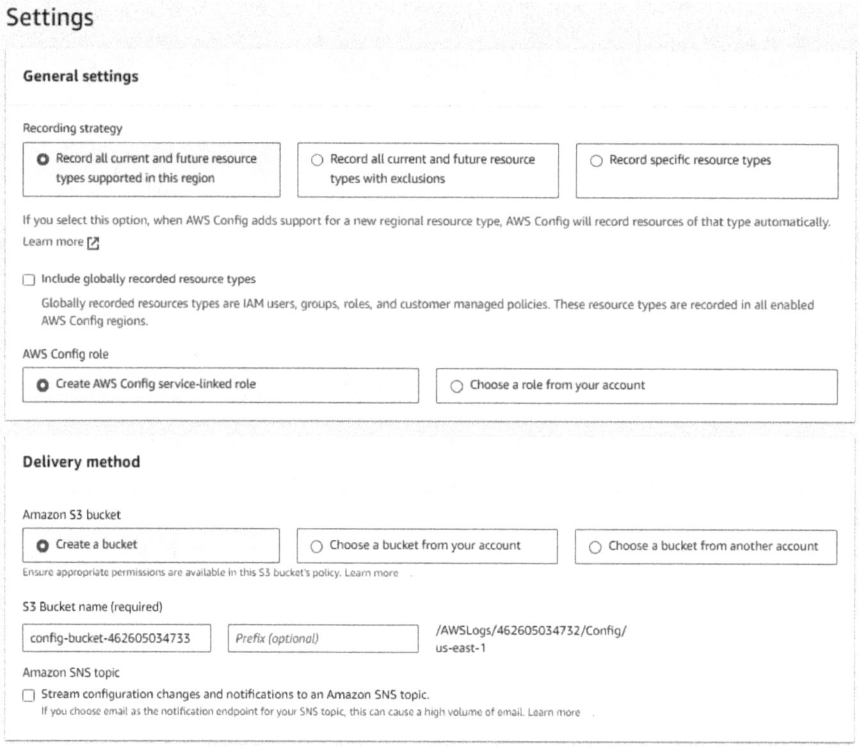

Figure 9.26 – Selecting the recording strategy for AWS config

How to do it...

1. Select AWS Config service; if this is your first time starting with AWS Config, you can select **1-click setup** as shown:.

Figure 9.27 – Selecting 1-click setup

2. In **Step 2 – Rules**, under **AWS Managed Rules**, select **s3-bucket-logging-enabled** and then click on **Review and create**:

Specify rule type

Add rules to define the desired configuration setting of your AWS resources. Customize any of the following rules to suit your needs, or create a custom rule. To create a custom rule, you must create an AWS Lambda function for the rule.

Select rule type

● Add AWS managed rule	○ Create custom Lambda rule	○ Create custom rule using Guard
Customize any of the following rules to suit your needs.	Create custom rules and add them to AWS Config. Associate each custom rule with an AWS Lambda function, which contains the logic that evaluates whether your AWS resources comply with the rule.	Create custom rules using Guard Custom Policy that evaluates whether your AWS resources comply with the rule.

AWS Managed Rules (338)

🔍 s3-bucket-logging-enabled   ✕   1 match   ‹ 1 › ⚙

	Name ▲	Labels	Supported evaluation mode	Description
●	s3-bucket-logging-enabled	S3	DETECTIVE, PROACTIVE	Checks if logging is enabled for your S3 buckets. The rule is NON_COMPLIANT if logging is not enabled.

Figure 9.28 – Selecting relevant rule name in AWS Managed Rules

Head to the **Rules** part in the AWS Config console, where you can see the buckets that are not compliant under **Detective compliance**:

Figure 9.29 – The list of buckets that do not align with the rules

3. Select **Manage remediation** to configure automated remediation actions:

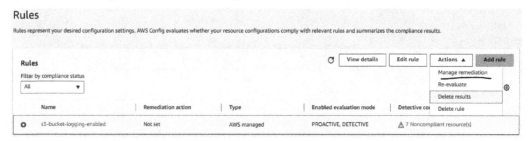

Figure 9.30 – Selecting Manage remediation to resolve the non-compliance

4. Select **Automatic remediation**, and in the **Remediation action details** section, choose **AWS-ConfigureS3BucketLogging**. More options exist, such as **Creating a Jira ticket** or **Publishing an SNS notification**. In this recipe, we will select **AWS-ConfigureS3BucketLogging**.

Edit: Remediation action

▼ **Select remediation method**

○ Automatic remediation The remediation action gets triggered automatically when the resources in scope become noncompliant.	**○ Manual remediation** You have to manually choose to remediate the noncompliant resources.

If a resource is still non-compliant after auto-remediation, you can set this rule to try again. Note, there are costs associated with running a remediation script.

Retries in   Seconds

`5`   `60`

▼ **Remediation action details**

The execution of remediation actions is achieved using AWS Systems Manager Automation

Choose remediation action

`AWS-ConfigureS3BucketLogging                                 ▼`

Enables Logging on S3 Bucket

▼ **Rate Limits**

You can specify the percentage of resources against which SSM documents are executed at a time and also the percentage of failed SSM executions for which the entire batch is marked as failed

Concurrent Execution Rate   Error Rate

`2`   `5`

Figure 9.31 – Editing the Remediation action section

5. Choose **BucketName** under **Resource ID parameter**:

▼ **Resource ID parameter**

Using the dropdown list, you can pass the resource ID of noncompliant resources to a remediation action by choosing a parameter that is dependent on the resource type. The parameters available in the dropdown list depend on the selected remediation action.

`BucketName                                                    ▼`

Figure 9.32 – Selecting BucketName for remediation action

6.   In the **Parameters** field, enter the values shown in the following image and then save the changes:

BucketName	>	RESOURCE_ID
GrantedPermission	>	FULL_CONTROL
GranteeType	>	Group
TargetBucket	>	config-bucket-4■■■■■■
GranteeEmailAddress	>	(optional)
GranteeId	>	(optional)
GranteeUri	>	(optional)
TargetPrefix	>	(optional)
AutomationAssumeRole	>	arn:aws:iam::■■■■■■■■role/SSMServiceRol

Figure 9.33 – Parameters for remediation action

How it works...

AWS Config rules can verify specific settings within your AWS environment, such as whether Amazon S3 buckets have logging enabled. These rules employ AWS Lambda functions to conduct compliance assessments, returning either compliant or non-compliant statuses for the inspected resources. If a resource is non-compliant, it can be corrected through a remediation action linked to the AWS Config rule. The auto-remediation feature of AWS Config rules allows this corrective action to be triggered automatically, immediately addressing any detected non-compliance.

There's more...

With AWS server access logs enabled, you can use Athena to analyze these logs.

In addition to using predefined AWS Config rules, you have the flexibility to create custom AWS Config rules to enforce your own corporate security policies. These custom rules are linked to an AWS Lambda function that you develop and manage. When a custom rule is triggered, it executes the associated Lambda function, enabling you to enforce specific configurations tailored to your organization's needs. Furthermore, AWS Config provides real-time notifications whenever a resource is misconfigured or violates the defined security policies, allowing for prompt remediation and enhancing the security posture of your Data Lake infrastructure.

See also

- Setting up AWS Config with the console: `https://docs.aws.amazon.com/config/latest/developerguide/gs-console.html`

- Remediating non-compliant AWS Config rules with AWS SSM automation runbooks: `https://aws.amazon.com/blogs/mt/remediate-noncompliant-aws-config-rules-with-aws-systems-manager-automation-runbooks/`

- Analyzing Amazon S3 server access logs using Athena: `https://repost.aws/knowledge-center/analyze-logs-athena`

- Amazon S3 server access log format: `https://docs.aws.amazon.com/AmazonS3/latest/dev/LogFormat.html`

Tracking AWS Data Lake cost per analytics workload

In an enterprise data lake, multiple teams often use the data lake to run several business campaigns. Tying the cost to the business value it provides is crucial to assess the return on investment later.

It is essential to have a cost allocation strategy. There are multiple ways to build the systems, such as implementing cost allocation tagging to reflect business units that utilize the resources as well as closely monitoring the cost through set budgets, and building Cloud Intelligence dashboards such as Cost and Usage report and Cost Intelligence Dashboards.

This recipe will look at how to track cost per analytics workload. First, we will look at how to create tags and then at how to bulk-add tags for multiple resources using Tag Editor. Finally, we will use the cost categories.

Getting ready

Ensure that you can access AWS billing and cost services and discuss with stakeholders to create a strategy for monitoring costs, such as using a tagging strategy.

How to do it...

1. First, tag your resources in a key-value pair. The **Name** tag is often used in the AWS GUI to identify resources. Each tag should have a single value. For example, if you want to remember that a particular redshift cluster is for a marketing test environment, you should tag as shown in the following figure instead of combining the tag such as **Name**: `marketing-test`.

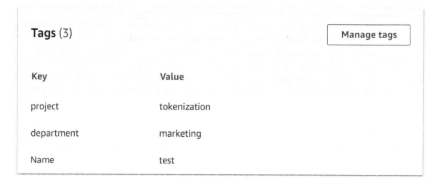

Figure 9.34 – Tagging an example of a key-value pair

2. Select **Resource Groups & Tag Editor** in the console.

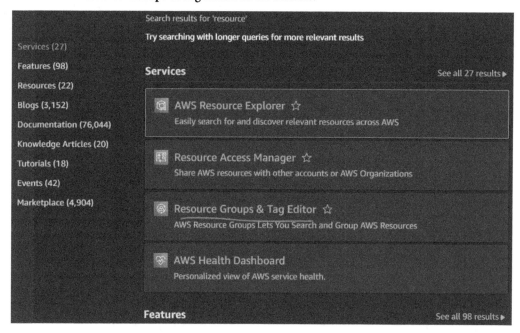

Figure 9.35 – Selecting the resource group and tag editor in AWS portal

3. Go to **Tag Editor** under **Tagging** and choose the AWS regions that you deployed the resources to. In the **Tag** key, choose department and tag value as **marketing** in the **Tags** section. Then click on the **Search resources** button. In this example, we created a project named tokenization; thus, we will filter the resources search box as Tag: project: tokenization and hit the *Enter* key.

Figure 9.36 – The tag editor for resource groups

4. Select all the resources. Then click on **Manage tags of selected resources**.

Figure 9.37 – Managing the tags of selected resources

5. In the **Edit tags of all selected resources** section as shown next, you can add additional tags besides the pre-existing ones. Add the cost_center tag and then click on **Review and apply tag changes** to bulk-apply this to the resources that you previously tagged for the project tokenization.

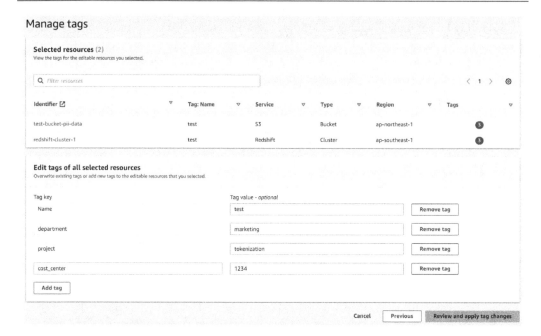

Figure 9.38 – Applying changes after editing tags

6. In the following steps, we will start creating cost categories. First, you need to go to **Cost allocation tags** and activate the tags that you want to categorize such as the name, department, or cost_ center tags, which we created in the previous steps. Click on the tag you want to activate, then click on the **Activate** button. Please note that sometimes tags take time to show up in **Cost allocation tags**.

Figure 9.39 – Creating cost allocation tags

7. Go to the cost category. In the **Cost category details** section, give a name for your categorization such as `Cost per environment` to reflect categorization per environment usage.

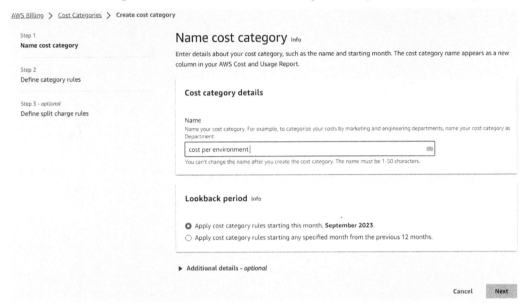

Figure 9.40 – Naming the cost category

8. Select the lookback period to apply the rules to the previous date.

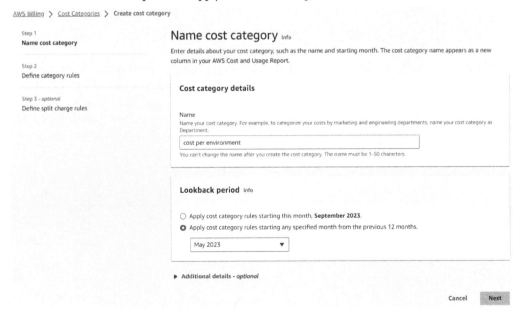

Figure 9.41 – Selecting the lookback period

9. On the **Define category rules** page, select **Inherited value**, then choose **Cost Allocation Tag** as **Dimension** and **Name** as **Tag key**.

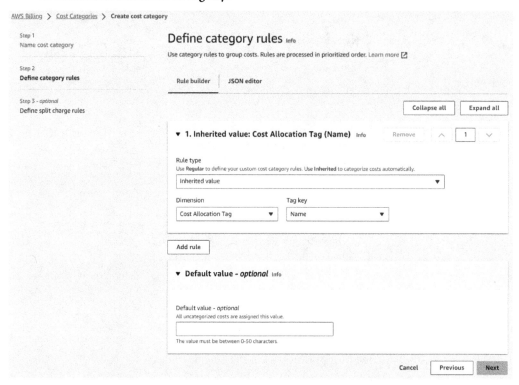

Figure 9.42 – Defining the category rules

10. Click on **Create cost category** and wait for the category to finish processing. The longer of a lookback period it has, the longer it will take to process the category.

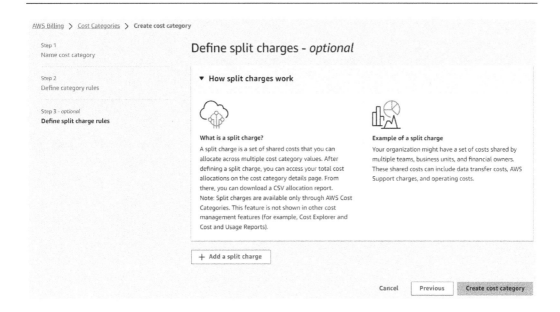

Figure 9.43 – Creating a cost category

How it works...

Cost allocation tags allow you to categorize your AWS resources as EC2 instances, EMR clusters, Glue jobs, and so on based on the workload they are processing. In addition to that, you can define a tag per **BusinessUnit**, **Workload**, and **Department**, and then use AWS Tag Editor or SDK/CLI to apply these tags to relevant resources in your Data Lake.

There's more...

Organizing your costs with cost categories is a foundational step for your organization to visualize these categories and apply split charges. You can further create cost visualizations, such as building a QuickSight dashboard or using cost explorer in **Cost Management**. To use cost explorer for dashboard visualization, simply use the **Cost category** option in the **Report** parameters. Make sure to save it to the report library for quick access later.

Split charges let you divide a single cost category's expenses across multiple target values, either proportionally, by fixed percentages, or as an even split. For example, you might have a general **Project** cost category. Still, you can apply split charges to allocate specific portions of this category to multiple teams based on their actual resource utilization or business objectives.

See also

- User-defined tag restrictions: `https://docs.aws.amazon.com/awsaccountbilling/latest/aboutv2/allocation-tag-restrictions.html`

- Cost modeling Data Lakes for beginners: `https://d1.awsstatic.com/whitepapers/cost-modeling-data-lakes.pdf`

- Best practices for tagging AWS resources: `https://docs.aws.amazon.com/whitepapers/latest/tagging-best-practices/tagging-best-practices.html`

10

Building a Serving Layer with AWS Analytics Services

In this chapter, we will explore how to manage a serving layer with Amazon Redshift and QuickSight. The consumption layer focuses on building solutions for data users to access, deriving data insights, and building dashboards to demonstrate the insights. Analysts must query this data quickly and efficiently to generate insights and reports. However, due to the sheer volume and complexity of the data, traditional querying methods are slow and cumbersome. Thus, implementing a robust consumption layer can address this challenge by enabling fast, efficient access and querying capabilities, thus empowering your analytics team to derive actionable insights without delay.

This chapter will walk you through the first step of managing a serving layer, from loading the data to Redshift, connecting client applications using a VPC endpoint, querying using Redshift Serverless, and using AWS SDK to manage QuickSight.

The recipes in this chapter are as follows:

- Using Redshift **workload management** (**WLM**) to manage workload priority
- Querying large historical data with Redshift Spectrum
- Creating a VPC endpoint to a Redshift cluster
- Accessing a Redshift cluster using JDBC to query data
- Using AWS SDK for pandas, the Redshift Data API, and Lambda to execute SQL statements
- Using the AWS SDK for Python to manage Amazon QuickSight

Technical requirements

Before going ahead with the recipes in this chapter, it would be useful to have an understanding of the data lake architecture and a basic knowledge of how data warehouses and data ingestion using Glue work.

The code files for this chapter are available on GitHub: `https://github.com/PacktPublishing/Data-Engineering-with-AWS-Cookbook/tree/main/Chapter10`.

Using Redshift workload management (WLM) to manage workload priority

Amazon Redshift's WLM feature is designed to help manage and prioritize queries and other database operations. By enabling and configuring WLM, users can ensure critical queries receive the necessary resources without being stalled by other less urgent processes. This recipe will guide you through setting up WLM, including the configuration of automatic WLM, to manage workloads in your Redshift environment efficiently.

Getting ready

Before configuring WLM in Amazon Redshift, you must have administrative access to your Redshift cluster and the AWS Management Console. Ensure your Redshift cluster is operational and you have familiarized yourself with the basic concepts of how Redshift handles queries and operations.

How to do it...

1. Log in to the Redshift management console.
2. Navigate to your cluster, click on **Configurations** as shown, and click on **Workload management**:

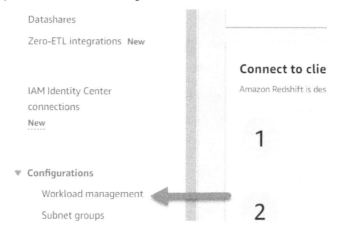

Figure 10.1 – Clicking on Workload management on the configuration

3.   On the **Workload management** screen, you can see the default parameter group that comes with the cluster creation. In this step, we will create a customized parameter group to help us:

Figure 10.2 – Clicking on Create in the Parameter groups section

4.   In the pop-up menu, fill in the **Parameter group name** and **Description** fields:

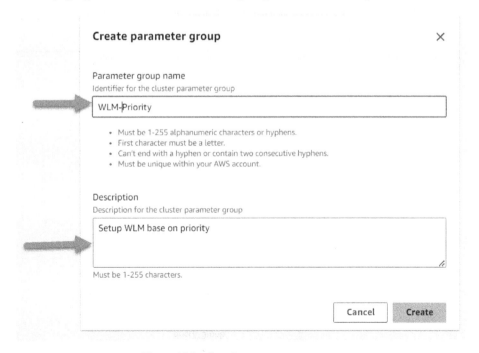

Figure 10.3 – Creating a parameter group

5. After you create the new WLM feature, there will be a new parameter group. On the right-hand side, you have two options: either using automatic WLM or creating workload queues. The default, **Automatic WLM**, is enabled. Click on **Switch WLM mode** to change the mode to **Manual WLM**:

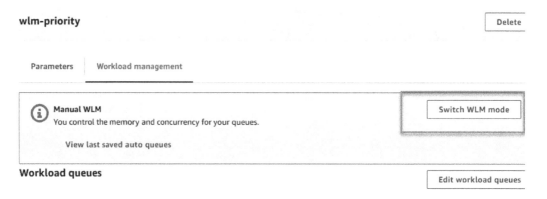

Figure 10.4 – Switching WLM to manual WLM

6. By switching to manual WLM, you will have full control of memory and concurrency for your queues:

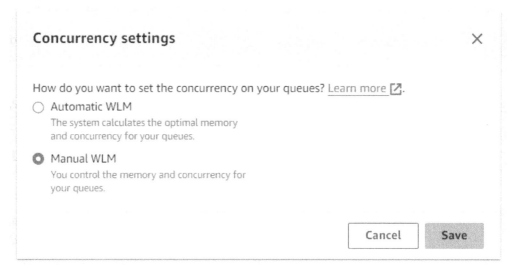

Figure 10.5 – Switching from automatic WLM to manual WLM

7. In the **Modify workload queues** screen, you can create or modify existing queues to define workload priorities and resource allocation. On this page, you can also set parameters to monitor each queue, such as memory allocation, concurrency, and timeout settings. The following screenshot is an example of a queue where you can configure memory and concurrency, create a monitoring rule, and use wildcards for the query group:

Figure 10.6 – Configuring workload queues

How it works...

Amazon Redshift WLM is designed to efficiently handle multiple workloads by ensuring that short, fast-running queries do not get delayed by longer, resource-intensive queries. By prioritizing queries, Redshift WLM helps maintain optimal performance for critical workloads and user groups.

Some workloads or user groups may require higher performance than others. Redshift allows you to set priorities for different workloads using a system of queues. Each queue can be assigned a priority level, which is then inherited by all queries associated with that queue. The priority levels, from highest to lowest, are the following:

- HIGHEST
- HIGH
- NORMAL
- LOW
- LOWEST

To manage the prioritization of queries, you can map user groups and query groups to specific queues. This ensures that queries from important users or critical workloads are processed with the appropriate priority. For instance, you might assign executive reports to a queue with the highest priority, while routine data loads could be assigned to a low-priority queue. Redshift allows up to 50 concurrent queries, so your concurrent queue needs to be a summary of 50. For an example JSON template for WLM config, please access the following GitHub repo: `https://github.com/PacktPublishing/Data-Engineering-with-AWS-Cookbook/blob/main/Chapter10/Recipe1`.

There's more...

Implementing effective WLM in Redshift goes beyond the basic configuration. Here are additional tips and advanced techniques:

- **Use short query acceleration (SQA)**: SQA is a feature in Amazon Redshift that prioritizes short-running queries over longer-running queries. This can help improve query performance and reduce query wait times. The default SQA maximum runtime is dynamic but can be manually set to a fixed value between 1-20 seconds. SQA is set automatically in automatic WLM to reduce overall query latency or through clicking on the **Enable short query acceleration for queries whose maximum runtime is** checkbox to a value of your choice:

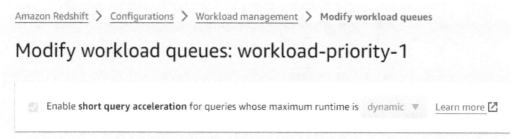

Figure 10.7 – Enabling SQA

- **Custom WLM rules**: Create custom rules to manage specific scenarios, such as automatically aborting long-running queries or reallocating resources during peak times. You can further monitor query performance and resource utilization using system tables such as `STL_WLM_QUERY`, `STL_QUERY`, and `STV_WLM_QUERY_STATE`.
- For this recipe, it might be helpful to know queries to get users grouped together. You can use the `pg_group` table from the system catalog table:

```
Select * from pg_group where groname='reporting_users'
```

See also

- *Configuring the WLM parameter using the AWS CLI*: `https://docs.aws.amazon.com/redshift/latest/mgmt/workload-mgmt-config.html#Configuring-the-wlm-json-configuration-Parameter`

- *WLM dynamic and static configuration properties*: `https://docs.aws.amazon.com/redshift/latest/dg/cm-c-wlm-dynamic-properties.html`

- *WLM queue assignment rules*: `https://docs.aws.amazon.com/redshift/latest/dg/cm-c-wlm-queue-assignment-rules.html`

Querying large historical data with Redshift Spectrum

Amazon Redshift Spectrum is a feature of Amazon Redshift that allows you to query exabytes of data stored in Amazon S3 directly without prior loading to Redshift tables. This can be useful for various reasons, such as querying historical datasets that have expanded to multiple years or having multiple Redshift workgroups to query the same dataset. You can directly query scalar data or nested data formats stored in Amazon S3.

By using Redshift Spectrum, you will launch a cluster that is independent of your existing cluster.

Getting ready

To use Amazon Redshift Spectrum, you need to have the following:

- **An SQL client**: This cluster is independent of your existing Redshift cluster.

- **At least three subnets**: Each subnet should be associated with different **Availability Zones (AZs)** for your Redshift Serverless workspace. Make sure you understand how the subnet mask and subnet **Classless Inter-Domain Routing** (**CIDR**) block work prior to creating these subnets. Fundamental knowledge of networking is required. Please see this link for more information: `https://docs.aws.amazon.com/vpc/latest/userguide/subnet-sizing.html`.

- **S3 bucket**: The data that you want to query from the S3 bucket must be in the same region as your Amazon Redshift cluster.

- **An Identity and Access Management (IAM) role for Redshift**: To create an IAM role for Redshift, follow the instructions at `https://docs.aws.amazon.com/redshift/latest/dg/c-spectrum-iam-policies.html`.

In this recipe, we will use a `default-namespace` namespace, so make sure your `default-namespace` namespace has the relevant IAM roles attached. You can check for the same under **Security and encryption** in your `default-namespace` namespace:

Figure 10.8 – Creating an IAM role for Redshift

How to do it...

1. Select **Redshift Serverless** from the Amazon Redshift console.

2. Select **Create workgroup** and follow the next page to fill in the workgroup name, capacity, network and security, and namespace. In **Network and security**, choose the subnets that you created for Redshift Spectrum. For the capacity, as this is a test recipe, make sure you only use the smallest **Redshift Processing Unit** (**RPU**) value, which is 8 RPUs:

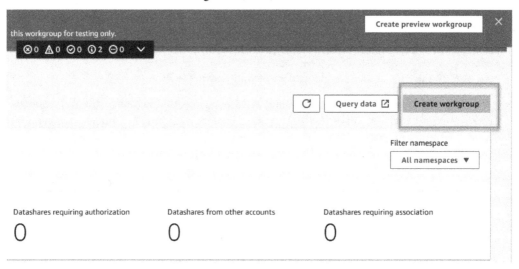

Figure 10.9 – Selecting Create workgroup

3. When you go to the Redshift console, there is a default namespace available. Make sure your `default-namespace` namespace already has the relevant permission to access Glue and S3, as mentioned in the *Getting ready* section of this recipe:

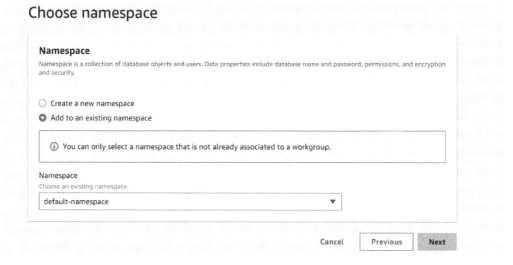

Figure 10.10 – Choosing a relevant namespace

4. Once your Amazon Redshift Spectrum namespace is created, click on **Query editor v2** as shown:

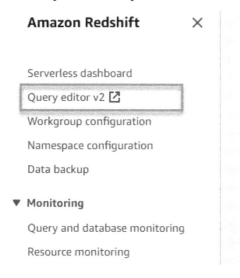

Figure 10.11 – Selecting Query editor v2

5. When you first start using the query editor, you can choose **Federated user** to start with. Once you select it, you will see the data that exists in your Glue data catalog:

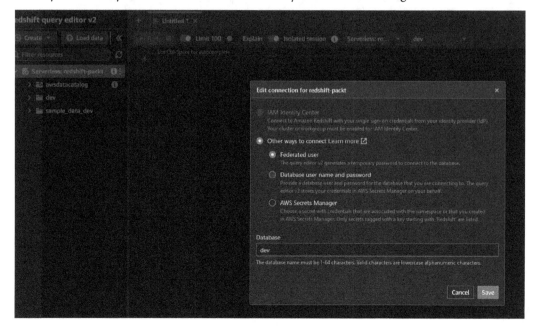

Figure 10.12 – Selecting Federated user

6. The next step is to create an external schema and external tables. You can choose from various external databases, such as an Amazon Athena Data Catalog, an AWS Glue Data Catalog, or an Apache Hive Metastore:

```
create external schema spectrum_data from data catalog
database 'sample_db'
iam_role 'arn:aws:iam::xxxxxxxx:role/xxxxx'
create external table if not exists;
```

7. Once you have created an external schema, you can create an external table using the following SQL statement. The LOCATION clause specifies the location of the data in Amazon S3. The bucket_name parameter is the name of the S3 bucket where the data is stored:

```
CREATE EXTERNAL TABLE spectrum_data.your_table_name (
    first_and_last_name VARCHAR,
    email VARCHAR,
    id BIGINT,
    id_2 BIGINT,
    gender VARCHAR,
    country VARCHAR,
```

```
        fax VARCHAR
    )
    ROW FORMAT DELIMITED
    FIELDS TERMINATED BY ','
    LOCATION 's3://sample-test-wf09/'
    TABLE PROPERTIES ('skip.header.line.count'='1');LOCATION 's3://
    bucket_name/prefix/';
```

8. Once you have created the external table, you can use standard SQL queries to query the data in Amazon S3. For example, the following SQL statement will query the top three rows in the table you just created:

```
select top 3 *
from spectrum_data.your_table_name
```

How it works...

Amazon Redshift Spectrum is a feature of Amazon Redshift that allows you to run SQL queries on data stored directly in Amazon S3. Redshift Spectrum supports the same data format as Athena.

The process begins with creating an external schema and an external table to reference the data stored in Amazon S3. Once the external table is set up, you can run SQL queries on this data directly from Amazon Redshift. For more information on the CREATE EXTERNAL TABLE statement command, refer to the *Creating external tables for Redshift Spectrum* blog (https://docs.aws.amazon.com/redshift/latest/dg/c-spectrum-external-tables.html). Once you create an external table, you can use SQL to query data similar to how you would use Athena.

To summarize, the following table compares Redshift Spectrum, Amazon Athena, and S3 Select:

COMPARISON	AWS ATHENA	S3 SELECT	AMAZON REDSHIFT SPECTRUM
Main purpose	Serverless SQL querying of S3 data	Built-in S3 feature for simple queries	Serverless SQL querying of S3 data
Allowed queries	SQL, based on Presto	A limited subset of SQL commands. S3 select has a maximum length of 256 KB for SQL expressions.	SQL, based on PostgreSQL
Managed/serverless	Serverless	Serverless	Managed service requires Amazon Redshift cluster
Table config	Virtual tables (for example, using AWS Glue)	N/A	Virtual tables, manually configured with external tables

COMPARISON	AWS ATHENA	S3 SELECT	AMAZON REDSHIFT SPECTRUM
Use case	Ad hoc querying; interactive analytics; exploring data in S3	Retrieving a subset of data from a single object; simple querying of an S3 object	Enterprise reporting; **business intelligence** (**BI**); complex analytical use cases from apps that require a low latency

Table 10.1 – Comparison between Athena, Redshift Spectrum, and S3 Select

There's more...

Amazon Redshift Spectrum also supports federated query access, which allows you to query data stored in Amazon Aurora directly from the Amazon Redshift query editor. To use federated query access, follow these steps:

1. You need to create a database user in Amazon Redshift that has the **SUPERUSER** role. You then need to create a database user in Amazon Aurora that has the same name and password as the database user in Amazon Redshift. The LOCATION clause specifies the location of the data in Amazon Aurora. The HOST and PORT clauses specify the hostname and port of the Amazon Aurora database. The DATABASE, SCHEMA, and TABLE clauses specify the name of the database, schema, and table, respectively, in Amazon Aurora. Once you have created the database users, you can use the following SQL statement to create a federated table in Amazon Redshift:

```
CREATE TABLE table_name (
  column_name1 data_type1,
  column_name2 data_type2,
  ...
)
LOCATION (
  TYPE = aurora
  HOST = 'aurora_host_name'
  PORT = 'aurora_port'
  DATABASE = 'aurora_database_name'
  SCHEMA = 'aurora_schema_name'
  TABLE = 'aurora_table_name'
)
```

2. Once you have created the federated table, you can use standard SQL queries to query the data in Amazon Aurora directly from the Amazon Redshift query editor. For example, the following SQL statement will query the `table_name` table and return the `column_name1` and `column_name2` columns:

```
SELECT column_name1, column_name2
FROM table_name;
```

3. To estimate the cost of a query executed using Redshift Spectrum, you can use the `SVL_S3QUERY_SUMMARY` system view. The `s3_scanned_bytes` column is the number of bytes scanned; it divides it by 1024^4 to convert it to bytes. Multiply by 2.5 to account for the cost per GB of data scanned in Amazon Redshift Spectrum:

```
SELECT s3_scanned_bytes / 1024^4 * 2.5 AS spectrum_cost
FROM SVL_S3QUERY_SUMMARY
WHERE query = <your_query_id>;
```

4. The `STL_ALERT_EVENT_LOG` table monitors performance thresholds that are exceeded. You can join `STL_ALERT_EVENT_LOG` with `STL_QUERY` to get detailed information about queries that trigger performance alerts. The following SQL example demonstrates how to filter the results based on a specific job run ID (to be replaced from your system):

```
SELECT A.*, B.*
FROM STL_ALERT_EVENT_LOG A
LEFT JOIN STL_QUERY B
ON A.QUERY = B.QUERY
WHERE A.PID IN
  (SELECT PID
   FROM STL_QUERY
   WHERE QUERYTXT LIKE '%<job_run_id>%'
   GROUP BY 1
  );
```

5. To further optimize the performance of Amazon Redshift Spectrum queries, it's essential to pre-filter the data before joining it with other tables and ensure that filtering is done on partition keys. This practice helps in reducing the data scanned. Consider the following example where a temporary table is created to pre-filter the data:

I. First, create a temporary table to pre-filter data from an external source:

```
CREATE TEMP TABLE TEMP_HITS_A
DISTKEY (MARKER_ID)
COMPOUND SORTKEY (MARKER_ID)
AS
(SELECT MARKER_ID, COUNT(1) AS TOTAL_HITS
```

```
FROM EXTERNAL_SCHEMA.SOURCE_TABLE_A
WHERE ENTITY_ID = 101
AND MARKET_ID = 1
AND EVENT_DAY AND IS_EVENT = 1
GROUP BY MARKER_ID);
```

II.   Then, create another temporary table to pre-filter data from a different source:

```
CREATE TEMP TABLE TEMP_HITS_B
DISTKEY (MARKER_ID)
COMPOUND SORTKEY (MARKER_ID)
AS
SELECT MARKER_ID, COUNT(1) AS TOTAL_HITS
FROM DATA_WAREHOUSE.SOURCE_TABLE_B
WHERE REGION_ID = 1
AND MARKET_ID = 1
AND EVENT_DAY AND IS_EVENT = 2024-04-20'::DATE
GROUP BY MARKER_ID;
```

III.   Lastly, join the pre-filtered temporary tables and calculate the difference:

```
SELECT A.MARKER_ID,
       A.TOTAL_HITS AS HITS_A,
       B.TOTAL_HITS AS HITS_B,
       A.TOTAL_HITS - B.TOTAL_HITS AS HIT_DIFFERENCE
FROM TEMP_HITS_A A
LEFT OUTER JOIN TEMP_HITS_B B
ON A.MARKER_ID = B.MARKER_ID;
```

See also

- *Best Practices for Amazon Redshift Spectrum | AWS Big Data Blog*: https://aws.amazon.com/blogs/big-data/10-best-practices-for-amazon-redshift-spectrum/

- *How do I calculate the query charges in Amazon Redshift Spectrum?* https://repost.aws/knowledge-center/redshift-spectrum-query-charges

- SVL_S3QUERY_SUMMARY: https://docs.aws.amazon.com/redshift/latest/dg/r_SVL_S3QUERY_SUMMARY.html

- *Monitoring queries and workloads with Amazon Redshift Serverless*: https://docs.aws.amazon.com/redshift/latest/mgmt/serverless-monitoring.html

- *Improving Amazon Redshift Spectrum query performance*: https://docs.aws.amazon.com/redshift/latest/dg/c-spectrum-external-performance.html

- *Issues of overlapping CIDR ranges for subnets*: https://repost.aws/questions/QUjrakcRqgRTGwD4pD0K8v8Q/overlapping-cidr-ranges-for-subnets

Creating a VPC endpoint to a Redshift cluster

This recipe will guide you in creating a Redshift VPC endpoint to ensure secure, private connectivity between your Redshift cluster and client applications within your VPC. A VPC is an isolated virtual network within the cloud that provides private access to your resources. By creating this endpoint, you enable direct, secure access to your Redshift cluster from BI tools such as Tableau Online, Qlik Sense, and Looker without exposing your data cluster to a public IP address or routing traffic through the internet.

There are two types of VPC endpoints:

- Interface endpoint
- Gateway endpoint

In this recipe, we will demo two concepts:

- Show you step-by-step how to create an S3 gateway endpoint so that the Redshift cluster can communicate with S3. You can use it later when you create a Glue connection.
- Create a Redshift-managed VPC endpoint so that you can further integrate your BI application. This scenario would be useful when you have your Redshift cluster in a VPC that is different from your BI applications.

It's important to note that Redshift-managed VPC endpoints are only available for Redshift clusters that are running on the RA3 node type and have either cluster relocation or Multi-AZ enabled. Additionally, the cluster or workgroup must be available within the valid port ranges, and the VPC endpoint quota limits the number of endpoints.

In the code, **AWS Systems Manager** (**SSM**) will be used to help automate operational tasks across your AWS resources. The Parameter Store capability within SSM is useful for storing sensitive data such as database connection details, credentials, and configuration settings in a secure and structured manner, with optional encryption using AWS **Key Management Service** (**KMS**).

Getting ready

Before starting, ensure you have the following prerequisites:

- An active AWS account with an IAM role that has permission to manage Redshift, VPC, Glue, and SSM
- Knowledge of SSM
- An active Redshift cluster

How to do it...

1. In the Glue service console, select **Data connections** and then **Create connection**:

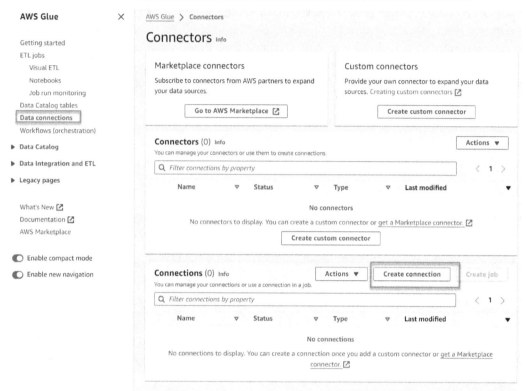

Figure 10.13 – Selecting Data connections and creating a connection

2. This step assumes that you have an existing Redshift cluster that you created in the previous recipe. Select Amazon Redshift as the data source, then follow the flow of **Step 2**, **Step 3**, and **Step 4**, as highlighted on the left, to finish the setup:

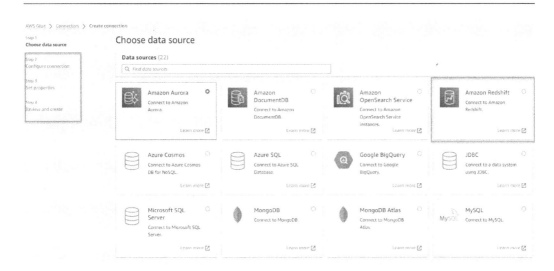

Figure 10.14 – Selecting Amazon Redshift as a data source

3. In **Step 2**, you need to select the relevant database instances, database name, and credential type. The best practice is to use AWS Secrets Manager for storing credentials. If you configure your Redshift cluster using username and password, you can head to the Redshift console, select **Namespace configuration**, and then **Edit admin credentials** to change the password:

Figure 10.15 – Heading to Namespace configuration to find out
your admin credentials or to change your password

4. In **Step 3**, set properties; if you want to reuse your connection later on in the Jupyter notebook, you should create a name without a space. For this recipe, we will use a connection name of `redshift_serverless` instead of the default name. Later on, if you want to reuse the connection in the Glue session, you should use the `%connections redshift_serverless` magic cell:

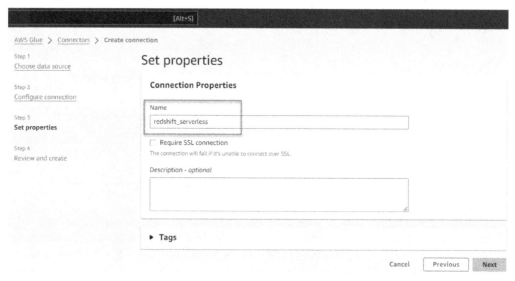

Figure 10.16 – Editing the connection name

5. In **Step 4**, when you click on **Create connection**, you will see a **Redshift connection successfully created** message. However, at this step, your connection won't work. You can find out why by clicking on **Test connection**:

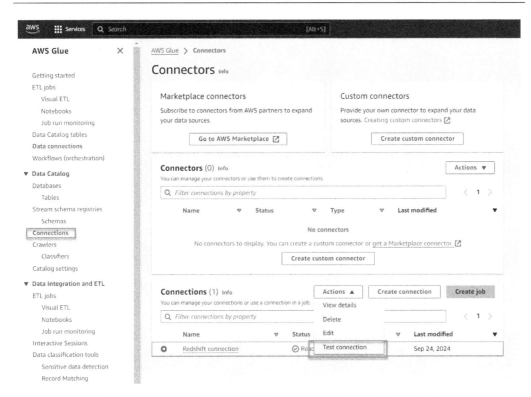

Figure 10.17 – Testing connection of the Redshift connection created

6. To test the connection, you need to select the relevant IAM role. You will receive an **InvalidInputException: VPC S3 endpoint validation failed** error message:

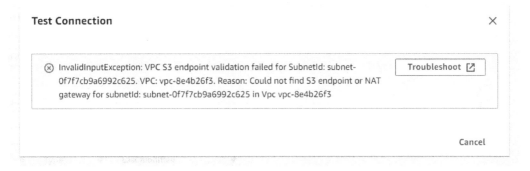

Figure 10.18 – Error message prior to creating VPC endpoint

7. Head to the VPC service, click on **Endpoints**, and then click on **Create endpoint**.

8. In **Endpoint settings**, select **AWS services**, and for the service name, select com.amazonaws. us-east-1.s3. Make sure you select the one marked as the **Gateway** type:

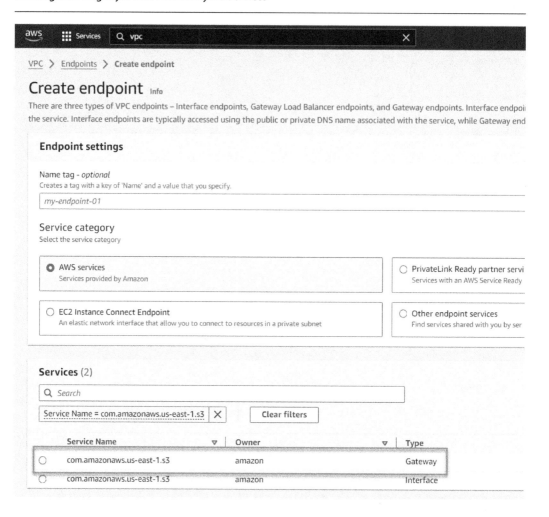

Figure 10.19 – Selecting an S3 gateway in Services

9.   Select the VPC that you want to create the endpoint and click on **Create**. Once you create the
endpoint, you can go back and test the connection, and you will see that after creating the
endpoint, your test connection will be successfully connected:

Figure 10.20 – Successfully created Redshift connection

How it works...

Creating connections to Amazon Redshift from AWS Glue requires an Amazon S3 VPC gateway endpoint. This is to ensure secure and private communication between the AWS Glue job and the Redshift cluster without exposing the data to the public internet.

To set up a VPC endpoint for Amazon S3, you would need to do the following:

1. Create a VPC endpoint for the S3 service in the same VPC as your Redshift cluster.

2. When you create the VPC endpoint, you can customize the level of access by configuring the route tables to allow the AWS Glue job to access the S3 bucket through the VPC endpoint.

There's more...

For secure, reliable connectivity, you can further integrate an AWS network firewall, an AWS shield, and a **Network Load Balancer** (**NLB**) instance to enhance security and performance. Additionally, you can employ the **AWS Cloud Development Kit** (**AWS CDK**) to define cloud **infrastructure as code (IaC)**.

An NLB is not required with a Redshift VPC endpoint, but it can be used to distribute incoming traffic to multiple Redshift clusters or workgroups. If you have multiple Redshift clusters or workgroups that need to be accessed by the same client tool, you can use an NLB to distribute incoming traffic to different Redshift endpoints. This can ensure that traffic is distributed evenly across the different Redshift clusters or workgroups. To set up an NLB with a Redshift VPC endpoint, you need to create a load balancer and configure it to distribute traffic to different Redshift endpoints. You will also need to configure security groups for the Redshift clusters or workgroups to allow traffic from the load balancer. It's important to note that the NLB will need to be in the same VPC as the Redshift clusters or workgroups, and the security groups for the Redshift clusters or workgroups will need to be configured to allow traffic from the load balancer.

The script that automates the creation of a Redshift VPC endpoint, enabling secure, private connectivity for your Redshift cluster within the VPC, can be found in the following GitHub repo: `https://github.com/PacktPublishing/Data-Engineering-with-AWS-Cookbook/blob/main/Chapter10/Recipe3`.

The steps involve the following:

1. Initializing Boto3 clients and defining environment variables.

2. Fetching necessary parameters from AWS SSM.

3. Checking if the VPC endpoint exists or if the cluster type is RA3.

4. Creating a managed VPC endpoint if one doesn't already exist.

The script uses `create_endpoint_access()` to create a VPC endpoint with the provided cluster ID, subnet, and security group. It will wait for 400 seconds after initiating the creation process to allow the endpoint to be fully set up, then use an API call in `describe_endpoint_access()` to get the newly created endpoint details. The VPC endpoint details are then stored back in SSM for future reference.

See also

- *Redshift-managed VPC endpoints*: `https://docs.aws.amazon.com/redshift/latest/mgmt/managing-cluster-cross-vpc.html`

- *Gateway endpoints for Amazon S3*: `https://docs.aws.amazon.com/vpc/latest/privatelink/vpc-endpoints-s3.html`

- *Different types of VPC endpoints*: `https://tutorialsdojo.com/vpc-interface-endpoint-vs-gateway-endpoint-in-aws/`

- *Enhance data security and governance for Amazon Redshift Spectrum with VPC endpoints*: `https://aws.amazon.com/blogs/big-data/enhance-data-security-and-governance-for-amazon-redshift-spectrum-with-vpc-endpoints/`

Accessing a Redshift cluster using JDBC to query data

Redshift has many connection options depending on the use case and downstream requirements. In this recipe, we will focus on how to connect Glue to Redshift using JDBC. The main benefits of using a JDBC connection are the following:

- You can connect to Redshift using a wide variety of tools

- Depending on your use case, you can access Redshift data within your ETL job without going through the Glue crawler, which will help you save on cost

Getting ready

To connect with Amazon Redshift through JDBC, you'll need the following:

- **AWS Glue notebook**: Make sure you know how to use Glue notebooks.

- **Redshift cluster**: You should have a working Redshift cluster with the necessary data. If you don't have one, create it before proceeding.

- **Access credentials**: You should have access credentials to your cluster.

- **Configure security group and VPC endpoint**: You should complete the setup of relevant security rules to allow inbound and outbound traffic. Please follow the steps mentioned at `https://docs.aws.amazon.com/glue/latest/dg/aws-glue-programming-etl-connect-redshift-home.html`. You also need to create a VPC endpoint to connect your application. Please see the *Creating a VPC endpoint to a Redshift cluster* recipe to create a managed VPC endpoint or reuse the `redshift_serverless` connection using the `%connections` syntax. For more information, please see `https://docs.aws.amazon.com/glue/latest/dg/aws-glue-programming-etl-connect-redshift-home.html`.

- **Sample data**: Have sample data to query. For this recipe, we will use a `users` table in Redshift, as shown:

```
CREATE TABLE public.users (
    user_id integer identity(1, 1) ENCODE az64,
    first_name character varying(50) ENCODE lzo,
    last_name character varying(50) ENCODE lzo,
    email character varying(100) ENCODE lzo,
    created_at timestamp without time zone DEFAULT ('now'::
text):: timestamp with time zone ENCODE az64
) DISTSTYLE AUTO;
```

How to do it...

1. Open the AWS Management Console and type `Redshift` in the search bar at the top. Click on the result under **Services** to access the Amazon Redshift management console:

Figure 10.21 – Selecting Amazon Redshift

2. On the Redshift page, click on **Provisioned clusters dashboard** and then click on the cluster you would like to connect:

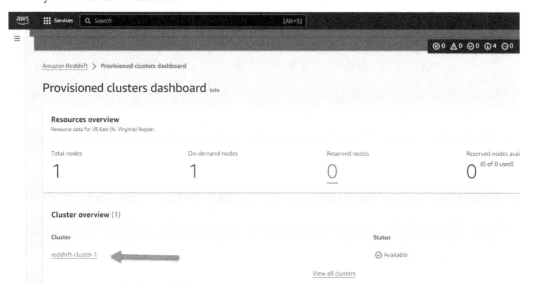

Figure 10.22 – Selecting Amazon Redshift cluster to connect

3. On the right side, you will see options to connect to Redshift, such as **Endpoint**, **JDBC URL**, and **ODBC URL**:

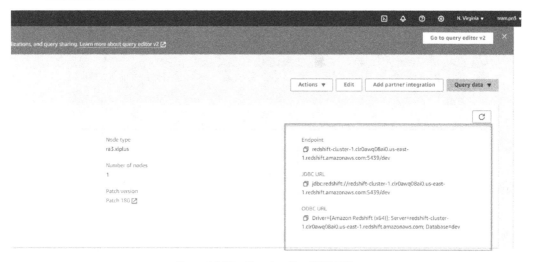

Figure 10.23 – Copying the JDBC URL

Depending on your application usage, you could consider a JDBC or **Open Database Connectivity (ODBC)** endpoint. The following code is an example of how to use Glue to connect to the Redshift cluster using the JDBC URL. Thus, for this recipe, you will need the JDBC URL as shown:

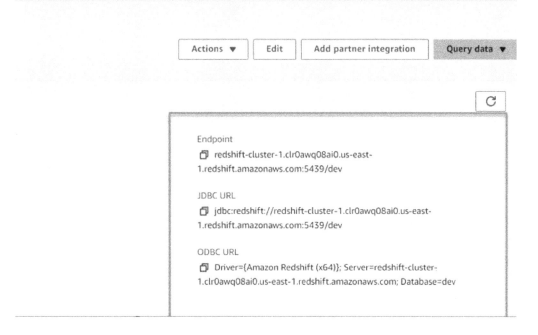

Figure 10.24 – Noting down Redshift JDBC URL

4. Next, we need to enter the code in the Glue crawler.

  I.   We will start with installing the necessary libraries:

```
from awsglue.context import GlueContext
from awsglue.dynamicframe import DynamicFrame,
DynamicFrameWriter
from awsglue.job import Job
from awsglue.transforms import *
from awsglue.utils import getResolvedOptions
from awsglue.context import GlueContext
from awsglue.dynamicframe import DynamicFrame
from pyspark.context import SparkContext
from pyspark.sql import *
import sys
import boto3
```

II.   Then, we create a Spark session:

```
sc = SparkContext.getOrCreate()
glueContext = GlueContext(sc)
spark = glueContext.spark_session
job = Job(glueContext)
```

III.   Set up your Redshift credentials:

```
redshift_properties = {
    "user": "your_username",
    "password": "your_password",
    "driver": "com.amazon.redshift.jdbc.Driver"
}


# Define Redshift connection URL
redshift_url = ""
```

IV.   Read the Redshift table into a Spark DataFrame:

```
df = spark.read.jdbc(url=redshift_url, table="your_table",
properties=redshift_properties)
```

V.   Create a temporary view for running SQL queries:

```
df.createOrReplaceTempView("users")
spark.sql("""
INSERT INTO public.users (first_name, last_name, email)
    SELECT 'John' AS first_name, 'Doe' AS last_name, 'john.
doe@example.com' AS email
    UNION ALL
    SELECT 'Jane', 'Smith', 'jane.smith@example.com'
    UNION ALL
    SELECT 'Bob', 'Johnson', 'bob.johnson@example.com'
    """)
```

VI.   Write it back to Redshift:

```
df.write.jdbc(url=redshift_url, table="your_data",
mode="overwrite", properties=redshift_properties)
""")
```

There's more...

You need to analyze requirements clearly to understand whether you need a Redshift cluster or if Redshift Spectrum would be sufficient. We will learn how to use Redshift Spectrum and the pros and cons of using Redshift Spectrum in the next chapter.

Amazon Redshift provides several system tables that can be used to monitor and troubleshoot the performance of your cluster. These tables are divided into six categories: STL, STV, SVV, SYS, SVCS, and SVL. When you migrate to Redshift Serverless, some SVV views do not directly apply because they are designed to work with the architecture and management features specific to provisioned clusters. Instead, Redshift Serverless uses SYS system views.

For this recipe, we have created a Glue notebook on GitHub at `https://github.com/PacktPublishing/Data-Engineering-with-AWS-Cookbook/blob/main/Chapter10/Recipe4.ipynb`.

See also

- *Top 10 performance tuning techniques for Amazon Redshift*: `https://aws.amazon.com/blogs/big-data/top-10-performance-tuning-techniques-for-amazon-redshift/`

- *Best practices to optimize your Amazon Redshift and MicroStrategy deployment*: `https://aws.amazon.com/blogs/big-data/best-practices-to-optimize-your-amazon-redshift-and-microstrategy-deployment/`

- *How to retain system tables' data spanning multiple Amazon Redshift clusters and run cross-cluster diagnostic queries*: `https://aws.amazon.com/blogs/big-data/how-to-retain-system-tables-data-spanning-multiple-amazon-redshift-clusters-and-run-cross-cluster-diagnostic-queries/`

- *Types of system tables and views*: `https://docs.aws.amazon.com/redshift/latest/dg/cm_chap_system-tables.html#c_types-of-system-tables-and-views`

- *How can I troubleshoot high or full disk usage with Amazon Redshift?*: `https://repost.aws/knowledge-center/redshift-high-disk-usage`

Using AWS SDK for pandas, the Redshift Data API, and Lambda to execute SQL statements

The AWS SDK for pandas library (previously known as AWS Data Wrangler) is a powerful tool that provides a pandas interface to various AWS services, including Glue, Redshift, Athena, and more using pandas syntax. This library is handy for data scientists, and analysts already using the pandas library and want to interact with AWS data and analytics services. One of the use cases of AWS SDK for pandas is to simplify the querying and manipulating data stored in AWS data stores.

In this recipe, we will demonstrate how to use AWS SDK for pandas and the Redshift Data API. In this flow, we will demonstrate a Lambda function with two steps:

1. Execute an SQL statement to retrieve the data.

2. Save the DataFrame to S3.

Getting ready

Before starting, ensure you have the following prerequisites:

- AWS Lambda set up with the necessary execution role permissions to interact with Redshift and S3. Your Lambda instance needs to be at the same VPC subnet as your Redshift instance unless you set the VPC endpoint. This recipe assumes that you created a VPC endpoint.

- Able to use AWS Lambda managed layers. Check out https://docs.aws.amazon.com/lambda/latest/dg/chapter-layers.html for more information.

- Have an Amazon Redshift cluster and associate it with an AWS IAM role.

How to do it...

1. Create a policy in IAM named LambdaRedshiftDataAPIRole and add the following permission. Make sure to change the Resource value so that it is relevant to the Redshift cluster that you want the Lambda instance to interact with:

```
{
  "Version": "2012-10-17",
  "Statement": [
    {
      "Sid": "VisualEditor0",
      "Effect": "Allow",
      "Action": [
        "logs:CreateLogStream",
        "redshift-data:GetStatementResult",
        "redshift:GetClusterCredentials",
```

```
          "redshift-data:DescribeStatement",
          "logs:CreateLogGroup",
          "logs:PutLogEvents",
          "redshift-data:ExecuteStatement",
          "redshift-data:ListStatements"
        ],
        "Resource": "*"
      }
    ]
  }
```

2. Log in to the Lambda service in your AWS portal. Click on the **Create a function** button to create a new Lambda function:

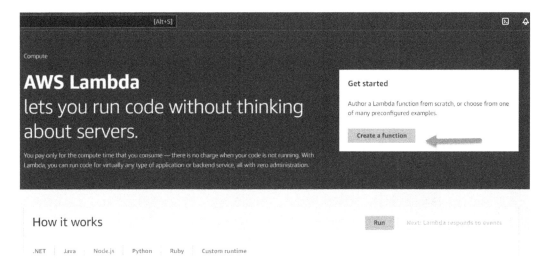

Figure 10.25 – Clicking on Create a function

3. Fill in the details shown in the following screenshot. Make sure you create a new role or choose an existing relevant IAM role in the **Change default execution role** section. If you create a new role, make sure to note it down for the next step. As of the time of writing this book, AWS Data Wrangler version 2.13.0 works with Python version 3.9, so make sure you use Python version 3.9 when creating the function. Fill in the information and then click on the **Create function** button:

Figure 10.26 – Filling in the Runtime and Function name fields

4.   After the function is successfully created, go to the **Lambda | Roles** section in IAM and attach the policy we created in *step 1*:

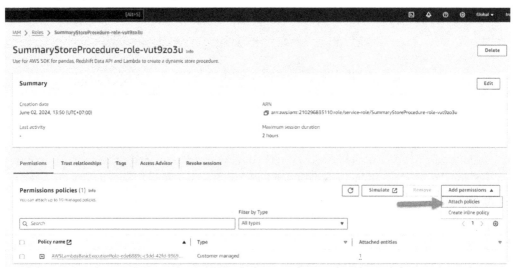

Figure 10.27 – Attaching policies to your Lambda function

5. Go back to the created Lambda function, scroll down to the **Layers** section, and click on **Add a layer**:

Figure 10.28 – Adding a layer to your Lambda function

6. Add a layer under **Specify an ARN**. You should select the **Amazon Resource Name (ARN)** that matches the runtime setting. Make sure to fill in your Lambda region in the ARN. After that, click on **Add** to add the layer:

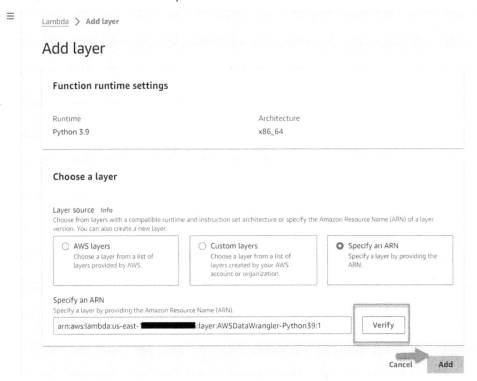

Figure 10.29 – Verifying the ARN then clicking on Add

7. Remember to change the **Timeout** and **Memory** values so that they fit your use case. In this exercise, the timeout will be 10 minutes (maximum allowed timeout):

Figure 10.30 – Making sure to edit the timeout and memory

8. Paste the following script in the **Code** tab. Edit the SQL to fit your use case:

```python
import logging
import awswrangler as wr
import json

statement = """
        UNLOAD (SELECT * FROM redshift.{table_name}')
        TO '{S3_location}'
        IAM_ROLE '{'arn:aws:iam:your-account-id:role/your-role'
}'
        PARQUET PARALLEL ON
        PARTITION BY (product_category)'
    """

def lambda_handler(event, context):
    logger.info(json.dumps(event))
    query_id = event['Input'].get('query_id')
    con_redshift = wr.data_api.redshift.connect(
        workgroup_name="aws-sdk-pandas",
        database="test_redshift",
        secret_arn="arn:aws:secretsmanager:us-east-1:your-
account-id:secret:your-secret-name",
    )


    wr.data_api.redshift.read_sql_query(
        sql=statement,
        con=con_redshift,
    return {
        'statusCode': 200,
        'body': json.dumps("finished")
    }
```

How it works...

With Lambda, you create a serverless, highly available computing architecture that can be further elevated by integrating with EventBridge for **event-driven architecture (EDA)** or **Simple Notification Service (SNS)** for error notification. It can help to trigger a stored procedure based on specific events or conditions. You need to carefully plan your trigger and outcome to fit the use case's goal and outcome.

The Data API enables you to access Redshift data without maintaining a persistent connection. This API can execute queries from any application or service that invokes HTTPS requests, including AWS Lambda and web applications. The Data API returns the result asynchronously, so ideally, for a complex job, you should integrate with Step Functions to get the job status using the `DescribeStatement` API and get the `GetStatementResult` API to see if the execution is finished. To use in a web application, it is better to parse the response to JSON instead of a pandas DataFrame.

In a real-world application, the S3 bucket that stores the DataFrame from Lambda execution could be the storage for the analytical workbench where you store relevant data for your team to explore. Due to the limitations of Lambda timeout and memory, you should consider appropriate tools such as Glue or SageMaker notebooks if the dataset is large.

There's more...

When working with AWS Lambda, you need to pay attention to the selected Python version. To take a step further to integrate into a web application, you can add an API gateway to leverage WebSocket API and REST API capability for your application.

You need to make sure your Python version in runtime matches with AWS SDK for pandas. If you make a mistake while choosing the runtime version, you can go to the runtime settings to change the version.

See also

- *Managing Lambda dependencies with layers*: https://docs.aws.amazon.com/lambda/latest/dg/chapter-layers.html

- *AWS Lambda Layer*: https://aws-sdk-pandas.readthedocs.io/en/2.14.0/install.html#aws-lambda-layer

- *Stored procedure limitations*: https://docs.aws.amazon.com/redshift/latest/dg/stored-procedure-constraints.html

- *Security and privileges for stored procedures*: https://docs.aws.amazon.com/redshift/latest/dg/stored-procedure-security-and-privileges.html

- *Configure Lambda function timeout*: https://docs.aws.amazon.com/lambda/latest/dg/configuration-timeout.html

Using the AWS SDK for Python to manage Amazon QuickSight

The AWS SDK for Python (Boto3) could be used to create, configure, and manage Amazon QuickSight. There are various use cases for it, from quickly granting permission to exporting QuickSight assets from one account to another account. The following script in this recipe will demonstrate how to update permissions for data sources, datasets, analyses, dashboards, and themes in Amazon QuickSight. It grants a specific user or role the necessary permissions to access and manage these resources.

Getting ready

Before starting, ensure you have the following prerequisites:

- An active AWS account with a QuickSight subscription
- Python 3 and Jupyter Notebook installed on your system
- Boto3 and PyYAML libraries installed (`pip install boto3 PyYAML`)
- AWS credentials and profile configured

How to do it...

1. From your AWS CLI, retrieve the Amazon QuickSight Principal ARN:

    ```
    aws quicksight list-users --region <aws-region> --aws-account-id
    <account-id> --namespace <namespace-name>
    ```

2. In your Jupyter notebook, start with `import boto3`, set up the client, and input your Principal ARN. Replace the principal value with the value from *step 1*:

    ```
    import boto3

    client = boto3.client(
        "quicksight",
    )
    principal = 'arn:aws:quicksight:<aws-region>:<account-
    id>:user/<namespace-name>/<quicksight-user-name>'
    ```

3. Input the recipe code from the *Using AWS SDK for pandas, the Redshift Data API, and Lambda to execute SQL statements* recipe; please check the *How it works...* section.

How it works...

The AWS SDK for QuickSight provides a set of APIs that you can use to manage your QuickSight resources programmatically. These APIs allow you to perform various tasks, such as creating and deleting users, managing datasets and analyses, and configuring permissions and security settings. In this recipe, we went through several important terms such as dataset, analysis, , and dashboard. Let's elaborate more on these terms and how the API of these groups works:

- A **dataset** is a collection of data used to create visualizations and dashboards. Datasets can be shared across users and groups and can be used to create multiple analyses and dashboards. With the API, you can specify the data source and configure data refresh schedules. You can also retrieve a list of datasets in your QuickSight account and search for datasets based on various criteria, such as dataset name or data source.

- An **analysis** is a workspace where you can create and customize visualizations and dashboards. It contains a dataset, which is the source of the data used in the visualizations, and a set of visualizations that you can arrange and customize to suit your needs. You can create new visualizations, add filters, remove visualizations, and apply formatting to the data.

- A **template** is a reusable set of metadata that defines the structure and settings of an analysis or dashboard. It includes the dataset, visualizations, filters, and formatting that you have defined in the analysis or dashboard and can be reused to create a new dashboard.

- A **dashboard** is a read-only version of your data visualization. Dashboards can be scheduled to refresh at regular intervals and can be delivered to users via email or other communication channels. They can also be embedded in external websites and applications, allowing you to share your data with a wider audience.

When you use the AWS SDK for QuickSight, you can use the programming interface to manage users, datasets, analyses, templates, and dashboards. This solution can help you to manage the access of your visualization in a programmatic way. It is a good way to apply DevOps practices such as version control and **continuous integration / continuous deployment (CI/CD)** to your QuickSight artifacts.

There's more...

With the `boto3` library for QuickSight, you can transfer dashboard ownership and create a CI/CD pipeline to move the dashboard from one environment to another or from one region to another region.

Besides the previously mentioned API, two frequently useful APIs can help you quickly create and download your QuickSight template:

- The `create_template` API helps you create a template from an existing analysis or template.

- The `describe_template` API allows you to retrieve metadata about a specific template, including its name, ARN, creation and last updated times, description, sheets, source entity ARN, status, theme ARN, and version number. This API is useful for getting information about a template that you have created or that has been shared with you.

Please note that not all visuals are supported over APIs, so you should check the *API Reference Index* page in the *See also* section for more details.

See also

- *Data Types*: https://docs.aws.amazon.com/quicksight/latest/APIReference/API_Types.html

- *API Reference Index*: https://docs.aws.amazon.com/quicksight/latest/APIReference/API_Reference.html

- Boto3 documentation – create_template: https://boto3.amazonaws.com/v1/documentation/api/1.26.85/reference/services/quicksight/client/create_template.html

- Boto3 documentation – describe_template: https://boto3.amazonaws.com/v1/documentation/api/1.26.93/reference/services/quicksight/client/describe_template.html

11

Migrating to AWS – Steps, Strategies, and Best Practices for Modernizing Your Analytics and Big Data Workloads

In this chapter, we'll embark on the journey of migrating a data warehouse from an on-premises environment to AWS, a transformative move that brings numerous benefits and a few challenges. This migration process represents a crucial step for organizations seeking to leverage the scalability, flexibility, and cost-effectiveness of cloud computing.

We'll explore the key considerations, strategies, and best practices for a successful transition, including data migration methods, security concerns, performance optimization, and leveraging AWS-specific services such as Amazon Redshift.

Our focus will be on ensuring a seamless migration with minimal disruption, while also optimizing the data warehouse to harness the full potential of AWS's cloud capabilities. This process not only involves physically transferring data but also a strategic shift in data management and operations, aiming to enhance accessibility, analytics, and overall **business intelligence** (**BI**) in the cloud environment.

The following recipes will be covered in this chapter:

- Reviewing the steps and processes for migrating an on-premises platform to AWS
- Choosing your AWS analytics stack – the re-platforming approach
- Picking the correct migration approach for your workload
- Planning for prototyping and testing

- Converting ETL processes with big data frameworks
- Defining and executing your migration process with Hadoop
- Migrating the existing Hadoop security authentication and authorization processes

Technical requirements

Before initiating a data warehouse migration to AWS, it's essential to set a robust foundation by addressing a set of technical prerequisites. Initially, a comprehensive infrastructure assessment is necessary so that you can evaluate the existing data warehouse's specifications, network setup, and data volume to ensure the AWS environment is provisioned appropriately. Compatibility is important; hence, in-depth analysis to ensure data types, schemas, and procedures align with AWS services such as Amazon Redshift is crucial. Employing tools such as AWS **Schema Conversion Tool** (**SCT**) for assessment and conversion planning can significantly aid in this assessment and conversion planning process. The key considerations are as follows:

- **Network readiness** must be ensured. For this, you may potentially require a high-bandwidth, secure, and possibly dedicated connection such as AWS Direct Connect to manage the substantial data transfer process.

- **Security considerations** are non-negotiable, demanding a thorough review and planning for encryption, compliance, and governance standards. A detailed resource inventory, including databases, ETL jobs, and connected applications, should be documented and accompanied by a performance baseline of the current system to measure post-migration efficiency.

- **A robust backup and recovery strategy** is imperative, ensuring data integrity during and after the transition. Operational readiness, which involves setting up monitoring, logging, and maintenance mechanisms in AWS, will ensure smooth, ongoing operations.

Addressing these prerequisites diligently sets the stage for a successful, seamless migration where you can leverage AWS's scalability, performance, and extensive feature set.

Reviewing the steps and processes for migrating an on-premises platform to AWS

Migrating from an on-premises data warehouse to AWS is a strategic move that can yield significant benefits such as cost savings, enhanced scalability, and improved performance. This recipe provides a comprehensive guide to the migration process while covering key considerations, strategies, best practices, and actionable steps to ensure a successful transition.

Getting ready

Migration consists of five stages, as shown in *Figure 11.1*:

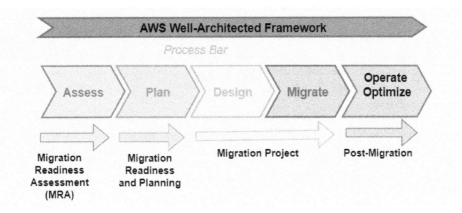

Figure 11.1 – The five phases of migration

Let's take a closer look at these phases in detail:

1. **Assessment or opportunity evaluation**: Begin by taking stock of your current on-premises environment. This involves cataloging your infrastructure (servers, storage, and networking), applications, data, and the dependencies between them. Understand your business goals and performance requirements to identify which workloads are suitable for migration and which strategy aligns best with your objectives.

2. **Planning or portfolio discovery and planning**: This phase is where the blueprint for your migration takes shape. Develop a comprehensive migration plan that outlines the chosen strategy, detailed timeline, resource allocation (both personnel and financial), and a thorough risk assessment. A well-crafted plan is your compass, guiding you through the complexities of the migration process.

3. **Migration or application design and migration**: This is the heart of the process and is where you execute your plan. It can involve various strategies:

    *   **Rehosting (lift-and-shift)**: This involves moving your applications and data to AWS with minimal changes. It's the quickest approach but may not fully leverage cloud benefits.

    *   **Replatforming**: This involves making minor modifications to your applications so that you can take advantage of cloud-native features such as managed databases or autoscaling.

    *   **Refactoring/re-architecting**: This is a more comprehensive approach where you redesign your applications so that they're cloud-native, often while utilizing microservices and serverless technologies.

4. **Testing and validation**: After migrating your workloads, rigorous testing is paramount. This involves functional testing to ensure applications behave as expected, performance testing to validate that they meet your requirements under load, and security testing to identify and mitigate any vulnerabilities.

5. **Optimization and cutover:** Once testing is complete, it's time to optimize your AWS resources for cost efficiency and performance. This can involve right-sizing instances, using reserved instances or savings plans, and employing autoscaling to adapt to varying workloads. Finally, you'll transition your production environment to AWS, either through a phased approach or a complete cutover.

You can read more at `https://aws.amazon.com/blogs/apn/tips-for-becoming-an-aws-migration-consulting-and-delivery-competency-partner/`.

How to do it...

Follow these steps to perform a detailed migration:

1. **Discovery and planning (assessment phase)**:

   - **Infrastructure inventory**: Start by documenting all your infrastructure components, including servers, storage systems, network equipment, software licenses, and any other essential assets.

   - **Application inventory**: List all your applications, their dependencies, and how they interact with infrastructure and data.

   - **Data assessment**: Analyze your data in terms of volume, type, structure, and sensitivity. This will help you determine the optimal migration and storage strategy.

   - **Business case development**: Clearly define the objectives for the migration, whether they're cost reduction, scalability, performance improvements, or faster time to market.

   - **Cloud-readiness assessment**: Evaluate the organization's readiness for cloud adoption by considering technical skills, security practices, and governance protocols.

2. **Migration planning (planning phase)**:

   - **Migration strategy selection**: Choose the most appropriate strategy (rehosting, replatforming, or refactoring) based on the workload's requirements

   - **AWS account structure design**: Set up your AWS accounts according to your organizational needs, such as separate accounts for production, development, and testing environments

   - **Network architecture design**: Plan your **virtual private cloud** (**VPC**), including subnets, security groups, and routing configurations

   - **Resource provisioning**: Estimate the AWS resources that are required (EC2 instances, RDS databases, S3 buckets, and so on) based on your workload and capacity needs

- **Cost estimation**: Use the AWS Pricing Calculator to predict costs and create a budget for your cloud infrastructure

- **Migration timeline**: Develop a timeline outlining key tasks, dependencies, and deadlines

- **Risk assessment**: Identify potential risks such as downtime, data loss, or security vulnerabilities, and put mitigation plans in place

- **Communication plan**: Create a communication strategy to keep stakeholders informed and involved during the migration process

3. **Migration execution (migration phase)**:

- **AWS environment setup**: Build your AWS environment by setting up VPCs, subnets, and security groups. Establish network connectivity (such as Direct Connect or VPN) between your on-premises infrastructure and AWS.

- **Data migration**:

    - Use AWS **Database Migration Service (DMS)** to move databases from on-premises to AWS (for example, Oracle to Amazon RDS)

    - Implement AWS DataSync for fast and secure file transfers to Amazon S3 or EFS

    - For large-scale transfers, consider AWS Snowball

- **Application migration**:

    - **Rehosting**: Use AWS **Server Migration Service (SMS)** or AWS **Application Migration Service (MGS)** to replicate VMs from your data center to AWS

    - **Replatforming**: Shift applications to AWS-managed services, such as Amazon RDS, ElastiCache, or Amazon MQ

    - **Refactoring**: Redesign applications so that they leverage cloud-native services such as microservices, containers (Amazon ECS or EKS), or serverless functions (AWS Lambda)

4. **Testing and validation (testing phase)**:

- **Functional testing**: Develop and run test cases to ensure that all application features and business processes function as expected in AWS

- **Performance testing**: Simulate real-world workloads to verify that your applications meet performance benchmarks in the cloud

- **Security testing**: Conduct vulnerability scans and penetration tests to detect and fix potential security risks

- **Data validation**: Confirm data accuracy and integrity between the on-premises and AWS environments

5. **Optimization and cutover (final phase)**:

- **Resource optimization**: Continuously monitor AWS resource usage (CPU, memory, and storage) and adjust it to optimize performance and cost-efficiency

- **Cost optimization**: Regularly review your AWS billing to identify cost-saving opportunities, such as Reserved Instances or Savings Plans

- **Performance tuning**: Fine-tune applications and databases to enhance performance in the AWS cloud

- **Cutover**: There are two cutover approaches we can use:

    - **Phased migration**: Gradually migrate applications and data in stages, allowing for validation and testing at each phase.

    - **Big bang migration**: Migrate all applications and data in one go. While faster, this method carries more risk.

By precisely following these steps and understanding the five key phrases of migration, you can confidently and successfully transition your on-premises platform to AWS.

Choosing your AWS analytics stack – the re-platforming approach

In this recipe, we'll focus on the re-platforming approach, which involves migrating to a new platform while making modifications to your architecture so that you can take advantage of cloud benefits. We'll examine the rich ecosystem of AWS services that can power your modern cloud data warehouse.

Getting ready

Before you initiate your migration, ensure you have the following in place:

- **Clear business objectives**: Clearly define your motivations for migrating to AWS. Are you seeking cost savings, scalability, improved performance, or increased agility? These goals will guide your decision-making throughout the migration process.

- **An AWS account**: If you don't already have one, you'll need to create an AWS account. This will be the foundation for all your cloud resources.

- **Network connectivity**: Establish secure connectivity between your on-premises data center and AWS. This can be achieved through AWS Direct Connect (dedicated private network connection) or a site-to-site VPN.

- **Inventory and assessment**:

  - **Infrastructure**: Catalog your existing on-premises data warehouse infrastructure (servers, storage, and network devices)

  - **Applications**: Identify and document all applications that interact with your data warehouse

  - **Data**: Analyze your data assets (volume, type, structure, and quality) to understand the scope of the migration

  - **Dependencies**: Map the relationships between applications, data, and infrastructure to identify potential migration challenges

  - **Performance requirements**: Determine your desired performance metrics for the cloud data warehouse

- **Skills and resources**: Ensure you have the necessary technical expertise in-house or have access to external consultants with experience in AWS data warehousing technologies.

Why re-platform on AWS?

Re-platforming on AWS offers a compelling middle ground between the speed of rehosting and the full transformation of refactoring. Here's why it's a popular choice:

- **Reduced operational overhead**: AWS takes care of infrastructure management, allowing you to focus on your data and analytics

- **Scalability and elasticity**: You can dynamically scale your resources up or down to match demand, paying only for what you use

- **Managed services**: You can leverage AWS's fully managed services, such as Amazon Redshift, to offload maintenance tasks and accelerate innovation

- **Performance and cost optimization**: You can choose from a variety of compute and storage options to fine-tune performance and cost-effectiveness

- **A rich ecosystem of services**: You can integrate with a wide range of AWS services for data ingestion, transformation, analytics, and visualization purposes

The AWS analytics ecosystem

AWS offers a comprehensive suite of services tailored for building a modern data warehouse:

- **Amazon Redshift**: A fast, fully managed, petabyte-scale data warehouse optimized for analytics. It offers excellent performance for complex queries and supports a wide range of data types and SQL functions.

- **Amazon S3**: A scalable, high-durability object storage service for storing vast amounts of structured and unstructured data. It's a cost-effective option for data lakes and can be queried directly by Redshift using Redshift Spectrum.

- **AWS Glue**: A serverless data integration service that simplifies discovering, preparing, and combining data for analytics and machine learning. Glue offers crawlers for schema discovery, **extract, transform, and load** (ETL) capabilities, as well as a data catalog.

- **Amazon Athena**: A serverless interactive query service that allows you to analyze data in S3 using standard SQL. Athena is a great option for ad hoc queries and exploration.

- **AWS Lake Formation**: This is a fully managed service that simplifies the process of creating, securing, and managing data lakes. It automates data ingestion, cataloging, and access control, allowing you to organize, secure, and analyze large-scale data efficiently in a centralized repository. By integrating with other AWS services, Lake Formation enables seamless data governance and analytics in a scalable, cost-effective manner.

- **AWS DynamoDB**: This is a serverless, NoSQL database service that enables you to develop modern applications at any scale.

- **AWS Lambda**: AWS Lambda is a serverless computing service that lets you run code without the need to provision or manage servers. You pay only for the compute time you consume. This eliminates the need for upfront infrastructure costs and allows you to focus on building applications.

- **AWS Managed Workflows for Apache Airflow (MWAA)**: This is a fully managed service that simplifies the deployment and operation of Apache Airflow, an open source workflow orchestration platform. It allows you to build, schedule, and monitor data pipelines and workflows without the need to manage the underlying infrastructure.

- **Amazon Kinesis**: This is a fully managed service that processes and analyzes real-time streaming data at scale. It can capture and process millions of events per second from various sources, making it ideal for applications such as clickstream analysis, IoT data processing, and real-time analytics.

- **Amazon EMR**: This service provides the cloud big data platform for processing vast amounts of data using open source tools.

- **Amazon QuickSight**: A scalable, serverless BI service for creating visualizations, dashboards, and reports from your data warehouse data.

If you're aiming to re-platform your analytics stack and wish to leverage the AWS analytics ecosystem, here's a guide on choosing the right tools and services:

1. **Data ingestion and integration**:

   - **AWS Glue**: This is a serverless data integration service that simplifies the process of discovering, preparing, and combining data for analytics. It automates many of the tasks involved in ETL, allowing you to focus on building data-driven applications.

   - **AWS Data Migration Service (DMS)**: This service helps migrate on-premises databases to AWS easily and securely.

   - **Amazon Kinesis or MSK**: For real-time streaming data.

2. **Data storage**:

   - **Amazon S3**: Scalable storage for data lakes and analytics. S3's integrations and data protection features make it ideal for a data lake.

   - **Amazon Redshift**: A petabyte-scale data warehouse service. For complex analytics workloads, you can analyze all your data using your existing BI tools.

   - **Amazon RDS and Aurora**: For relational data storage needs.

3. **Data processing and analysis**:

   - **AWS Glue**: With this service, you can visually clean and normalize data

   - **Amazon Elastic MapReduce (EMR)**: You can process large amounts of data with popular distributed frameworks such as Apache Spark and Hadoop

   - **AWS Lambda**: You can run code in response to events without provisioning servers, which can be useful for lightweight data processing tasks

4. **Data querying**:

   - **Amazon Athena**: A serverless interactive query service that allows you to analyze data in Amazon S3 using SQL

   - **Amazon Redshift Spectrum**: With this service, you can examine huge datasets in S3 without having to load or transform them

5. **Machine learning and advanced analytics**:

   - **Amazon SageMaker**: Build, train, and deploy machine learning models at scale

   - **Amazon Comprehend and Recognition**: For **natural language processing (NLP)** and video/image analysis, respectively

6. **Data visualization and BI**:

   - **Amazon QuickSight**: A BI service with native machine learning integrations

   - **Integration with third-party tools**: AWS supports integration with popular BI tools such as Tableau, Looker, and more

7. **Data security**:

   - **Amazon Macie**: Discover, classify, and protect sensitive data

   - **AWS Key Management Service (KMS)**: Create and manage cryptographic keys

   - **AWS Identity and Access Management (IAM)**: Manage user permissions and access

8. **Data governance and cataloging**:

   - **AWS Glue Catalog**: A central metadata repository integrated with a wide range of AWS services

   - **Lake Formation**: Simplifies and automates many of the complex tasks associated with setting up, securing, and managing data lakes

9. **Monitoring and Management**:

   - **Amazon CloudWatch**: Monitor resources and applications

   - **AWS CloudTrail**: Provides governance, compliance, operational auditing, and risk auditing

10. **Optimization and cost management**:

    - **AWS Cost Explorer**: View and analyze your costs and usage

    - **AWS Trusted Advisor**: Offers real-time guidance to provision resources while following AWS best practices

How to do it...

When re-platforming, consider these steps:

1. **Assess**: Understand your current analytics stack and identify areas of improvement.

2. **Select AWS services**: Based on your needs, pick the right services from the AWS analytics ecosystem.

3. **Data migration**: Move your data to AWS, either through batch transfers or real-time streams.

4. **Re-architect and develop**: Modify or redesign your analytics workflows and processes so that they fit within the AWS ecosystem.

5. **Testing**: Ensure that the new system works correctly and meets performance standards.

6. **Deployment and monitoring**: Roll out the new system and continuously monitor its performance, optimizing as necessary.

AWS analytics services architecture

Harnessing the power of AWS's comprehensive data analytics services, the following diagram outlines a scalable and flexible solution for processing, storing, and analyzing massive datasets efficiently.

The following mind map diagram shows a use case map for analytics services:

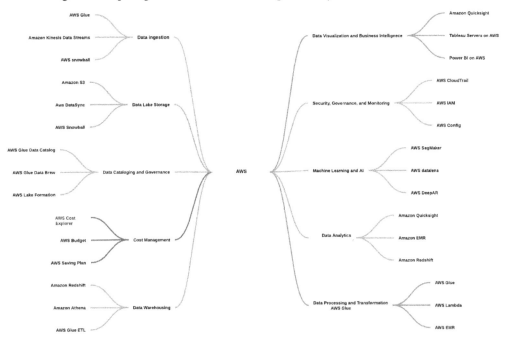

Figure 11.2 – Analytics service mind map

A modern data analytics architecture on AWS involves integrating various AWS services so that you can ingest, store, process, analyze, and visualize data in a scalable, secure, and cost-effective manner. Here's a breakdown of the components that are typically involved in such a reference architecture:

1. **Data ingestion**: Data comes from multiple sources, such as applications, databases, logs, and third-party systems. AWS provides several services for handling different types of data ingestion:

    *   **Amazon Kinesis or AWS Managed Streaming for Apache Kafka (MSK)**: For real-time streaming data such as IoT sensor data or log files

    *   **AWS DMS**: For migrating or replicating databases to AWS

    *   **AWS IoT Core**: For ingesting data from IoT devices

    *   **Amazon AppFlow**: For integrating data from SaaS applications such as Salesforce, ServiceNow, or Google Analytics

- **AWS DataSync**: For moving large amounts of unstructured data, such as files and logs

- **AWS Snowball**: For transferring petabyte-scale data from on-premises environments

2. **Data lake storage**: After data ingestion, the raw and processed data is stored in a scalable, durable, and secure data lake.

- **Amazon S3**: Serves as the primary data lake storage platform. It provides cost-effective, scalable storage with high durability.

- **AWS Lake Formation**: Helps in building, managing, and securing the data lake on S3.

3. **Data cataloging and governance**: Metadata management, data cataloging, and data governance are crucial for managing a data lake at scale:

- **AWS Glue**: For automatic schema discovery, data cataloging, and ETL workflows

- **AWS Glue DataBrew**: For data preparation and cleaning without writing code

- **AWS Lake Formation**: For unified governance, which helps manage security, access controls, and audit logs centrally

4. **Data processing and transformation**: Once the data has been ingested and cataloged, various processing methods can be applied, depending on the data type and latency requirements:

- **AWS Glue**: For running ETL jobs that transform and load data into a format suitable for analysis

- **Amazon EMR**: For big data processing using Apache Spark, Hive, and other open source frameworks

- **Amazon Redshift Spectrum**: Enables querying directly from the data lake using SQL

- **AWS Lambda**: For serverless data processing, especially for real-time or event-driven processing

5. **Data warehousing**: For structured, high-performance, and scalable querying, data is often transformed into a data warehouse optimized for analytics:

- **Amazon Redshift**: AWS's fully managed data warehouse service, optimized for querying large datasets and integrating with the data lake using **Redshift Spectrum**

- **Amazon RDS/Aurora**: For structured data storage with relational databases such as MySQL, PostgreSQL, and Aurora

6. **Data analytics**: With data ingested, processed, and stored, analytics can be performed using various tools to derive insights and perform machine learning:

- **Amazon Athena**: An interactive query service for analyzing data in S3 using standard SQL

- **Amazon OpenSearch**: For full-text search, log analytics, and real-time monitoring

- **AWS Kinesis Data Analytics**: For real-time analytics on streaming data

- **AWS QuickSight**: AWS's BI tool for generating dashboards and reports

7. **Machine learning and AI**: Data science and machine learning workflows are essential for predictive analytics, recommendation systems, and anomaly detection:

    - **Amazon SageMaker**: A fully managed machine learning service for building, training, and deploying machine learning models

    - **AWS AI services**: Pre-built AI services such as Amazon Rekognition (image recognition), Amazon Comprehend (natural language processing), and Amazon Forecast (time series forecasting)

8. **Security, governance, and monitoring**: Security and governance play a critical role in modern analytics architecture, ensuring that data is protected and compliant with industry standards:

    - **AWS IAM**: For controlling access to services and resources

    - **AWS KMS**: For encrypting data at rest and in transit

    - **AWS CloudTrail**: For monitoring and logging all activities on AWS accounts for security audits

    - **Amazon CloudWatch**: For monitoring the performance of applications, services, and infrastructure

    - **AWS Config**: For monitoring compliance and security posture

9. **Data visualization and BI**: Visualizing insights is key for decision-making processes, and AWS offers a variety of services to accomplish this:

    - **Amazon QuickSight**: AWS's scalable BI tool that offers machine learning and AI-driven insights, visualizations, and dashboards

    - **Third-party tools**: Integration with popular BI tools such as Tableau, Power BI, or Looker

10. **Cost management**: With all services in place, managing costs is critical to ensure the analytics platform remains efficient:

    - **AWS Cost Explorer**: For visualizing and understanding costs

    - **AWS Trusted Advisor**: Provides real-time guidance to help optimize the AWS infrastructure in terms of cost, performance, security, and fault tolerance

Choosing your architecture

The right architecture for your AWS data warehouse will depend on your specific needs and use cases. Here are a few popular architectural patterns:

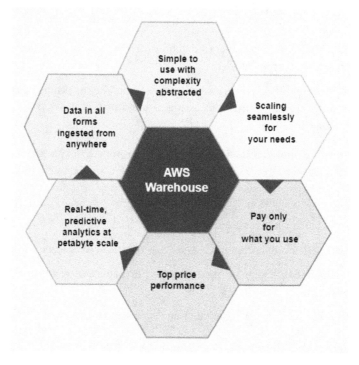

Figure 11.3 – AWS data warehouse benefits

This Redshift-centric architecture leverages Amazon Redshift's **massively parallel processing (MPP)** capabilities to process and analyze large datasets efficiently while supporting advanced BI and reporting needs.

Let's explore the diverse range of AWS architectures that are available to support your unique cloud computing needs:

- **Redshift-centric architecture**: This is ideal for high-performance analytics workloads. All your data is loaded into Redshift, and you use its built-in capabilities for analysis.

  In a Redshift-centric architecture, data from various operational systems, marketing data sources, and other systems is initially moved to Amazon S3. AWS Glue is then used to perform ETL processes, transforming and cleansing the data before it's loaded into Amazon Redshift. Redshift acts as the central data warehouse, serving as a producer that consolidates and stores this processed data. The consolidated data is then made accessible to multiple consumers, including Consumer 1, Consumer 2, and Consumer 3, each leveraging the data for analytics, reporting, and BI purposes. This setup ensures a streamlined, scalable, and efficient data pipeline that supports diverse analytical needs across the organization:

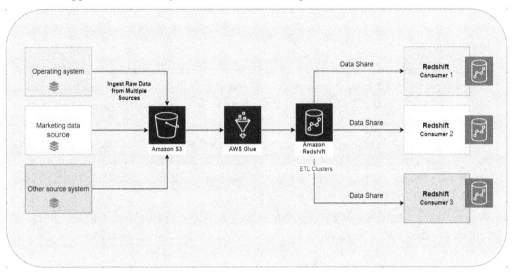

Figure 11.4 – AWS Redshift-centric architecture diagram

- **Data lakehouse architecture**: Combines the best of data lakes and data warehouses. Raw data is stored in S3, and Glue is used for ETL and transformation. You can use Redshift, Athena, or both to query the data.

  The initial layer, known as the landing layer, serves as the entry point for all source files in their native formats. Subsequently, these files undergo conversion and are stored in a standardized Parquet format within the raw layer. The stage layer maintains historical records of dimensional tables using **Slowly Changing Dimension Type 2 (SCD2)**, which is facilitated by Apache Hudi in Amazon S3 and AWS Glue jobs. AWS Glue orchestrates ETL tasks across layers, ensuring seamless data movement, cleansing, validation, and transformation. In the presentation layer, data is meticulously refined so that it aligns with business requirements through AWS Glue jobs. Finally, the data warehouse layer, powered by Amazon Redshift, houses the curated and cleansed data, accessible either through direct copying via AWS Glue or by creating Spectrum tables linked to the S3 location.

- **Serverless analytics architecture**: Leverages Athena and S3 for a fully serverless data lake. Glue can be used for ETL if needed. This is a cost-effective option for ad hoc or intermittent workloads.

- **AWS data mesh architecture**: This is a decentralized approach to data management that empowers business domains to own and manage their data. It contrasts with traditional data management, where data is centralized and managed by a dedicated team. In data mesh, each domain is responsible for the entire life cycle of its data, from ingestion to consumption. This includes ingesting data from various sources, processing it to make it usable, and serving it to consumers.

  Data mesh is similar to microservices in that it breaks down data management into smaller, independent units. This makes it more scalable and flexible than traditional data management approaches. It also makes it easier for business domains to innovate and experiment with new data use cases.

  AWS Lake Formation and AWS Glue are two services that can be used to implement a data mesh architecture on AWS. Lake Formation is a service that makes it easy to create and manage access and permissions for the data lake. AWS Glue is a serverless ETL service that can be used to extract, transform, and load data into a data lake.

  Data mesh is a relatively new concept, but it's gaining traction as organizations look for ways to improve their data management practices. AWS Lake Formation and AWS Glue are two services that can be used to implement a data mesh architecture on AWS.

Here are some key considerations when you're choosing your architecture:

- **Workload types**: Are your workloads primarily batch processing, interactive querying, or a mix of both?

- **Data volume**: How much data do you need to store and process?

- **Data variety**: Do you need to handle structured, semi-structured, or unstructured data?

- **Performance requirements**: What are your latency requirements for queries and reporting?

- **Cost optimization**: What is your budget, and how can you balance cost with performance?

The key considerations for data migration services and tools are as follows:

- **Schema conversion**: Use the AWS SCT to convert your on-premises data warehouse schema into a format compatible with Redshift.

- **Data migration**: Use AWS DMS, S3 Transfer Acceleration, or other tools to move your data to S3.

- **ETL with Glue**: Develop ETL jobs in Glue to transform and load the data into Redshift (or prepare it for querying with Athena).

- **Optimize and validate**: Fine-tune your Redshift cluster (or Athena queries) for performance. Validate data integrity and functionality.

By carefully considering your requirements and choosing the right architecture, you can use the full power of the AWS analytics ecosystem to build a scalable, flexible, and cost-effective data warehouse that meets your business needs.

See also

- *Modern Data Analytics Reference Architecture on AWS Diagram*: `https://docs.aws.amazon.com/architecture-diagrams/latest/modern-data-analytics-on-aws/modern-data-analytics-on-aws.html`

- *Use a reusable ETL framework in your AWS lake house architecture*: `https://aws.amazon.com/blogs/architecture/use-a-reusable-etl-framework-in-your-aws-lake-house-architecture/`

- *Design a data mesh architecture using AWS Lake Formation and AWS Glue*: `https://aws.amazon.com/blogs/big-data/design-a-data-mesh-architecture-using-aws-lake-formation-and-aws-glue/`

- *Serverless data lake centric analytics architecture*: `https://docs.aws.amazon.com/whitepapers/latest/aws-serverless-data-analytics-pipeline/serverless-data-lake-centric-analytics-architecture.html`

Picking the correct migration approach for your workload

Choosing the right migration approach depends on various factors, including the type of workload, existing infrastructure, performance needs, future scalability requirements, and budget constraints. The choice becomes even more nuanced when you're considering specific workloads, such as data warehouses or Hadoop clusters. This recipe will guide you through the process of picking the correct migration approach for different workloads.

In this recipe, we'll explore several migration approaches and discuss the factors to consider when you're choosing the best approach for your workload. We'll also provide a step-by-step guide to planning and executing a migration.

Each migration strategy comes with its pros and cons. The choice depends on the current state of your systems, the nature of the workload, the desired benefits, and any other constraints. A thorough assessment of the existing environment, coupled with a clear understanding of business and technical objectives, will guide the decision-making process.

Getting ready

Selecting the optimal migration strategy is crucial. Here are some common approaches and when to consider them.

Data warehouse migration to AWS offers several approaches, allowing you to choose the one that best fits your needs:

- **Lift and shift**: Lift and shift, also known as rehosting, is a migration strategy that involves moving workloads from on-premises infrastructure to the cloud without making any significant changes to the application code or architecture. This approach is often used for applications that are running well on-premises but that could benefit from the scalability, elasticity, and cost-effectiveness of the cloud:

  - **When to use**: Your current on-premises data warehouse meets performance and scalability requirements, but you desire the benefits of cloud infrastructure, such as operational efficiency, elasticity, and cost savings

  - **Advantages**: Faster migration, lower initial effort, and minimal changes to existing systems

  - **Example**: Migrating an on-premises Teradata or Oracle data warehouse to Amazon Redshift without major transformations

- **Lift and rewrite/re-architect**: Lift and rewrite/re-architect is a cloud migration strategy that involves rewriting or re-architecting applications to take full advantage of the cloud's capabilities. This approach is often used for applications that aren't running efficiently on-premises or that need to be modernized to meet new business requirements:

  - **When to use**: Your current data warehouse has inherent limitations, or you wish to leverage cloud-native features, optimized performance, and scalability

  - **Advantages**: Fully utilizing cloud capabilities, improved performance, and potentially reduced operational costs in the long run

  - **Example**: Transforming an on-premises data warehouse into a serverless architecture using Amazon Athena and S3

- **Hybrid**: A hybrid cloud environment combines on-premises infrastructure with a public cloud, such as AWS. This approach allows organizations to take advantage of the benefits of both environments, such as the scalability and elasticity of the cloud, while still maintaining control over their sensitive data on-premises:

  - **When to use**: You need to maintain some components on-premises due to data sovereignty, latency, or other business requirements, but you also want to leverage the cloud for scalability and flexibility

  - **Advantages**: You can balance on-premises control and cloud scalability

  - **Example**: Keeping a subset of sensitive data on-premises while bursting compute-intensive analytics tasks to Amazon Redshift

When choosing a migration approach, it's important to consider the following factors:

- **The size and complexity of the workload**: The size and complexity of the workload will determine the time and effort required to migrate it to the cloud. Larger and more complex workloads will typically require more time and effort to migrate than smaller and simpler workloads.

- **The skills and experience of the team**: The skills and experience of the team will determine how much work they can do on their own and how much help they will need from external consultants. A team with strong cloud skills and experience will be able to migrate workloads more quickly and efficiently than a team with limited cloud skills and experience.

- **The desired timeframe for migration**: The desired timeframe for migration will determine the urgency with which the migration needs to be completed. If the migration needs to be completed quickly, then a lift-and-shift approach may be the best option. If the migration can be completed more slowly, then a refactoring or re-platforming approach may be a better option.

- **Budget**: The budget for the migration will determine the resources that are available to complete the migration. A larger budget will allow for more resources to be allocated to the migration, which may allow for a more complex migration approach to be used.

- **Risk tolerance**: The risk tolerance of the organization will determine how much risk is acceptable when migrating workloads to the cloud. A risk-averse organization may prefer a lift-and-shift approach, while a more risk-tolerant organization may be willing to consider a more complex migration approach.

How to do it...

Once you've chosen a migration approach, you can start planning and executing the migration. The following is a step-by-step guide to planning and executing a migration:

1. **Define the scope of the migration**: The first step is to define the scope of the migration, which includes identifying the workloads that will be migrated, the target cloud environment, and the timeframe for the migration.

2. **Assess the readiness of the organization for migration**: The next step is to assess the readiness of the organization for migration. This includes assessing the skills and experience of the team, the availability of resources, and the organizational culture.

3. **Design the migration plan**: The design of the migration plan will include detailed specifications for the migration, such as the timeline for each step, the tools and technologies that will be used, and the procedures for testing and cutover.

4. **Execute the migration**: The execution of the migration will involve carrying out the steps in the migration plan. This includes migrating the workloads to the cloud, testing the migrated workloads, and cutting over to the cloud environment.

5. **Monitor and optimize the migration**: Once the migration is complete, it's important to monitor the migrated workloads to ensure that they're performing as expected.

Planning for prototyping and testing

When planning for a migration, especially a significant migration, such as moving to the cloud, prototyping and testing are critical phases. These steps allow organizations to validate assumptions, identify potential issues, and ensure smooth operations when the migration is executed at scale.

Getting ready

Prototyping and testing are crucial phases in the on-premises to AWS cloud migration process. These phases ensure that the migration is well-planned, executed seamlessly, and meets the organization's objectives.

Prototyping involves creating a small-scale replica of the on-premises environment in the AWS cloud. This allows for thorough testing and validation of the migration plan before it's applied to the entire production environment. Here are the benefits of prototyping:

- Identifying and resolving potential issues early in the migration process
- Validating the migration plan and ensuring it aligns with the organization's requirements
- Gaining hands-on experience with AWS services and tools

Testing is an ongoing process that encompasses both functional and non-functional testing. Functional testing verifies that the migrated applications are working as expected, while non-functional testing assesses aspects such as performance, security, and scalability. Here are some of the benefits of testing:

- You can ensure that the migrated applications are fully compatible with the AWS environment
- You can identify and address performance bottlenecks or security vulnerabilities
- You can optimize the applications for maximum efficiency and scalability in the cloud

By incorporating prototyping and testing into the on-premises to AWS cloud migration process, organizations can minimize risks, enhance the migration's success rate, and achieve a smooth transition to the cloud.

How to do it...

Planning for prototyping and testing from on-premises to AWS involves several key steps. Here's how to approach it:

1. **Define the objective**:

   - **Understand the primary goals of the migration**: Performance enhancement, cost-saving, scalability, and so on
   - **Clearly state what you hope to achieve with the prototype**: Validate migration tools, test performance, ensure data integrity, and so on

2. **Pick a prototype scope**:

   - **Representative subset**: Choose a subset of your systems/data that's a good representative of the whole. This should include various data types, applications, and workloads.

   - **Complexity**: Include both simple and complex components to ensure a comprehensive test.

   - **Size**: The dataset should be large enough to simulate real-world scenarios but not so vast that it becomes cumbersome.

3. **Develop a prototype plan**:

   - **Setup**: Define the tools (for example, AWS SCT and DMS) and resources needed

   - **Execution**: Outline the step-by-step process of migration for the prototype

   - **Validation**: Describe how you'll validate the success of the migration prototype

4. **Execute the prototype migration**:

   - **Migrate data**: Use the chosen tools to migrate the data

   - **Configure environment**: Ensure the target environment has been set up similarly to the planned final setup

   - **Document**: Keep detailed notes of any challenges or issues that you faced

5. **Test the prototype**:

   - **Functionality tests**: Ensure all applications and services are working as expected post-migration

   - **Performance tests**: Compare the performance of the prototype system to the original and your performance goals

   - **Security and compliance tests**: Ensure that data remains secure and that all compliance requirements are met post-migration

   - **Failover and recovery tests**: Check the reliability and resilience of the new setup

6. **Gather feedback**: Involve end users and stakeholders. Their feedback on application performance, accessibility, and any potential issues is invaluable.

7. **Refine the migration process**: Based on feedback and test results, refine your migration strategy. Address any issues or challenges that were noted during the prototyping phase.

8. **Plan for mass migration**:

   - **Iterate**: If major issues are found during the prototype phase, consider running additional prototypes or tests before committing to a mass migration.

   - **Scale up**: Once confident in the prototype's success, plan out the migration for the entire system. Use the prototype as a blueprint.

   - **Back up**: Ensure you have backups of everything before the mass migration.

Here are a few best practices:

- **Iterative approach**: It's okay to run multiple prototypes if necessary. It's better to identify issues now than during the mass migration.

- **Stakeholder communication**: Keep all stakeholders in the loop. Their insights can be invaluable and ensure everyone is aligned on expectations.

- **Documentation**: Maintain detailed documentation throughout the prototyping and testing phase. This will be invaluable during the mass migration.

- **Safety first**: Always ensure data integrity and security. Any signs of data loss or potential breaches should be addressed immediately.

Here are a few key takeaways:

- **Prototyping is crucial**: It offers a glimpse into the potential challenges and successes of a mass migration

- **Detailed planning**: Detailed and careful planning for the prototype phase increases the chances of a successful mass migration

- **Feedback and iteration**: Using feedback and iterating on the migration strategy ensures fewer issues during the actual migration

By following this guide, you'll be well-prepared to tackle the challenges of migration with a well-tested prototype, minimizing risks and ensuring a smoother transition.

Converting ETL processes with big data frameworks

As the volume and complexity of data continue to grow, traditional ETL processes are struggling to keep pace. Big data frameworks, such as Apache Hadoop, Apache Spark, and AWS, offer a powerful solution for migrating and managing big data workloads, enabling organizations to process, analyze, and extract valuable insights effectively from their vast data repositories.

Getting ready

Let's discover how AWS can help you overcome the limitations of traditional ETL processes and unlock new possibilities for data analysis:

- **Challenges of traditional ETL in big data**: Traditional ETL processes face several limitations in handling the massive scale and complexity of big data:

  - **Scalability**: Traditional ETL tools aren't designed to handle the massive scale of big data, leading to performance bottlenecks and slow processing times

  - **Flexibility**: Traditional ETL processes are often rigid and inflexible, making it difficult to adapt to the ever-changing nature of big data

  - **Real-time processing**: Traditional ETL processes are batch-oriented, making it challenging to handle real-time data streams and perform real-time analytics

- **Introducing big data frameworks for ETL migration**: Big data frameworks such as Apache Hadoop, Apache Spark, and AWS offer several advantages for migrating ETL processes:

  - **Scalability**: Big data frameworks are designed to handle massive datasets horizontally by distributing data across multiple nodes, enabling scalability and parallel processing

  - **Flexibility**: Big data frameworks provide flexible data processing capabilities, allowing for data ingestion, transformation, and analysis in a variety of formats and structures

  - **Real-time processing**: Big data frameworks enable real-time data processing and analytics, making it possible to respond to data changes and events in real time

- **Migrating ETL processes to big data frameworks on AWS**: Converting ETL processes into a big data framework often involves transitioning from traditional, monolithic ETL tools to scalable, distributed processing systems, such as those provided by big data platforms. The goal is to handle larger volumes of data more efficiently, ensure scalability, and reduce processing times.

How to do it...

When migrating ETL processes to a big data framework, you'd typically follow these steps:

1. **Assessment and planning**:

   - **Identify current ETL processes**: Understand current data sources, transformations, and load processes.

   - **Determine volume and velocity**: Estimate data volume and the rate of incoming data to choose the right big data tools.

   - **Set goals**: What do you want to achieve with the migration? It could be faster processing, handling larger datasets, cost savings, and so on.

2. **Choose a big data framework**:

   - **Apache Hadoop**: An open source framework for distributed storage and processing. Its ecosystem (for example, Hive for SQL-like operations and Pig for scripting) can aid in ETL processes.

   - **Apache Spark**: It offers fast, in-memory data processing and is particularly suited for ETL tasks. It supports various languages, including Python, Scala, and Java.

3. **Redesign ETL for scalability**:

   - **Parallel processing**: Instead of sequential processing, design ETL jobs so that they run in parallel, distributing the workload across clusters

   - **Optimize transformations**: Some transformations might be optimized or restructured to suit the distributed nature of big data frameworks

   - **Incremental loads**: Rather than full loads, consider incremental approaches, processing only new or changed data

4. **Data ingestion tools**: Use tools such as Apache Kafka for real-time data streaming or Apache Flume and Sqoop for batch data ingestion from various sources.

5. **Data transformation**:

   - **Leverage big data tools**: Use Spark's DataFrame or Dataset API for transformations. If you're using Hadoop, tools such as Hive and Pig can help.

   - **User-defined functions (UDFs)**: For complex transformations, you can write UDFs tailored to your needs.

6. **Data loading**:

   - **Storage options**: Depending on the nature and usage of the data, choose storage options such as **Hadoop Distributed File System (HDFS)**, cloud storage (for example, Amazon S3), or NoSQL databases

   - **Batch versus real time**: Depending on your needs, you can load data in batches (using tools such as Sqoop) or in real time (using Kafka or Spark Streaming)

7. **Optimization and performance tuning**:

   - **Tune cluster configuration**: Adjust configurations such as memory allocation, number of executors, and so on for optimal performance

   - **Partitioning and bucketing**: Partitioning divides data into subsets, whereas bucketing divides data based on column values, optimizing query performance

8. **Testing and validation**:

- **End-to-end testing**: Ensure that the new ETL process captures data correctly, transforms it as expected, and loads it without issues

- **Performance testing**: Confirm that the new process meets the performance goals that were set at the beginning

9. **Monitoring and maintenance**:

- **Monitoring tools**: Use tools such as Apache Ambari or Cloudera Manager to monitor health, performance, and failures

- **Automate failover and recovery**: Ensure that if nodes fail or processes crash, the system can recover automatically without data loss

10. **Documentation**: Ensure every part of the new ETL process is well-documented, from data sources and transformations to load processes and optimizations.

Migrating traditional ETL processes to a big data framework is a significant endeavor that can offer immense benefits in terms of scalability, performance, and flexibility. While the process requires careful planning, design, and testing, the result can lead to more efficient and cost-effective data processing and analysis.

Defining and executing your migration process with Hadoop

Defining and executing a Hadoop migration process from on-premises to AWS involves a complex transition that requires careful planning and execution. This process encompasses transferring data, applications, and infrastructure to AWS cloud-based Hadoop clusters. In this recipe, we'll explore the essential steps and considerations involved in successfully migrating your Hadoop ecosystem to AWS, helping you unlock the benefits of cloud scalability, flexibility, and cost-efficiency.

Getting ready

Before embarking on your migration journey, it's essential to lay a strong foundation with the following prerequisites and technical requirements:

- **Prerequisites**:

- **Clear objectives**: Define your migration goals and expected outcomes. What are you trying to achieve with this migration (cost savings, improved scalability, increased agility, and so on)?

- **Stakeholder alignment**: Ensure buy-in and collaboration from key stakeholders across IT, business, and management.

- **AWS account**: Ensure you have an active AWS account with the necessary permissions to create and manage resources.

- **Skilled team**: Assemble a team with expertise in AWS services, networking, security, and your existing on-premises infrastructure.

- **Technical requirements**:

  - **Inventory and assessment**: Thoroughly document your current on-premises environment, including servers, applications, databases, storage, and network configurations. Identify dependencies and performance requirements.

  - **Migration strategy**: Choose the most suitable migration approach for each application or workload.

  - **Rehost (lift and shift)**: Migrate applications without major changes, often using tools such as AWS SMS:

    - **Replatform**: Make minor modifications to take advantage of cloud-native features

    - **Refactor/rearchitect**: Redesign applications so that you can fully leverage AWS services and optimize for the cloud

  - **Connectivity**: Establish secure and reliable network connectivity between your on-premises environment and AWS using options such as AWS Direct Connect or a VPN.

  - **Migration tools**: Select and configure appropriate AWS migration tools, such as AWS Migration Hub, DMS, and Application Discovery Service.

How to do it...

When defining migration processes for a Hadoop framework, follow these steps:

1. **Planning phase**:

   - **Assessment**:

     - Analyze existing Hadoop infrastructure, data size, workloads, and dependencies

     - Evaluate AWS services for compatibility and cost-effectiveness (for example, EMR, S3, EC2, and Glue)

     - Consider performance, scalability, security, and governance requirements

- **Design**:

    - Choose appropriate AWS services and architecture (for example, multi-cluster EMR, hybrid setup, and so on)

    - Plan data transfer and storage strategies (for example, S3 buckets and data partitioning)

    - Address network connectivity and security measures

    - Determine monitoring and logging tools for post-migration analysis

    - Create a detailed migration timeline and rollback plan

2. **Execution phase**:

    - **AWS environment setup**:

        - Create AWS accounts and configure IAM roles for access control

        - Set up the necessary VPCs, subnets, security groups, and network configurations

        - Provision EMR clusters with appropriate hardware and software configurations

    - **Data migration**:

        - Transfer data from on-premises Hadoop storage to S3 using tools such as DistCp or S3DistCp

        - Optimize transfer speed and validate data integrity during the transfer

    - **Job migration**:

        - Refactor or modify Hadoop jobs so that they run on EMR while considering any syntax or API changes

        - Test modified jobs with sample data in the AWS environment

    - **Testing and validation**:

        - Thoroughly test migrated workloads on EMR to ensure functionality and performance

        - Validate data integrity and consistency post-migration

        - Monitor resource usage and costs for optimization

3. **Post-migration phase**:

    • **Monitoring and optimization**:

        • Continuously monitor EMR clusters for performance, resource usage, and costs

        • Implement cost optimization strategies (for example, spot instances, auto-scaling, and so on)

    • **Maintenance and security**:

        • Apply security patches and updates to AWS services and EMR clusters

        • Regularly review security configurations and access controls

    • **Troubleshooting**: Address any issues that arise while leveraging AWS support and documentation

    • **Data security**: Encrypt sensitive data in transit and at rest while adhering to compliance requirements

    • **Cost optimization**: Leverage AWS cost management tools and explore spot instances and reserved pricing

    • **Performance optimization**: Optimize EMR configurations and consider caching and data partitioning techniques

    • **Governance**: Implement data access control and auditing mechanisms for compliance

---

Note

Every migration is unique, so tailor your approach so that it fits your specific requirements and challenges. Don't hesitate to seek assistance from AWS experts or certified partners if needed. With the right strategy and execution, your AWS migration can be a smooth and successful journey.

Migrating the existing Hadoop security authentication and authorization processes

Migrating your security domain from on-premises to the AWS cloud is a pivotal transformation that can enhance the resilience, scalability, and manageability of your security infrastructure. This process involves transitioning critical security components and protocols to the cloud environment, ensuring that your organization can harness the power of AWS while maintaining robust security measures. In this context, we'll explore the crucial steps and considerations involved in successfully migrating your security domain to AWS, enabling you to safeguard your digital assets effectively in the cloud.

The steps to migrate security authentication and authorization from Hadoop to AWS are as follows:

1. **Understand your current setup**: Identify the security mechanisms currently in place in your Hadoop environment. This could include Kerberos for authentication, Apache Ranger or Apache Sentry for authorization, and encryption methods.

2. **Perform authentication and authorization**: Assess your Hadoop ecosystem's components (such as HDFS, Hive, Spark, and so on) and their security configurations:

   - **Authentication mechanisms**: Identify the methods being used (Kerberos, LDAP, and so on)

   - **Authorization mechanisms**: Determine how permissions are granted and enforced (HDFS ACLs, Ranger, and so on)

   - **Integrations**: Map any external systems that interact with Hadoop security

3. **Choose AWS services**:

   - **IAM**: Centrally manages users, groups, and permissions for AWS services and resources

   - **S3 access points**: Create access points with distinct permissions for S3 buckets, simplifying access management

   - **Lake Formation**: Optionally, create a secure data lake with fine-grained access control and auditing

   - **Encryption**:

     - Understand how encryption is handled in AWS using services such as AWS KMS and AWS Certificate Manager.

     - Plan how to migrate data securely. AWS offers services such as AWS DataSync, AWS Transfer Family, and AWS Snowball for large-scale data migrations.

     - Ensure data is encrypted during transit and at rest in AWS.

4. **Integrate with AWS security services**:

   - Integrate your Hadoop applications with AWS security services. This might require some reconfiguration or code changes.

   - Use AWS IAM for managing users and permissions. You can create IAM roles and policies that align with your existing Hadoop permissions.

5. **Testing and validation**:

   - Thoroughly test the security configurations in a staging environment before moving to production

   - Validate that authentication and authorization work as expected and that data is accessed securely

   - Utilize AWS CloudTrail and AWS Config for monitoring and auditing security configurations

   - Continuously monitor for any security threats or vulnerabilities

6. **Migrate authentication**:

   - **AWS credential providers**: Configure Hadoop so that it uses AWS credentials for S3 access via the following methods:

     - Access keys and secret keys

     - IAM roles for EC2 instances

     - Temporary security credentials

   - **Instance profile credentials provider**: For EC2 instances, use IAM roles for authentication

7. **Migrate authorization**:

   - **IAM policies**: Define permissions for S3 buckets and objects using IAM policies

   - **S3 access points**: Simplify permission management with access points for specific use cases

   - **Lake Formation**: For a data lake, manage permissions centrally with Lake Formation

8. **Map existing permissions**:

   - Translate existing Hadoop permissions into IAM policies or S3 access points

   - Consider using tools such as Apache Ranger for policy management across environments

9. **Migrate data**:

   - Transfer data to S3, preserving existing permissions or mapping them to IAM policies

   - Encrypt sensitive data at rest and in transit

10. **Others**:

- **Encryption**: Use S3 encryption features for data protection

- **Auditing**: Enable AWS CloudTrail for logging and auditing of S3 access

- **Third-party tools**: Consider tools for managing authentication and authorization across Hadoop and AWS (for example, Apache Ranger and Cloudera Navigator)

- **Access patterns**: Analyze access patterns to optimize S3 access policies and access points

- **Compliance requirements**: Adhere to regulatory or industry-specific compliance requirements

Getting ready

To ensure a successful migration, it's essential to understand the technical and security requirements outlined here.

Thoroughly document your on-premises Hadoop security setup, including the following aspects:

- **Authentication methods**: How users and services authenticate (for example, Kerberos, LDAP, and Active Directory).

- **Authorization mechanisms**: How access to data and resources is controlled (for example, Sentry and Ranger).

- **Data encryption**: Any encryption methods in place for data at rest and in transit.

- **AWS account and IAM**: An active AWS account with necessary IAM permissions to create and manage resources. Ensure you have familiarity with IAM roles and policies for fine-grained access control.

- **Target AWS services**: Knowledge of relevant AWS services for Hadoop migration (for example, EMR, S3, Glue, and Lake Formation). You must also understand how security integrates with these services.

The technical requirements are as follows:

- **Network connectivity**: Secure network connectivity between your on-premises environment and AWS (for example, Direct Connect and VPN)

- **Migration tools**: Tools such as AWS SCT, DMS, and DataSync for data and metadata transfer

- **The AWS CLI or SDK**: For interacting with AWS services programmatically

- **Backup and recovery**: Ensure proper backups of your on-premises Hadoop security configurations and data

How to do it...

When migrating existing Hadoop security authentication and authorization, you'd typically follow these steps:

1. **Planning and design**:

   - **Map security controls**: Identify the equivalent AWS security services and configurations to replicate your on-premises security model

   - **Authentication**: Consider AWS IAM, AWS SSO, or integrating with existing identity providers using SAML or OpenID Connect

   - **Authorization**: Evaluate AWS Lake Formation for fine-grained access control on data stored in S3, or continue using tools such as Ranger if you're migrating them to EMR

   - **Encryption**: Utilize server-side encryption with Amazon S3 or KMS for data at rest and in transit

   - **Provide a migration strategy**:

     - **Phased approach**: Migrate components gradually, starting with non-critical data and workloads

     - **Parallel testing**: Run your migrated environment in parallel with the on-premises setup for testing and validation purposes

2. **Implementation**:

   - **IAM**:

     - **Create IAM roles**: Create IAM roles for users, services, and applications, granting appropriate permissions

     - **Configure authentication**: Set up authentication mechanisms such as IAM users, federated identities, or SSO integration

   - **AWS Lake Formation**:

     - **Fine-grained control**: Set up Lake Formation to manage permissions on data stored in S3 data lakes

     - **Ranger (optional)**: If you're using Ranger, migrate it to your EMR cluster, and configure your policies

- **Encryption**:

    - **S3 encryption**: Enable server-side encryption (SSE-S3 or SSE-KMS) for S3 buckets that store sensitive data

    - **EMR encryption**: Configure encryption for EBS volumes attached to EMR instances

    - **In-transit encryption**: Ensure secure communication channels between your on-premises environment and AWS

3. **Testing and validation**:

    - **Thorough testing**: Test authentication and authorization mechanisms in the AWS environment

    - **Security review**: Conduct security assessments and penetration testing to identify vulnerabilities

4. **Cutover and go-live**:

    - **User communication**: Communicate the migration plan and any changes in authentication or access procedures to users

    - **Gradual cutover**: Gradually shift workloads to the AWS environment, monitoring them closely for any issues

    - **Rollback plan**: Have a well-defined rollback plan in case of unforeseen problems during the cutover

5. **Post-migration**:

    - **Monitoring and auditing**: Continuously monitor access logs and security events

    - **Refinement**: Refine security configurations and policies based on real-world usage patterns and evolving threats

6. **Decommission on-premises**: Once you're confident in the AWS setup, decommission the on-premises Hadoop security infrastructure.

12

Harnessing the Power of AWS for Seamless Data Warehouse Migration

In the ever-evolving landscape of data management, the migration of your data warehouse to the cloud is an essential step toward unlocking scalability, agility, and cost efficiency. As we embark on this journey, we'll examine the powerful tools that AWS offers to streamline and optimize this transition. In this chapter, we'll explore three key services: the AWS **Schema Conversion Tool** (**SCT**) and AWS **Database Migration Service** (**DMS**), as well as the AWS Snow Family. Each of these tools plays a crucial role in ensuring a seamless and successful migration of your valuable data assets to the AWS cloud.

We will cover the following recipes:

- Creating SCT migration assessment report with AWS SCT
- Extracting Data with AWS DMS
- Live example – migrating an Oracle database from a local laptop to AWS RDS using AWS SCT
- Leveraging AWS Snow Family for large-scale data migration

Technical requirements

Make sure you have an active AWS account. If you don't already have one, sign up for an AWS account at `https://aws.amazon.com/resources/create-account/` before proceeding. You can access the code for this project in the GitHub repository at `https://github.com/PacktPublishing/Data-Engineering-with-AWS-Cookbook/tree/main/Chapter12`.

Creating SCT migration assessment report with AWS SCT

AWS SCT is a valuable utility for automating the migration of database schemas between different database engines. Whether you're moving from an on-premises database to a cloud-based solution or converting between different types of databases, SCT simplifies the process by analyzing the source schema and generating a compatible schema for the target database. SCT automates much of the process. It analyzes the source database schema and converts it into a format compatible with the target database as shown, ensuring a smooth transition with minimal manual intervention.

Understand the data migration journey with the following flow diagram outlining the use of AWS SCT tools:

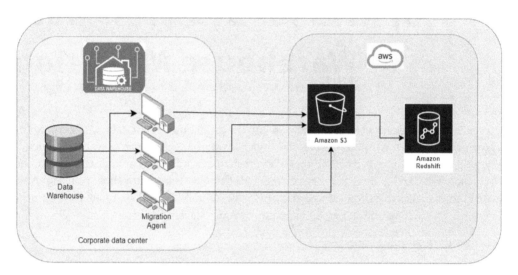

Figure 12.1 – SCT for data migration

This recipe demonstrates how to migrate an on-premises database (**Source**) to AWS (**Target**) using the AWS SCT. We'll focus on converting the database schema and generating comprehensive reports to streamline the migration process.

Figure 12.2 – SCT flow for data migration

When you run AWS SCT, it assesses your source database and provides a detailed report on what can be automatically converted and what needs manual intervention.

Getting ready

Before you begin generating your migration assessment report, ensure you have the following in place:

- **AWS account**:

  - **Active AWS account**: You need an active AWS account to use AWS SCT, as it's an AWS service.

- **Source and target database connectivity**:

  - Ensure that your source and target databases are reachable. You need to configure network settings (for example, VPC and firewall rules) to allow SCT to connect to both the source and target databases.

  - Supported databases include Oracle, Microsoft SQL Server, MySQL, MariaDB, PostgreSQL, Amazon Aurora, and many others.

- **AWS SCT installation**:

  - Download and install the AWS SCT application on your local machine (Windows, Linux, or macOS). It requires **Java Runtime Environment (JRE)** to run.

  - Ensure that your machine meets these system requirements:

    - At least 4 GB of RAM

    - At least 500 MB of disk space

- **AWS IAM role and permissions**:

  - Create or use an AWS **Identity and Access Management (IAM)** role that has the required permissions for SCT to access resources in AWS, such as AWS Glue, Amazon **Relational Database Service (RDS)**, and Amazon S3.

  - The IAM role should have permissions for AWS Glue, Amazon S3, Amazon Redshift, or any other target service you are using.

- **Schema conversion project setup**:

  - Define your source and target database types to configure the correct conversion project within SCT.

  - Make sure SCT is compatible with both the source and target database versions.

- **AWS DMS**: If you are planning to use AWS DMS for data migration after schema conversion, ensure that the AWS DMS endpoints are properly set up, and that your replication instance has access to both the source and target databases.

- **Sufficient disk space for SCT project files**: SCT generates project files during the schema conversion process, which can require significant storage space depending on the size of the schema and data to be converted.

- **Source database privileges**: Ensure that the user accounts for both the source and target databases have the necessary privileges to access schemas, tables, and other database objects for schema conversion.

- **Source and target databases**:

  - **Database details**: Have the connection details (hostname/IP, port, username, and password) ready for both your *source* and *target* databases.

  - **Compatibility**: Ensure that the source and target database platforms are supported by AWS SCT. Consult the AWS SCT documentation (`https://docs.aws.amazon.com/ SchemaConversionTool/latest/userguide/CHAP_Welcome.html`) for the latest compatibility information.

  - **Database access**: Make sure you have appropriate access permissions to the source and target databases to run queries, extract schemas, and potentially modify data (if necessary).

- **Software and drivers**:

  - **AWS SCT installation**: Download and install the latest version of AWS SCT on a machine that has network access to both your source and target databases:

    - **Download for Windows**: `https://s3.amazonaws.com/publicsctdownload/ Windows/aws-schema-conversion-tool-1.0.latest.zip`

    - **Download for Linux**: `https://s3.amazonaws.com/publicsctdownload/ Fedora/aws-schema-conversion-tool-1.0.latest.zip`

    - **Download for Ubuntu**: `https://s3.amazonaws.com/publicsctdownload/ Ubuntu/aws-schema-conversion-tool-1.0.latest.zip`

  - **JDBC drivers**: Install the JDBC drivers for both your source and target database platforms. These drivers are essential for AWS SCT to connect to and interact with your databases:

    - Download the JDBC driver for your Oracle database: `https://www.oracle.com/ jdbc`

    - Download the PostgreSQL driver: `https://jdbc.postgresql.org/download/ postgresql-42.2.19.jar`

    - For other JDBC drivers, please check the database driver page. All supported database drivers are listed in the AWS SCT user guide: `https://docs.aws.amazon. com/SchemaConversionTool/latest/userguide/CHAP_Installing. html#CHAP_Installing.JDBCDrivers`

- **Network connectivity**: Verify that the machine running AWS SCT can connect to both your source and target databases over the network. This may involve adjusting firewall rules or security group settings.

- **Schema understanding**:

  - **Source schema**: Have a good understanding of your source database schema, including tables, columns, relationships, data types, and any custom objects or stored procedures.

  - **Target schema (optional)**: If you're migrating to a specific target database, familiarize yourself with its schema as well to anticipate potential mapping challenges.

How to do it...

With all the necessary setup in place, we are now ready to begin configuring the AWS SCT tool.

Configuring JDBC drivers in AWS SCT

Configuring JDBC drivers in AWS SCT involves several steps. JDBC drivers are essential for AWS SCT to connect to your source and target databases. Here's a general guide to help you configure JDBC drivers in AWS SCT global settings:

1. **Download the JDBC drivers**: First, you need to download the appropriate JDBC drivers for your source and target databases. These drivers are typically available on the database vendors' websites. For example, if you are converting a schema from an Oracle database, you would download the Oracle JDBC driver from Oracle's website.

2. **Launch AWS SCT**: Open the AWS SCT on your computer. Once you have installed and opened SCT, you will see the following screen:

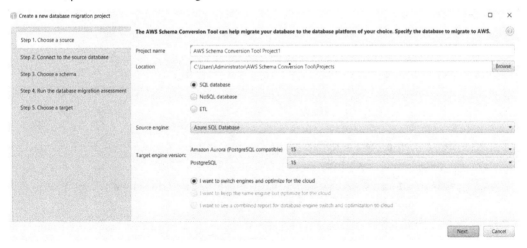

Figure 12.3 – Opening SCT after installation

3. **Access Global Settings**: In the AWS SCT, go to **Global Settings** on the **Settings** menu. This is usually found in the top menu bar. Locate the driver configuration. Inside **Global Settings**, look for a tab or section related to JDBC drivers. This might be labeled something such as **JDBC Drivers**, **Driver Configuration**, or similar.

4. **Add the JDBC drivers**: In the JDBC drivers section, you will likely see different categories for different database types (for example, Oracle, MySQL, and PostgreSQL). Click on the **Add** or **Install** button next to the database type for which you downloaded the JDBC driver. Navigate to the location where you downloaded the JDBC driver file, select it, and then open or install it as prompted.

5. **Verify or test the driver**: After adding the driver, there might be an option to test the driver to ensure that it's correctly installed and functional. Use this feature to verify the setup.

6. **Save the Configuration**: Once the drivers are added and tested, save your configuration settings.

7. **Restart AWS SCT (if required)**: In some cases, you may need to restart AWS SCT for the changes to take effect.

8. **Proceed with database connection**: Now, when setting up a new connection to a database in AWS SCT, it should use the JDBC driver you've configured.

Checking the compatibility

Ensure the JDBC driver version is compatible with your version of AWS SCT. Use AWS SCT documentation and your database's documentation for any specific instructions or requirements.

---

9. **Add source and target connections**: Adding source and target connections in AWS SCT is a critical step in the database migration process. This allows AWS SCT to analyze your source database and create a corresponding schema in your target database. Here's how you can add these connections:

I. **Adding a source connection**: Open AWS SCT and launch it on your computer:

Figure 12.4 – AWS SCT

i.   **Create a new project or open an existing one**: If you haven't already, create a new project by going to **File | New project** or open an existing project.

ii. **Initiate adding a source connection**: Go to **Database | Add source** or simply click on the **Add source** button if it's available on your toolbar.

iii. **Choose the source database type**: Select the type of your source database (for example, Oracle, Microsoft SQL Server, and MySQL) from the list provided.

iv. **Enter connection details**: Fill in the necessary details such as server name, port, database name, user, and password. These details are specific to your database environment.

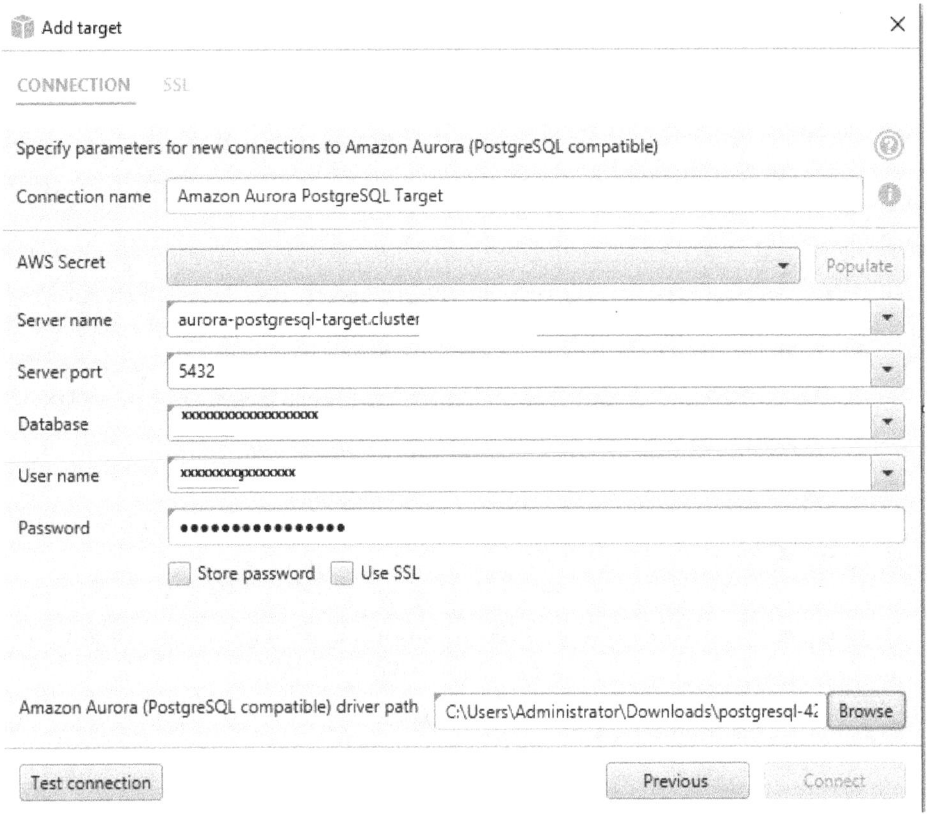

Figure 12.5 – Adding data source in SCT

Once you add the data source, it will be visible in the SCT editor.

- **Test the connection**: Click on the **Test connection** button to ensure that AWS SCT can successfully connect to your source database. Troubleshoot any connection issues if the test fails.

- **Finish and save**: Once the connection is successful, click on **OK** or **Save** to store the source database connection.

II. **Adding a target connection**:

i. **Initiate adding a target connection**: Go to **Database** | **Add target** or use the **Add target** button.

ii. **Select the target database type**: Choose the type of your target database. This could be an AWS service such as Amazon RDS, Amazon Redshift, Aurora, or another database type.

iii. **Input target connection details**: Just like with the source, input the required details for your target database: server name, port, database name, user, and password.

iv. **Test the target connection**: Use the **Test connection** feature to verify connectivity to the target database.

v. **Save the target connection**: Once the test is successful, save the connection by clicking on **Save**.

> **Note**
>
> Make sure you meet the following conditions:
>
> - **JDBC drivers**: Ensure you have the correct JDBC drivers installed for both the source and target databases before attempting to connect.
>
> - **Network accessibility**: Make sure that your source and target databases are accessible from where AWS SCT is running. This might involve configuring network settings or VPNs.
>
> - **Security and permissions**: Verify that the user credentials used for both source and target have the necessary permissions for the migration process.
>
> - **Firewall and security groups**: Adjust firewall settings or security groups as needed to allow AWS SCT to access the databases.

Once you have successfully added both the source and target connections, you can proceed with the schema conversion process and any other migration activities using AWS SCT.

Mapping source and target schemas

Mapping source and target schemas in the AWS SCT is a crucial step in the database migration process. This process involves aligning the schemas of your source database with those in your target database, which helps ensure a smooth and accurate transfer of data. Here's how you can map source and target schemas in AWS SCT:

1. **Open your project in AWS SCT**: Start by opening your project in the AWS SCT where you have already established the source and target connections.

    I. **Access the schema mapping interface**: Look for an option to view or manage your schemas, often found under a menu labeled something such as **Schema** or **Database**.

    II. **Select your source schema**: In the schema view, you'll see a list of schemas from your source database. Select the schema you want to migrate to.

    III. **Initiate the mapping process**: Right-click on the selected schema and look for an option to map it to a target schema. This might be labeled as **Map to target** or something similar.

IV.   **Choose the corresponding target schema**:

    i.   A list of available target schemas (from your target database) will be displayed. Select the schema in the target database that corresponds to or will host the schema from the source database.

    ii.   If the target schema does not exist, you may have an option to create a new schema.

V.   **Configure schema mapping options**: Depending on the complexity of your database, you might have additional options to configure during the mapping. This can include data type mappings, key conversions, and so on.

    i.   **Review and adjust mappings**: Carefully review the proposed schema mappings. AWS SCT might provide suggestions or automatic mappings, but it's essential to verify that these are correct and adjust as necessary.

    ii.   **Apply and save the mappings**: Once you are satisfied with the mappings, apply and save them.

    iii. **Repeat for additional schemas**: If you have multiple schemas to migrate, repeat this process for each one.

2. **Generate an assessment report**:

    I.   In AWS SCT, go to the **View** menu and select **Main view**.

    II.   In the left panel, choose the schema objects you want to include in the report. You can select individual objects, schemas, or the entire database.

    III. Go to the **View** menu and select **Assessment Report view**.

The report will provide a summary of the following:

- **Conversion success**: Objects converted successfully
- **Action items**: Objects requiring manual intervention or code changes
- **Complexity estimates**: Effort levels for addressing action items
- **Recommendations**: Suggestions for optimization and best practices

3. **Save the report**: You have the following options to save the report:

- **Export options**: You can save the assessment report as a PDF or a CSV file
- **PDF format**: This provides a comprehensive, formatted report suitable for sharing
- **CSV format**: This allows you to analyze the report data in a spreadsheet

4. **Review the assessment report**: In a multi-server project, the assessment report provides consolidated views of all the mapped schemas and targets from a single source database.

The following screenshot illustrates a sample assessment report from a multi-server project. This report consolidates information from a single source database, mapping various schemas to different target databases such as Aurora PostgreSQL and Amazon Redshift. The summary section offers a comprehensive overview of all target databases, allowing you to quickly select a specific server or scroll through detailed information for each.

Executive summary

Target Platform	Auto or minimal changes			Complex actions			
	Storage objects	Code objects	Conversion actions	Storage objects		Code objects	
				Objects count	Conversion actions	Objects count	Conversion actions
Amazon RDS for MySQL	54 (98%)	21 (17%)	3,508	1 (2%)	1	102 (83%)	1,077
Amazon Aurora (MySQL compatible)	54 (98%)	21 (17%)	4,921	1 (2%)	1	102 (83%)	1,385
Amazon RDS for PostgreSQL	54 (98%)	33 (27%)	5,493	1 (2%)	0	90 (73%)	65
Amazon Aurora (PostgreSQL compatible)	54 (98%)	33 (27%)	5,493	1 (2%)	0	90 (73%)	65
Amazon RDS for MariaDB	54 (98%)	34 (28%)	4,747	1 (2%)	1	89 (72%)	1,079

We completed the analysis of your Oracle source database and estimate that 98% of the database storage objects and 17% of database code objects can be converted automatically or with minimal changes if you select Amazon RDS for MySQL as your migration target. Database storage objects include schemas, tables, table constraints, indexes, types, collection types, sequences, synonyms, view-constraints, clusters and database links. Database code objects include triggers, views, materialized views, materialized view logs, procedures, functions, packages, package constants, package cursors, package exceptions, package variables, package functions, package procedures, package types, package collection types, scheduler-jobs, scheduler-programs, scheduler-schedules and queuing-tables. Based on the source code syntax analysis, we estimate 56% (based on # lines of code) of your code can be converted to Amazon RDS for MySQL automatically. To complete the migration, we recommend 4,586 conversion action(s) ranging from simple tasks to medium-complexity actions to complex conversion actions.

If you select Amazon Aurora (MySQL compatible) as your migration target, we estimate that 98% of the database storage objects and 17% of database code objects can be converted automatically or with minimal changes. Based on the syntax analysis we estimate that 27% of your entire database schema can be converted

Figure 12.6 – Viewing report in SCT

The details report is grouped by each target database. You can save this report in CSV and PDF format for viewing outside AWS SCT or sharing with others.

Figure 12.7 – Report in SCT

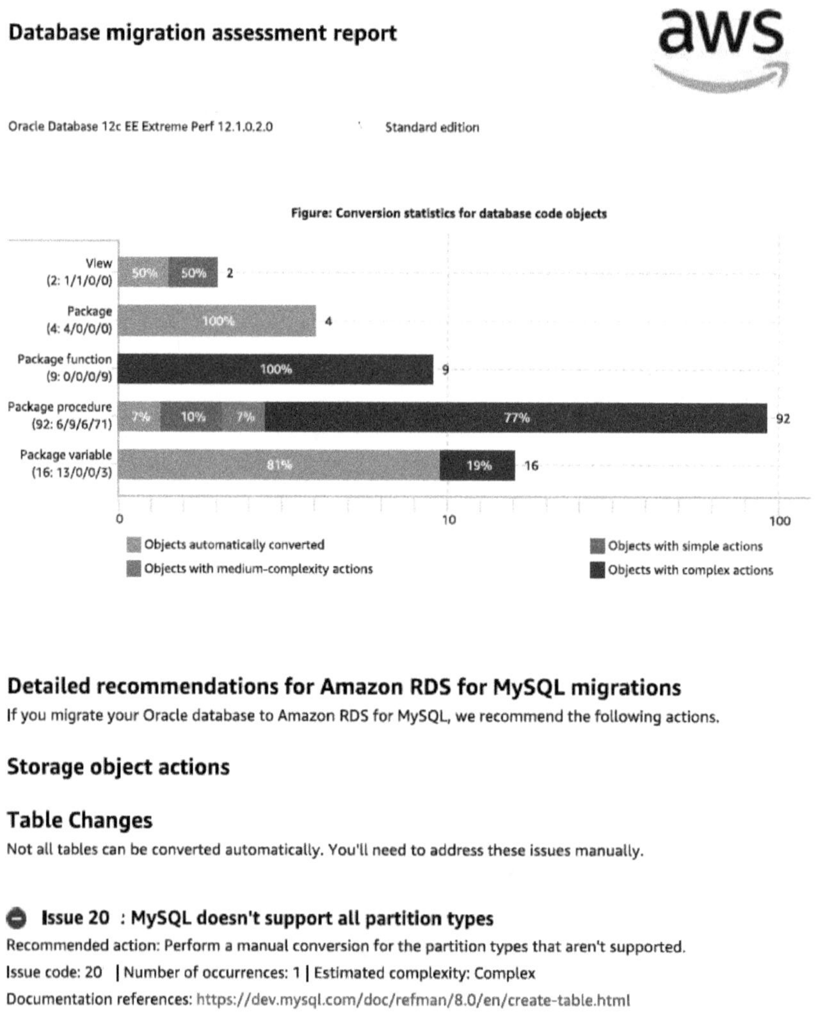

Database migration assessment report

aws

Oracle Database 12c EE Extreme Perf 12.1.0.2.0   Standard edition

Figure: Conversion statistics for database code objects

Figure 12.8 – Report details in SCT

5. **Convert schemas**: When you're ready to convert your schemas, complete the following steps:

    I.   Select the desired objects from the source tree (right-click).

    II.   Choose **Convert schema**.

    III. AWS SCT doesn't apply the converted code to the target database directly. You can view and edit the converted code inside AWS SCT or save it to a SQL script.

IV. To view or edit the converted code inside AWS SCT, choose an object in the source tree. The converted code is displayed in the lower-center panel.

V.   To save the converted code to a SQL script, choose the intended schema on the target tree (right-click), and choose **Save as SQL**.

VI. To apply the converted code to the database, choose the intended schema on the target tree (right-click), and choose **Apply to database**. This option is only available for non-virtual targets.

See also

- *What is the AWS SCT?*: https://docs.aws.amazon.com/SchemaConversionTool/latest/userguide/CHAP_Source.htm

- *Category: AWS SCT*: https://aws.amazon.com/blogs/database/category/database/aws-schema-conversion-tool/

Extracting data with AWS DMS

AWS DMS is a versatile tool designed to simplify database migrations and data transformations. While it's primarily known for database migrations, it's also a powerful tool for extracting data from your source databases and loading it into other destinations, such as Amazon S3, for further processing, analysis, or archival.

In this recipe, you'll learn how to use AWS DMS to extract data and migrate it to the cloud, along with converting database schemas using AWS SCT. DMS handles the extraction and migration of your database data, while SCT helps convert the schema when moving between different database engines. By the end of this recipe, you'll have a streamlined approach to converting database structures and extracting data efficiently, ensuring a smooth transition to AWS services such as Amazon RDS.

In this recipe, we will explore how to extract data from an on-premises database and migrate it to AWS using DMS. This process is divided into two parts:

- AWS SCT for database schema conversion (we already covered it in the previous recipe, *Creating SCT migration assessment report with AWS SCT*)

- AWS DMS for data migration

The following diagram illustrates how on-premises data can be migrated to the AWS cloud using an internet gateway, AWS SCT, and AWS DMS, ultimately landing in an Amazon RDS instance:

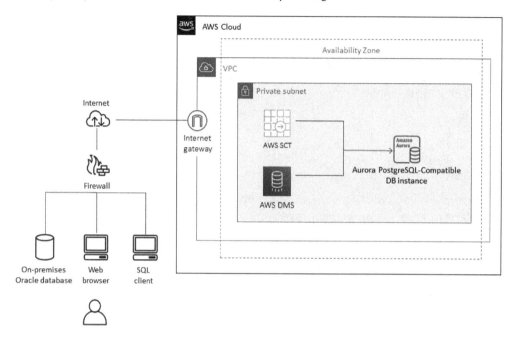

Figure 12.9 – AWS DMS architecture

Let's get started!

Getting ready

Before diving into data extraction with AWS DMS, ensure you have the following in place:

- **Active AWS account:** You'll need an active AWS account with sufficient permissions to create and manage DMS resources. If you don't have an account, you can create one for free at https://aws.amazon.com.

- **Source database:**

  - **Supported database:** Verify that your source database is compatible with AWS DMS. DMS supports a wide range of database engines, including popular options such as MySQL, PostgreSQL, Oracle, Microsoft SQL Server, and more. You can find the full list of supported engines in the AWS DMS documentation (https://docs.aws.amazon.com/dms/).

- **Connectivity**: Ensure that your DMS replication instance can connect to your source database. This may involve adjusting network settings or security groups to allow inbound connections.

- **Credentials**: Have valid database credentials (username and password) with sufficient permissions to read the data you want to extract.

- **Amazon S3 bucket**: Create an Amazon S3 bucket in the same AWS region as your DMS replication instance to store the extracted data.

  - **Permissions**: Ensure that your DMS replication instance has permission to write to the S3 bucket.

- **IAM roles**: To securely interact with your resources, you'll need to create and configure two IAM roles:

  - **DMS replication instance role**: This role allows your DMS replication instance to access your source and target resources. Attach the following managed policies to this role:

    - **AmazonS3FullAccess**: This grants full access to the S3 bucket for storing extracted data.

    - **AmazonDMSReplicationInstance**: This provides the necessary permissions for DMS to operate.

  - **DMS task role**: If you're accessing your source database through a VPC, create a custom IAM role and attach policies that grant access to your VPC and the source database.

How to do it...

1. **Create a DMS replication instance**: An AWS DMS *replication instance* is a key component in the DMS that performs the actual data migration tasks. It provides the necessary compute and memory resources for AWS DMS to migrate data from the source database to the target database. The replication instance handles the migration process, including extracting data, transforming it (if needed), and loading it into the target. The steps are as follows:

   I.   Open the AWS DMS console at `https://us-east-1.console.aws.amazon.com/dms/v2/home?region=us-east-1#firstRun`.

II.   Click on **Create replication instance**.

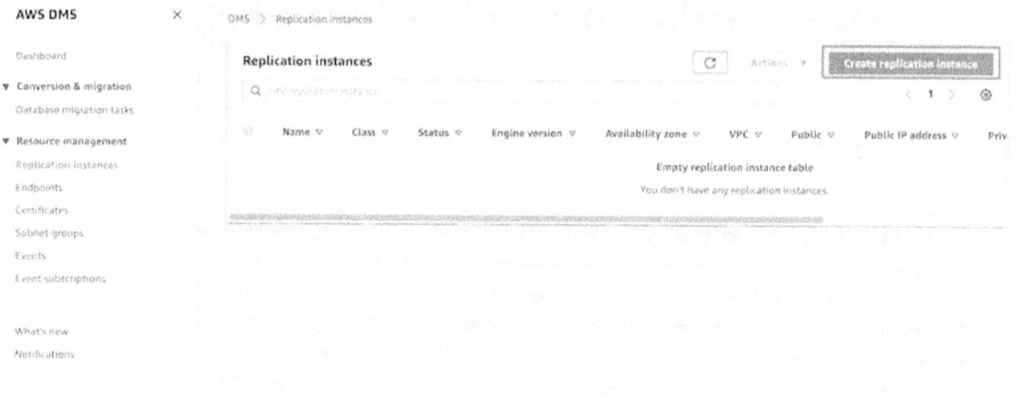

Figure 12.10 – DMS replication instance

III.   Provide a name, choose an instance class based on your data volume, and select your VPC (if applicable):

Parameter	Value
Name	Oracle-to-aurora-replication-instance
Description	Oracle to Aurora DMS replication instance
Instance Class	dms.c4.xlarge
Replication engine version	Latest
VPC	Your VPC id
Allocated storage (GB)	Leave default
Multi-AZ	Unchecked
Publicly accessible	Unchecked

Table 12.1 – Replication instance configuration

IV.   Select the **Publicly accessible** option if your source database is in a different network.

V.   Assign the DMS replication instance role you created.

2. **Create a source endpoint**: AWS DMS *source endpoints* are configurations that define the source database systems from which you want to migrate data. These endpoints specify the details of your on-premises or cloud-based databases, such as the database type, hostname, port number, and authentication credentials. It's easy to create your own source endpoint. In the DMS console, navigate to **Endpoints** and click on **Create endpoint**.

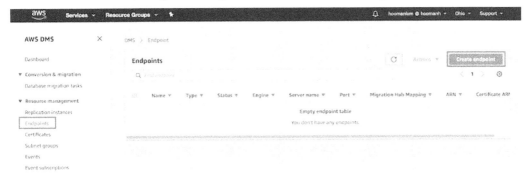

Figure 12.11 – DMS endpoints

3. **Select your source database engine**: When using AWS DMS to migrate data, the first critical step is selecting your *source database engine*. The source database is the system from which you're migrating data to AWS. AWS DMS supports a wide range of both commercial and open source database engines, allowing for migrations between heterogeneous or homogeneous systems. Here's what to do:

    I.   Provide the endpoint identifier, server name or IP address, port, username, and password.

    II. Test the connection to ensure it's successful.

4. **Create target endpoint**: In AWS DMS, the *target endpoint* refers to the destination database where your data will be migrated. It is the system or environment that receives the data from the source database during the migration process. Here are the steps:

    I.   Click on **Create endpoint** again.

    II. Select **S3** as the target endpoint type.

    III. Provide the endpoint identifier and S3 bucket name for storing extracted data.

IV.   Specify the output format (for example, CSV and Parquet) and configure any data format or compression settings:

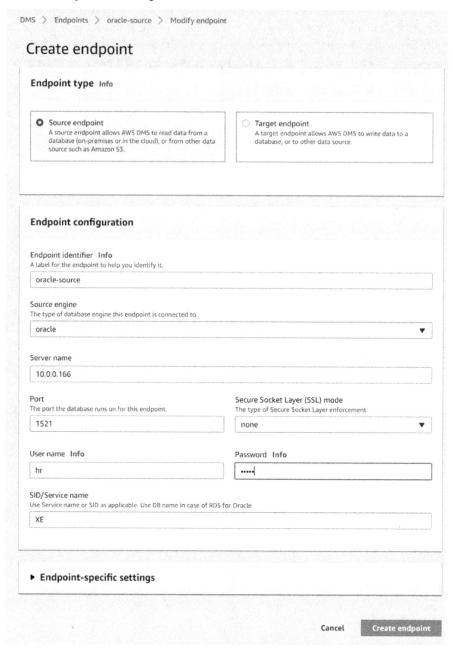

Figure 12.12 – Create endpoints

V.   Once the information has been entered, click on **Run test** under **Test endpoint connection** (optional). When the status turns to successful, click on **Create endpoint**.

5.   **Create DMS task:**

   I.   Go to **Tasks** and click on **Create task**.

   II.   Provide a task identifier and select the replication instance, source endpoint, and target endpoint you created.

   III. Choose **Migrate existing data** as the migration type.

   IV.   Select the tables or schemas you want to extract.

   V.   Optionally, configure task settings such as logging, error handling, and validation.

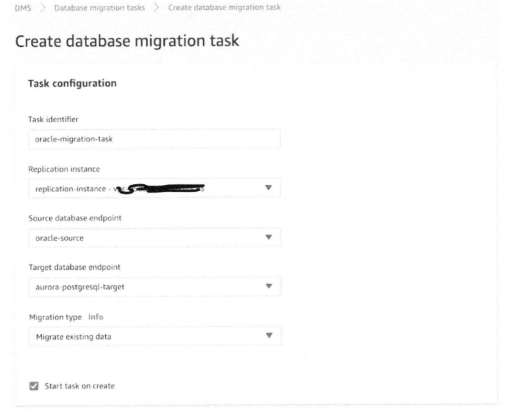

Figure 12.13 – DMS task

6. **Start task**:

   I.   Review the task settings and click on **Start task**.

   II.   AWS DMS will initiate the extraction process, transferring data from your source database to the S3 bucket.

Figure 12.14 – DMS task running

7. **Monitor and verify**:

   I.   Monitor the task progress in the DMS console.

   II.   Check the S3 bucket to ensure data is being extracted as expected.

   III. Review task logs for any errors or warnings.

Type of migrations

There are the following migrations types:

- **Large migrations**: For very large migrations, consider using AWS SCT's replication agent in conjunction with AWS DMS to stream data directly to the target database or a Snowball Edge device

- **Heterogeneous migrations**: If you're migrating between different database types, AWS SCT can assist with schema conversion before using DMS for data extraction

- **Data transformations**: AWS DMS supports various data transformations, such as filtering, masking, and converting data types, which can be applied during extraction

---

Throughout this recipe, we explored key components such as the replication instance, source and target endpoints, and various migration strategies, helping you understand how to successfully plan and execute database migrations to AWS. By leveraging DMS alongside complementary tools like the AWS SCT, you can ensure a smooth transition of both data and schema, while maintaining high availability and data integrity. As you plan your migration, AWS DMS empowers you to scale your data workloads effortlessly, optimizing performance and cost in the cloud.

See also

- *Migrating an On-Premises Oracle Database to Amazon Aurora MySQL*: `https://docs.aws.amazon.com/dms/latest/sbs/chap-on-premoracle2aurora.html`

Live example – migrating an Oracle database from a local laptop to AWS RDS using AWS SCT

In this recipe, we will walk through the process of migrating an Oracle database installed on your local laptop to AWS RDS using AWS SCT. This involves installing Oracle on your laptop, setting up the AWS RDS Oracle instance, using SCT for schema conversion, and completing the migration process with AWS DMS.

Getting ready

Before diving into data extraction with AWS DMS, ensure you have the following in place:

- **AWS account** with the necessary permissions (RDS, DMS, SCT, and networking)
- **Proper networking configurations** to ensure connectivity between your local machine and the AWS RDS instance

How to do it...

Installing the Oracle database on your laptop

1. **Download the Oracle database installer**:

    I.   Visit the Oracle website (`https://www.oracle.com/database/technologies/oracle-database-software-downloads.html`).

    II.   Download the database software (choose the version suitable for your OS such as Windows or macOS).

2. **Install Oracle database**: Follow the installation instructions provided by Oracle to install the database on your laptop. You'll need to do the following:

    I.   Set up a **System Identifier (SID)** for your Oracle instance.

    II.   Configure the listener for database connections (usually port 1521).

    III. Create a database username and password for access (such as system or another DBA account).

3. **Verify Oracle database**: After installation, verify that the Oracle Database is running correctly by logging into the database using Oracle SQL Developer. For detailed steps on how to configure SQL Developer, you can visit `https://docs.oracle.com/en/cloud/paas/exadata-express-cloud/csdbp/connect-sql-developer.html#GUID-00D45398-2BF3-48D5-B0E9-11979D5EAFFC`.

Setting up an AWS RDS Oracle instance

1. **Create an RDS instance**: Log into your AWS Management Console, go to **RDS Dashboard**, and click on **Create Database**.

2. **Select Oracle**: Choose **Oracle** as the database engine. Select the desired version of Oracle, keeping it compatible with your local installation.

3. **Configure instance settings**: Configure your database settings:

   - **DB instance class**: Choose the appropriate instance size (for example, `db.t3.medium`)

   - **Storage**: Choose the storage size (based on your data size)

   - **Master username/password**: Set up the admin credentials for the RDS instance

4. **Security and network**:

   I.   Ensure the RDS instance is accessible by configuring the **VPC** and **Security Group** options to allow incoming traffic on port **1521**.

   II. If you're migrating from your laptop, you'll need to configure the laptop's IP address in the security group to allow access.

5. **Launch RDS instance**: Click on **Create database** and wait for the RDS instance to be created. Note down the endpoint of your RDS Oracle instance for later use.

Installing and setting up AWS SCT

1. **Download and install SCT**:

   I.   Download SCT from the AWS SCT download page (`https://aws.amazon.com/dms/schema-conversion-tool/`) for your OS and install it on your laptop.

   II. You need to follow the detailed steps given in the previous recipe, *Creating SCT migration assessment report with AWS SCT*.

2. **Launch SCT**: Open SCT and create a new project. Go to **File** | **New Project**, name your project, and select **Oracle** as the source database.

3. **Connect to Oracle on Laptop**:

   I.   Add a connection to your local Oracle database:

      • **Endpoint**: localhost

      • **Port**: 1521

      • **Username**: System or another user with DBA privileges

      • **SID**: Use the SID or service name of your local Oracle database

   II.   Test the connection to ensure it is working correctly.

4. **Connect to AWS RDS Oracle**:

   I.   Add a target database connection to your RDS instance:

      • **Endpoint**: Use the RDS endpoint from the AWS console

      • **Port**: 1521

      • **Username/Password**: Use the credentials you set during RDS instance creation

   II.   Test the connection to ensure it is properly configured.

Schema conversion with SCT

1. **Convert the schema**:

   I.   After setting up connections to both your local Oracle and RDS Oracle, SCT will display the schemas in your local Oracle database.

   II.   Select the schema(s) or objects you wish to migrate (tables, views, procedures, and so on).

   III. Right-click and select **Convert schema**. SCT will attempt to convert the schema into a format compatible with the AWS RDS Oracle database.

2. **Analyze compatibility**: Review the compatibility report generated by SCT. If there are issues with converting certain features, SCT will highlight them. You may need to manually edit some parts of the schema if they are incompatible with RDS.

3. **Apply the schema to RDS**:

   I.   Once the conversion is complete, right-click on the converted schema and select **Apply to database**.

   II.   SCT will then apply the converted schema to your AWS RDS Oracle instance, creating tables, indexes, and other schema objects in the cloud.

Data migration with AWS DMS

1. **Set up a DMS replication instance**: Go to the AWS DMS Console and click on **Create replication instance**. This instance serves as the engine for migrating your data between the source (local Oracle) and target (RDS Oracle).

    For all configuration steps, please refer to the detailed steps in the *Extracting data with AWS DMS* recipe.

2. **Create source endpoint**: In the DMS console, create a source endpoint using the connection details of your local Oracle database:

    *   **Endpoint type**: Source

    *   **Database engine**: Oracle

    *   **Server name**: Your laptop's IP address or localhost (if configured)

    *   **Port**: 1521

    *   **Username/Password**: Oracle DB credentials

3. **Create target endpoint**: Create a target endpoint pointing to your AWS RDS Oracle instance:

    *   **Endpoint type**: Target

    *   **Database engine**: Oracle

    *   **Server name**: RDS Endpoint.

    *   **Port**: 1521

    *   **Username/Password**: RDS admin credentials.

4. **Create a migration task**: Create a new migration task in DMS to migrate data from your local Oracle database to your RDS Oracle instance:

    I.   Choose whether to perform a full load or continuous data replication.

    II.   Select the tables or schemas to migrate.

5. **Start the migration task**: Start the task and monitor its progress through the DMS console. DMS will transfer the data from your local Oracle database to the AWS RDS Oracle instance.

Verification and post-migration steps

1. **Verify schema and data**:

    I.   Once the migration task is complete, use an Oracle SQL client such as SQL*Plus or Oracle SQL Developer to log into your AWS RDS Oracle instance.

II.   Verify that all schema objects (tables, indexes, and so on) have been created and that the data has been migrated successfully.

2. **Update application configuration**: If your application was pointing to your local Oracle database, update it to point to the new AWS RDS Oracle endpoint.

3. **Monitoring and tuning**: Monitor your RDS instance performance and fine-tune parameters such as storage auto-scaling, backup policies, and performance metrics using Amazon CloudWatch.

By following these detailed steps, you have successfully migrated an Oracle database from your local laptop to AWS RDS using AWS SCT for schema conversion and AWS DMS for data migration. This approach minimizes downtime and ensures a smooth migration to the cloud.

Leveraging AWS Snow Family for large-scale data migration

In the dynamic landscape of data management and migration, businesses often face the challenge of moving massive volumes of data securely and efficiently. AWS offers a suite of physical devices under the Snow Family, designed to facilitate large-scale data transfers.

This recipe explores the scenarios in which AWS Snow Family products should be considered for data migration. Additionally, we will delve into the Snow Family Large Data Migration Manager, a valuable tool for managing migrations of 500 TB or more.

AWS Snow Family comprises a range of physical devices designed to transfer large amounts of data to and from AWS. These devices include Snowcone, Snowball Edge, and Snowmobile, each tailored for different scales and types of data migration.

Here's a comparison of the key features of Snowball, Snowball Edge, and Snowmobile:

- **Snowball**:
  - **Capacity**: 80 TB
  - **Networking**: 10 GB+ network connection
  - **Ideal for**: Medium-sized data transfers

- **Snowball Edge**:
  - **Capacity**: 100 TB
  - **Networking**: 10 GB+ network connection
  - **Ideal for**: Medium to large-scale data transfers and edge computing

- **Snowmobile**:

  - **Capacity**: Exabyte-scale
  - **Ideal for**: Extremely large data transfers

The Snow Family Large Data Migration Manager as a service

This service simplifies the process of migrating petabytes of data from your on-premises data centers to AWS. It eliminates the need for manual tracking and provides a centralized platform to manage multiple Snowball Edge or Snowmobile jobs simultaneously.

Let's talk about choosing the right device:

- **Consider data size**: Choose Snowball for medium-sized data, Snowball Edge for medium to large, and Snowmobile for exabyte-scale
- **Evaluate edge computing requirements**: Snowball Edge offers both storage and compute capabilities
- **Prioritize security**: Snowmobile provides the highest level of security

By understanding these differences, you can select the most appropriate Snow Family device for your data transfer needs.

A typical Snowball import job flow is as follows:

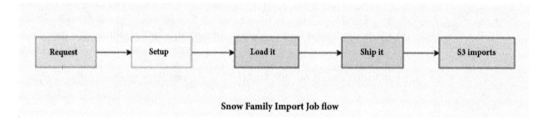

Snow Family Import Job flow

Figure 12.15 – Snow family job flow

Let's walk through the process of ordering AWS Snow Family devices:

1. **Request**: In the AWS web console, order one or more snowballs to your location.
2. **Setup**:

   I.   Install the Snowball client app on a workstation.

   II.   Do an import speed test.

   III. Connect the Snowball to your local network.

3. **Load it**:

   I.   Download the job manifest file from the AWS web console.

   II.   Connect the Snowball client to the snowball device.

   III. Start the import process.

4. **Ship it**:

   I.   Power off and close the snowball device.

   II.   Arrange for UPS pickup. Shipping is prepaid.

   III. UPS provides a tracking number.

5. **S3 import**:

   I.   AWS loads the Snowball contents into your S3 bucket.

   II.   S3 access is controlled via the IAM role and bucket policy.

Substeps involved in using the Large Data Migration Manager include the following:

1. **Create a data migration plan**:

   I.   Access the AWS Snow Family Management Console.

   II.   Provide details about your data migration goals, including the following:

   - Total data size to be migrated
   - Desired migration timeframe
   - Number of Snow devices you plan to use

   The Large Data Migration Manager will analyze your input and generate a projected schedule for your migration project. It will also recommend a job ordering schedule to optimize the process and meet your deadlines.

2. **Create and monitor jobs**:

   I.   Based on the recommended schedule, create individual jobs for each Snow device.

   II.   Specify the source and destination locations for your data.

   III. Track the status of each job in real time through the console. You can monitor progress, identify any issues, and receive notifications about important events.

3. **Manage and track devices:**

   I.   The service provides tools to manage the Snow devices associated with your migration plan.

   II.   Track the physical location of devices, their status (for example, in transit, at AWS, and so on), and any maintenance updates.

4. **Optimize and adjust:**

   I.   The Large Data Migration Manager allows you to adjust your plan as needed.

   II.   If your data size or timeframe changes, you can modify the plan and the service will update the projected schedule and job ordering recommendations.

Getting ready

Let's get started with the AWS Snow Family! First, here's what you'll need to have in place:

- **AWS account:**

  - **Active account**: An active AWS account is required to create and manage Snow Family jobs

  - **Permissions**: Ensure your IAM user or role has the necessary permissions to create Snow Family jobs, interact with S3 buckets, and use other relevant AWS services (for example, KMS for encryption)

- **Data preparation:**

  - **Identify data**: Determine the data you want to transfer and its location (on-premises or in the cloud)

  - **Estimate size**: Estimate the total data size accurately to order the appropriate Snow Family device or cluster

  - **Data format**: Ensure your data is in a format compatible with AWS S3 or other supported data formats

- **Network and logistics:**

  - **Network connectivity**: Ensure your location has adequate network connectivity to support data transfer to/from the Snowball device

  - **Shipping address**: Provide a valid shipping address that can receive and ship large packages

  - **Physical space**: Allocate sufficient space to accommodate the Snowball device, especially for larger Snowball Edge or Snowmobile options

- **Software and tools**:

  - **AWS Snow Family client**: Download and install the appropriate AWS Snow Family client software for your operating system

  - **Data transfer tools**: If you're not using the Snow Family client, ensure you have compatible tools for data transfer (for example, S3Cmd and AWS s3 sync)

How to do it...

1. **Plan your migration**: Before starting your migration, ensure you have the following prepared:

   I.   **Estimate data size**: Measure how much data you need to transfer. If it's 500 TB or more, AWS will recommend using multiple Snowball or Snowball Edge devices, or even Snowmobile for petabyte-scale transfers.

   II. **Identify source data**: Determine where the data is currently stored (on-premises, data center, and so on) and the size of each dataset.

   III. **Decide target location**: Decide where you want the data stored in AWS, typically in Amazon S3 or another AWS service.

   IV. **Choose your Snow Family device**:

        - **Snowball**: Ideal for data transfers ranging from 10 TB to multiple petabytes

        - **Snowball Edge**: Includes both storage and local compute capabilities, making it suitable for edge processing before migration

        - **Snowmobile**: The best option for migrations of **100 PB or more**, typically used in large-scale data center migrations

2. **Create a Large Data Migration Job in AWS Snow Family Console**:

   I.   **Log in to the AWS Snow Family Console**: Go to the AWS Snow Family section in the AWS Management Console.

   II. **Create a new job**: Click on **Create Job** to start the migration process.

III. **Select job type**: Choose between **Import into Amazon S3** (if you're transferring data to AWS) or **Export from Amazon S3** (if you're moving data from AWS to your location).

Figure 12.16 – AWS snow creates new job

IV. **Choose the Snowball device**: Under **Device options**, choose **Snowball Edge** devices, which are designed for large data migrations (such as Snowball Edge Storage Optimized or Compute Optimized).

V. **Select the Large Data Migration option**: In the **Job options** section, you will be asked to specify your use case. Choose **Large Data Migration** to indicate that you're using Snowball to transfer large volumes of data (for data migrations of 500 TB or more, choose **Large Data Migration Manager**).

VI. **Define job details**:

- **S3 bucket**: Select the S3 bucket where you want to transfer the data.

- **Encryption**: Select or create an **AWS KMS key** for encryption. This key ensures that all data is encrypted while in transit.

- **Add IAM**: Create and add IM policy to the job. Please check the sample policy in the following GitHub path: `https://github.com/PacktPublishing/Data-Engineering-with-AWS-Cookbook/blob/main/Chapter12/snowball-policy.json`.

- **Shipping details**: Provide the shipping address for where AWS will deliver the Snowball or Snowball Edge devices.

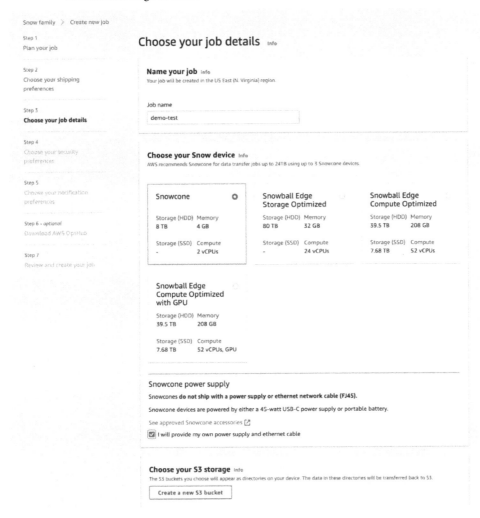

Figure 12.17 – AWS job details

3. **AWS ships the devices to you**:

I.   **Receive Snowball devices**: AWS will ship the required number of Snowball or Snowball Edge devices to your location based on your data size. You can track the shipment via the AWS console.

II. **Unpack and set up devices**: Follow the instructions provided with the devices to set them up in your data center. You will need to plug them into your network.

4. **Load data onto Snowball devices**:

    I.   **Install the AWS Snowball Client**:

        i.   AWS provides a *Snowball Client* that allows you to transfer data to the Snowball device using a **Command-Line Interface (CLI)**.

        ii.   Download the client from the *AWS Snow Family* documentation (`https://docs.aws.amazon.com/snowball/`).

    II.   **Connect to Snowball**: Use the Snowball Client to connect to the device. You will need the job ID and unlock code (provided in the AWS console).

    III.   **Start data transfer**: Begin loading data onto the Snowball using the CLI or APIs. For large datasets, consider segmenting the data to load it in parallel across multiple devices. Here's an example command for transferring data:

```bash
snowball cp /path/to/data snowball://snowball_device/data
--recursive
```

    IV.   **Verify data transfer**: Ensure all data has been transferred successfully. Use checksums or other data validation techniques to verify data integrity.

5. **Ship devices back to AWS**

    I.   **Pack the Snowball devices**: After completing the data transfer, pack the devices securely using the provided packaging materials.

    II.   **Return shipment**: AWS provides pre-paid shipping labels. Attach them to the Snowball devices and schedule a pickup or drop them off at the specified shipping carrier.

    III.   **Track the return**: Track the shipment via the AWS console to monitor the device's return to AWS data centers.

6. **Data import to Amazon S3**:

    I.   **AWS receives the devices**: Once AWS receives the devices, they will automatically begin transferring the data to the specified Amazon S3 bucket.

    II.   **Monitor progress**: Use the AWS Snow Family Console to monitor the progress of the data transfer. You can check whether the import is complete or whether any issues have occurred.

    III.   **Validation**: AWS will validate the integrity of the data and notify you once the transfer is successful.

7. **Post-migration operations**

I. **Access data in Amazon S3**: Once the data transfer is complete, you can access your data in Amazon S3. This data can now be used for further processing, analysis, or storage.

II. **Edge compute results (optional)**: If you used Snowball Edge for edge computing, you can also access the results of any compute tasks that were performed locally during the migration.

III. **Data life cycle management**: Configure S3 storage classes, life cycle policies, and versioning to optimize your storage costs.

Migrating *500 TB or more* of data using the AWS Snow Family is a secure, efficient way to handle large-scale data migrations where network transfer is impractical. By leveraging the Large Data Migration Manager, you can streamline the process of managing multiple Snowball devices, ensuring that your data transfer is seamless and secure. Whether you're moving to Amazon S3, performing edge compute tasks, or managing long-term storage, AWS Snow Family provides the tools needed to perform large-scale data migrations with minimal impact on your operations.

Key considerations

Here are a few key considerations:

- **Planning**: Accurate data estimation is crucial to avoid ordering too few or too many devices

- **Data preparation**: Ensure your data is organized and ready for transfer before the devices arrive

- **Security**: Use strong encryption keys and IAM roles to protect your data during transit and in the cloud

- **Network**: Adequate network bandwidth is essential for efficient data transfer

- **Tracking**: Use the Large Data Migration Manager's dashboard to monitor the progress of your migration and address any issues promptly

---

See also

- *AWS Snowball Edge Primer*: `https://explore.skillbuilder.aws/learn/course/external/view/elearning/45/aws-snowball-edge-getting-started`

- *AWS Snowball Edge Logistics and Planning*: `https://explore.skillbuilder.aws/learn/course/external/view/elearning/115/aws-snowball-edge-logistics-and-planning`

13

Strategizing Hadoop Migrations – Cost, Data, and Workflow Modernization with AWS

Migrating large-scale data and machine learning processes to cloud platforms such as **Amazon Web Services (AWS)** brings multiple benefits. AWS and similar services offer a wide variety of on-demand computing resources, cost-effective and durable storage solutions, and managed environments that are up to date and user-friendly for big data applications. This setup allows data engineers, developers, data scientists, and IT staff to concentrate on data preparation and insight extraction.

Tools such as Amazon **Elastic MapReduce (EMR)**, AWS Glue, and Amazon **Simple Storage Service (S3)** facilitate the separation and independent scaling of computing and storage resources while ensuring a cohesive, robust, and well-managed environment. This significantly reduces many of the challenges associated with on-premises methods. Adopting this cloud-based approach results in quicker, more flexible, user-friendly, and cost-effective solutions for big data and data lake projects.

In this chapter, we'll guide you through various strategies for migration, covering how to transfer your cluster data, catalog metadata, **extract, transform, and load (ETL)** jobs, and workflow services. You'll also discover methods to incorporate quality checks to ensure your migration is successful.

The recipes we'll explore include the following:

- Calculating total cost of ownership (TCO) using AWS TCO calculators
- Conducting a Hadoop migration assessment using the TCO simulator
- Selecting how to store your data
- Migrating on-premises HDFS data using AWS DataSync
- Migrating the Hive Metastore to AWS
- Migrating and running Apache Oozie workflows on Amazon EMR
- Migrating an Oozie database to the Amazon RDS MySQL
- Setting up networking – establishing a secure connection to your EMR cluster
- Performing a seamless HBase migration to AWS
- Migrating HBase to DynamoDB on AWS

Gaining insight into these areas will equip you with a comprehensive understanding of the crucial factors to consider during cloud migration planning. Additionally, you'll learn about the necessary modifications to adapt your existing applications for a cloud-native architecture.

Technical requirements

Make sure you have an active AWS account. If you don't already have one, sign up for an AWS account at `https://aws.amazon.com/resources/create-account/` before proceeding.

You can access the code for this project in the GitHub repository: `https://github.com/PacktPublishing/Data-Engineering-with-AWS-Cookbook/tree/main/Chapter13`.

Calculating total cost of ownership (TCO) using AWS TCO calculators

Navigating the financial landscape of cloud migration can be complex. AWS **total cost of ownership (TCO)** calculators simplify this process by offering an estimate of your potential cost savings when transitioning from on-premises infrastructure to AWS cloud services. These tools allow you to compare the expenses associated with maintaining your current infrastructure to the projected costs of running similar workloads on AWS.

By factoring in variables such as server costs, storage, networking, software licenses, and even personnel expenses, TCO calculators paint a comprehensive picture of your current IT expenditure. They then contrast this with the potential costs of AWS, considering various pricing models, usage patterns, and optimizations available in the cloud.

While a TCO calculator provides a useful starting point, it's crucial to consider it as an estimate. Your actual costs will depend on your unique configuration and how you use the service and configuration. Nonetheless, these tools serve as a crucial guide in your cloud migration journey, helping you make informed decisions based on potential cost savings and a clearer understanding of the financial implications of moving to AWS.

In this recipe, we aim to guide you through the process of evaluating the cost of migrating an on-premises Hadoop cluster to Amazon EMR using the **AWS TCO calculator**. By following a step-by-step approach, you will learn how to input your current infrastructure details, compare on-premises costs with AWS services, and assess potential savings.

In this migration from an on-premises Hadoop cluster to Amazon EMR, we assume a 10-server Hadoop setup with 200 TB of **Hadoop Distributed File System** (**HDFS**) storage. The EMR cluster is configured with `r5.2xlarge` instances for master, core, and task nodes. Amazon S3 will replace HDFS for data storage, with 200 TB of data transferred during the initial migration. Additionally, 500 GB of **Amazon Elastic Block Store** (EBS) storage is allocated per EMR instance for local storage needs.

Getting ready

To facilitate the migration decision-making process, ensure that you collect the following metrics from your current Hadoop clusters:

- Total count of physical CPUs
- CPU clock speed and core counts
- Total memory capacity
- HDFS storage volume (excluding replication)
- Maximum aggregate network throughput
- Utilization graphs span at least one week for the resources
- **Current Hadoop cluster inventory**: Detailed information on the existing Hadoop cluster, including the following:
  - Hardware specifications (servers, storage, and network)
  - Software versions (Hadoop distribution, related tools, and so on)
  - Current utilization and performance metrics
- **AWS cost estimation tools**: Access to AWS Pricing Calculator, AWS Cost Explorer, or similar tools to estimate the cost of running equivalent workloads on AWS
- **TCO analysis framework**: A methodology or tool to compare on-premises costs (hardware, software, maintenance, and personnel) against estimated AWS costs

How to do it...

Here's a step-by-step guide on how to use the TCO calculator to assess the cost of moving your Hadoop cluster to AWS:

1. **Access the AWS TCO calculator:**

    I.   Navigate to `https://calculator.aws/#/`.

    II. Click on **Create estimate**.

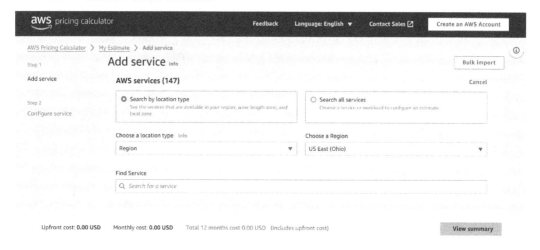

Figure 13.1 – The pricing calculator

2. **Input on-premises Cloudera Hadoop details:**

    I.   **Select the workload:** Choose **Hadoop** (for example, Cloudera or Hortonworks) as the current workload.

    II. **Add server details:** Let's assume you have 10 physical servers running your Cloudera Hadoop cluster. Enter the following server details in the calculator:

    *   **Number of servers:** 10
    *   **CPU cores:** 64 cores per server
    *   **Memory:** 256 GB RAM per server
    *   **Storage:** 20 TB per server (SSD or HDD)
    *   **Utilization:** Choose **High** if your servers are running at high capacity

    III. **Storage details:** Let's assume that the HDFS storage on your Hadoop cluster amounts to 200 TB in total. Specify the type of storage used for Hadoop, whether SSD or HDD.

IV. **Network costs**: Estimate the network bandwidth used for data transfers between your cluster nodes and other systems. In this example, assume a bandwidth usage of 10 Gbps.

3. **Add additional on-premises costs**:

I. **Data center costs**: For electricity, cooling, and physical space, assume the following estimates:

- **Power and cooling**: $100 per month per server
- **Data center space**: $500 per month for rack space

II. **Labor costs**: If you have an IT team managing the Hadoop cluster, input labor costs. For example:

- **IT staff**: 2 full-time employees, each with a salary of $100,000/year
- **Cloudera licensing**: $50,000/year for Cloudera Manager licenses

III. **Hardware refresh costs**: If you refresh hardware every 4-5 years, input an average of $10,000 per server for hardware replacement.

4. **Configure the AWS cloud settings for EMR**:

I. **Amazon EMR cluster setup**: Let's estimate that you will use Amazon EMR with r5.2xlarge instances for your new setup:

- **Number of instances**: You need 10 **Amazon Elastic Compute Cloud (EC2)** instances, similar to your current on-premises setup
- **Instance type**: Select r5.2xlarge (8 vCPUs, 64 GB of RAM)

Add a configuration for one master node, six core nodes, and three task nodes

II. **Amazon S3 for storage**: You plan to move your HDFS storage (200 TB) to Amazon S3. Add the estimated cost of S3 storage based on your data size. In this example, you'll be using standard S3 storage for 200 TB of data.

III. **EBS storage for EMR nodes**: Specify the amount of EBS storage needed for the EMR nodes, assuming 500 GB per instance.

IV. **Network and data transfer costs**: Estimate the costs of transferring your data from on-premises to AWS using AWS Direct Connect or public internet. Assume you'll transfer 200 TB of data initially.

V. **Additional AWS services**: You may use AWS Glue for ETL processing or Amazon CloudWatch for monitoring. Add these costs based on your projected usage.

VI. **Auto-scaling for EMR**: If you plan to use auto-scaling, configure your scaling policy (for example, scaling up to 20 instances during peak times).

Once you've entered all the required information, the TCO calculator will generate the estimated total cost:

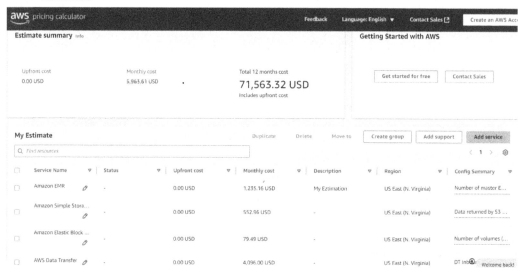

Figure 13.2 – The pricing calculator estimate summary

5. **Run the TCO calculation**: After inputting the required details for both your on-premises Cloudera Hadoop cluster and your Amazon EMR configuration, click on **Calculate TCO**:

    I.   **View the results**: The AWS TCO calculator will provide a comparison of the cost of running your Hadoop cluster on-premises versus running it on Amazon EMR. You will see a breakdown of the costs, such as the following:

         •   Compute costs (for EMR EC2 instances)

         •   Storage costs (for S3 and EBS)

         •   Networking costs (for data transfer)

         •   Labor and licensing savings (since Cloudera licensing and IT staff costs may decrease)

6. **Review and analyze the results**:

    I.   **Estimated savings**: The TCO report will show how much you could potentially save by moving to Amazon EMR. For example, it might show that your on-premises Cloudera cluster costs $500,000/year, while running Amazon EMR could cost $300,000/year, representing a 40% savings.

II. **Download the report**: You can export the results as a PDF or Excel file to share with stakeholders:

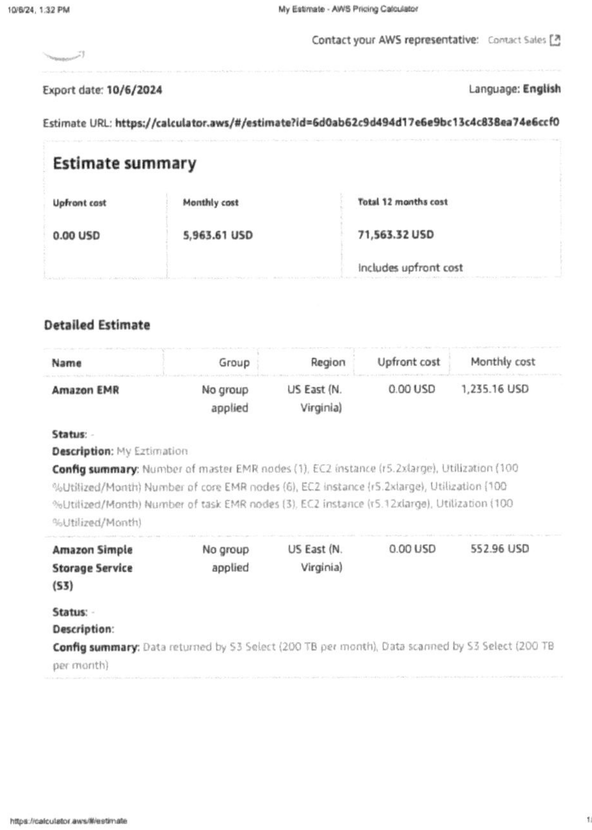

10/6/24, 1:32 PM My Estimate - AWS Pricing Calculator

Contact your AWS representative: Contact Sales

Export date: **10/6/2024** Language: **English**

Estimate URL: **https://calculator.aws/#/estimate?id=6d0ab62c9d494d17e6e9bc13c4c838ea74e6ccf0**

Estimate summary

Upfront cost	Monthly cost	Total 12 months cost
0.00 USD	5,963.61 USD	71,563.32 USD Includes upfront cost

Detailed Estimate

Name	Group	Region	Upfront cost	Monthly cost
Amazon EMR	No group applied	US East (N. Virginia)	0.00 USD	1,235.16 USD

Status: -
Description: My Eztimation
Config summary: Number of master EMR nodes (1), EC2 instance (r5.2xlarge), Utilization (100 %Utilized/Month) Number of core EMR nodes (6), EC2 instance (r5.2xlarge), Utilization (100 %Utilized/Month) Number of task EMR nodes (3), EC2 instance (r5.12xlarge), Utilization (100 %Utilized/Month)

Name	Group	Region	Upfront cost	Monthly cost
Amazon Simple Storage Service (S3)	No group applied	US East (N. Virginia)	0.00 USD	552.96 USD

Status: -
Description:
Config summary: Data returned by S3 Select (200 TB per month), Data scanned by S3 Select (200 TB per month)

https://calculator.aws/#/estimate 1/2

Figure 13.3 – The pricing calculator estimate summary

Summary:

- You have a 10-node on-premises Cloudera Hadoop cluster with 200 TB of data and a total cost of $500,000/year
- Moving to Amazon EMR with r5.2xlarge instances, S3 storage for 200 TB, and auto-scaling configured, the estimated annual cost is around $300,000/year
- This migration could potentially save you around 40% in operational and hardware costs

By following these steps and inputting similar values into the AWS TCO calculator, you can create your own estimates for migrating from Cloudera Hadoop to Amazon EMR. This hands-on approach helps in understanding the potential cost benefits of moving workloads to the cloud.

See also

- *Create and configure an estimate*: https://docs.aws.amazon.com/pricing-calculator/latest/userguide/create-configure-estimate.html#create-estimate

Conducting a Hadoop migration assessment using the TCO simulator

Using AWS's TCO assessment tool to evaluate the cost of migrating your current Hadoop cluster to AWS is an effective approach to understanding and comparing the expenses involved. The TCO assessment helps you get a comprehensive view of the financial impact of the migration. Here's how you can use the AWS TCO assessment for your Hadoop cluster migration:

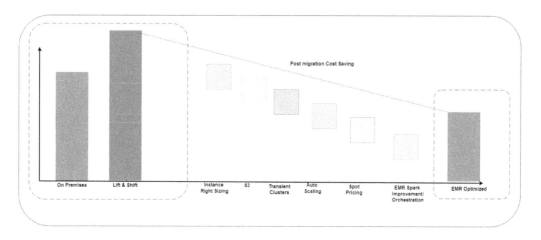

Figure 13.4 – The TCO cost-saving summary

Hadoop to Amazon EMR TCO simulator

The AWS ProServe **Hadoop Migration Delivery Kit** (**HMDK**) TCO tool is designed to assist organizations in migrating from on-premises Hadoop clusters to Amazon EMR. This tool helps in assessing the TCO by simulating the resource usage of future EMR clusters based on historical Hadoop job data. It provides insights into workload patterns, job timelines, and resource utilization, enabling organizations to design cost-effective and optimized EMR clusters.

For a more in-depth analysis of migration, AWS professional services offer a valuable asset: the Hadoop migration assessment TCO tool. This tool is now integrated into the AWS ProServe HMDK. We'll explore this further in this recipe.

This step-by-step recipe provides a comprehensive approach to assessing and optimizing the migration of Hadoop workloads to Amazon EMR using the TCO simulator tool.

Getting ready

Before using the TCO simulator, ensure that you have the following prerequisites in place:

- **Access to Hadoop cluster logs**: You need access to the YARN logs from your existing Hadoop cluster. These logs provide essential data on job execution and resource usage, which the TCO tool will analyze.

- **Python environment**: A Python environment is required to run the log collector and analyzer scripts. The tool supports execution through both native Python environments and Docker containers. You can download it from Python's official website (https://www.python. org/downloads/).

- **AWS account**: An active AWS account with necessary permissions to create and manage EMR clusters and to use services, such as Amazon QuickSight for visualization.

- **Excel**: The final TCO calculations are done using an Excel template with macros, so you need a working installation of Excel that supports macro execution.

Technical requirements are as follows:

- **YARN log collector**: This tool collects logs from the Hadoop YARN Resource Manager. The logs are securely transported using HTTPS and converted into CSV format, which serves as input for further analysis.

- **YARN log analyzer**: After collecting the logs, this analyzer processes the data to extract key metrics such as job timelines, user activity, and resource usage. These metrics are visualized using Amazon QuickSight dashboards.

- **Optimized TCO calculator**: This calculator processes the aggregated log data to estimate the costs of running equivalent workloads on Amazon EMR. The calculator uses inputs such as instance types, storage costs, and resource allocation to provide detailed cost estimates.

How to do it...

1. **Set up the environment**:

    I.   **Clone the repository**: Start by cloning the TCO simulator repository from GitHub using the following command:

    ```
    git clone https://github.com/awslabs/migration-hadoop-to-
    emr-tco-simulator.git
    ```

II.   **Install Python dependencies**: Navigate to the cloned directory and install the required Python packages. If using a virtual environment, you can do the following:

```
cd migration-hadoop-to-emr-tco-simulator
```

- Under the `migration-hadoop-to-emr-tco-simulator` folder, you will find `requirements.txt`:

```
pip install -r requirements.txt
```

III. **Set up Docker (optional)**: If you prefer using Docker, ensure Docker is installed on your system. You can follow the instructions from Docker's official installation guide: `https://www.docker.com/get-started/`.

2. **Collect YARN logs**:

I.   **Run the YARN log collector**: Use the provided script to collect logs from your Hadoop cluster. Ensure you have the necessary permissions to access the YARN ResourceManager API. The command looks like this:

```
python yarn-log-collector.py --resource-manager <YARN_
RESOURCE_MANAGER_ENDPOINT> --output-dir ./logs
```

You can find details about commands and how to run them in the README file at `https://github.com/awslabs/migration-hadoop-to-emr-tco-simulator/tree/main/yarn-log-collector`.

II.   **Convert logs to CSV**: The script will convert the collected JSON logs into CSV format, ready for analysis.

3. **Analyze logs**:

I.   **Run the log analyzer**: Use the `yarn-log-analysis` script to analyze the collected logs. This will generate detailed metrics that you can use to understand your current Hadoop workload.

```
python yarn-log-analysis.py --input-dir ./logs --output-dir
./analysis
```

II.   **Visualize with Amazon QuickSight**: Set up Amazon QuickSight using the provided CloudFormation template to visualize the analysis results.

4. **Run the TCO calculator**:

I.   **Aggregate log data**: Use the `tco-input-generator.py` script to aggregate the log data:

```
python tco-input-generator.py --input-dir ./analysis
--output-file tco-input.csv
```

II.   **Calculate TCO in Excel**: Open the provided Excel template, enable macros, and input your data. Use the green cells to enter parameters such as instance types, data size, and other relevant configurations.

5.   **Interpret results**: Use the insights from the QuickSight dashboards and Excel-based TCO estimates to design your future EMR architecture. Adjust your cluster design based on workload types, peak usage times, and resource needs to optimize for both performance and cost.

See also

- *Amazon EMR pricing*: `https://aws.amazon.com/emr/pricing/`

- *Introducing the AWS ProServe Hadoop Migration Delivery Kit TCO tool*: `https://aws.amazon.com/blogs/big-data/introducing-the-aws-proserve-hadoop-migration-delivery-kit-tco-tool/`

- *Hadoop Migration Assessment (TCO-Simulator)*: `https://github.com/awslabs/migration-hadoop-to-emr-tco-simulator`

Selecting how to store your data

When migrating from an on-premises Hadoop cluster to AWS, one of the crucial decisions to be made is selecting the appropriate storage solution for your data. Amazon S3 and HDFS both offer robust data storage capabilities, but they differ in their architecture, features, and use cases. This recipe will help you navigate this choice by comparing S3 and HDFS, examining their technical requirements, and offering guidance on how to make an informed decision based on your specific needs.

Choosing between Amazon S3 and Hadoop HDFS depends on your specific use case, performance requirements, and long-term goals. Amazon S3 offers unmatched scalability and integration with AWS services, making it ideal for cloud-native workloads and data lakes. HDFS, on the other hand, is well suited for high-throughput big data processing within a Hadoop ecosystem. By carefully evaluating your needs against the capabilities of each storage solution, you can select the most appropriate technology for your environment.

Here's a brief overview of HDFS and S3:

- AWS presents **Amazon S3** as an object storage service. S3 boasts industry-leading attributes, including remarkable scalability, data availability, security, and performance. This comprehensive feature set empowers customers to securely store and safeguard vast volumes of data, catering to a wide spectrum of use cases. Amazon S3 is best for cloud-native applications, multi-tenant environments, and when leveraging other AWS analytics and machine learning services.

- **HDFS** serves as a distributed filesystem specifically crafted for the efficient management of extensive data sets spread across multiple nodes. Within the Apache Hadoop ecosystem, HDFS plays a pivotal role, ensuring robust and high-throughput access to application data while gracefully managing failures. HDFS is ideal for on-premises deployments, traditional data warehousing, and when using a Hadoop-centric processing tool.

When deciding between Amazon S3 and Hadoop HDFS for storing your data, it's crucial to consider various factors, such as data size, access patterns, cost, performance, scalability, and your specific use case requirements. Here's a step-by-step guide to help you make an informed choice.

In an HDFS versus S3 comparison, it's vital to recognize that each storage system possesses its unique strengths and weaknesses, tailored to distinct scenarios and objectives. HDFS excels in environments prioritizing data locality and replication, while S3 stands out due to its exceptional scalability, durability, and seamless integration with various AWS services:

- **Understand the basics of Amazon S3 and Hadoop HDFS:**

  - **Amazon S3:**

    - **Type**: Object storage service

    - **Scalability**: Highly scalable, designed for 99.999999999% durability

    - **Access**: Accessible via HTTP/HTTPS; supports a wide range of APIs

    - **Cost**: Pay-as-you-go pricing; no upfront costs; different storage classes (Standard, Intelligent-Tiering, Glacier) depending on access frequency

    - **Performance**: Optimized for high availability and scalability, with the ability to manage large volumes of data

    - **Hard limits**:

      - **Number of buckets**: You can have up to 100,000 buckets per AWS account

      - **Object size**: Individual objects can be up to 5 TB in size

      - **Number of objects per bucket**: There is no strict limit on the number of objects per bucket, but performance and scalability might be affected by extremely large numbers

    - **Quotas**:

      - **Storage capacity**: The amount of storage you can use depends on your AWS account plan and usage. AWS provides flexible storage options, including Standard, **Infrequent Access (IA)**, Glacier, and Glacier Deep Archive.

      - **Data transfer**: There are limits on the amount of data you can transfer in and out of S3. These limits can vary based on your region and usage patterns.

- **API requests**: The number of API requests you can make to S3 is subject to quotas. These quotas are generally adjusted based on your usage patterns.

- **Hadoop HDFS**:

  - **Type**: Distributed filesystem designed for large-scale data processing

  - **Scalability**: Scales across multiple nodes; provides high throughput

  - **Access**: Typically accessed within a Hadoop ecosystem; supports POSIX-like file operations

  - **Cost**: Costs are tied to the underlying hardware, including storage devices and network infrastructure

  - **Performance**: Optimized for batch processing with high throughput; can be configured for low-latency access depending on the hardware

  - **Hard limits**:

    - **Cluster size**: The number of nodes in your HDFS cluster will determine the overall storage capacity and processing power. As you add more nodes, your HDFS cluster can scale to handle larger datasets.

    - **Data volume**: While HDFS can handle massive amounts of data, there are practical limits based on the hardware resources of your cluster. Factors such as disk space, network bandwidth, and processing power will influence how much data your HDFS cluster can effectively store and process.

    - **File size**: Individual files in HDFS can be very large, but there might be limitations based on the underlying filesystem and hardware.

  - **Quotas**: HDFS doesn't have predefined quotas like AWS S3. However, there are practical considerations and limitations that can affect your HDFS cluster's performance and capacity:

    - **Storage capacity**: The total storage capacity of your HDFS cluster is determined by the combined storage of all nodes. While there's no fixed limit, you might need to add more nodes to accommodate larger datasets.

    - **Network bandwidth**: The network bandwidth between nodes in your HDFS cluster can impact data transfer speeds. If your network is congested, it can affect performance.

    - **Processing power**: The processing power of your HDFS nodes will influence how quickly data can be processed and analyzed. If your workloads are demanding, you might need to upgrade your hardware.

- **Assess your use case requirements**:

  - **Data size and growth**:

    - **Amazon S3**: Ideal for storing vast amounts of data that can scale infinitely without concern for the underlying infrastructure.

    - **HDFS**: Suitable for large datasets that are part of a Hadoop ecosystem. However, scaling requires adding more hardware nodes.

  - **Data access patterns**:

    - **Amazon S3**: Best for scenarios where data is accessed infrequently or requires global access. Suitable for object storage, backups, logs, and data lakes.

    - **HDFS**: Ideal for high-throughput data processing tasks, such as big data analytics, where data is processed in large batches within a cluster.

  - **Performance requirements**:

    - **Amazon S3**: This provides scalable performance for read and write operations but is typically slower than HDFS for high-speed, sequential access required by some big data workloads

    - **HDFS**: This is optimized for sequential reads and writes, making it highly efficient for data processing tasks such as Spark and MapReduce jobs in an EMR Hadoop cluster, where large-scale data is processed in parallel.

  - **Cost considerations**:

    - **Amazon S3**: This is cost-effective for storing large volumes of data with different storage classes to optimize costs based on access patterns.

    - **HDFS**: Costs are higher if you need to manage and scale the physical hardware and associated infrastructure. However, HDFS can be more cost-effective for on-premises deployments if you already have the hardware.

  - **Scalability and management**:

    - **Amazon S3**: Automatically scales to accommodate growing data volumes without manual intervention

    - **HDFS**: This requires manual management to scale by adding more nodes to the cluster, which increases operational complexity

- **Consider data security and compliance**:

  - **Amazon S3 security**:

    - This offers advanced security features such as encryption at rest (**Server-Side Encryption with Amazon S3-Managed Keys (SSE-S3)** and **Server-Side Encryption with AWS Key Management Service-Managed Keys (SSE-KMS)**) and in transit (**Secure Sockets Layer/Transport Layer Security (SSL/TLS)**)

    - This supports fine-grained access control with **Identity and Access Management (IAM)** policies, bucket policies, and **Access Control List (ACLs)**

    - Compliance certifications such as **Health Insurance Portability and Accountability Act (HIPAA)**, **General Data Protection Regulation (GDPR)**, and others make it suitable for storing sensitive data

  - **HDFS security**:

    - This provides encryption at rest using Hadoop's transparent encryption features

    - This supports Kerberos for authentication and Apache Ranger for fine-grained access control

    - Security depends on the underlying environment; additional configurations may be needed for compliance

- **Evaluate integration with your ecosystem**:

  - **Amazon S3**:

    - This integrates seamlessly with AWS services such as AWS Lambda, Amazon Athena, Amazon EMR, and AWS Glue

    - Suitable for building serverless architectures, data lakes, and integrating with cloud-native applications

    - This supports native integration with big data tools via S3 connectors (for example, Spark and Hive)

  - **HDFS**:

    - Tight integration with the Hadoop ecosystem and tools such as Apache Hive, Apache HBase, and Apache Spark

    - Best for environments where Hadoop is central to data processing, and you rely on its ecosystem for analytics

- **Match the solution to your specific use case**:

  - **Use cases for Amazon S3**:

    - **Data lakes**: Store structured and unstructured data for analytics and reporting
    - **Backup and archive**: Cost-effective storage for long-term backups and archival
    - **Web content hosting**: Serve static content such as images, videos, and HTML files globally
    - **Big data analytics**: Store raw data for processing using AWS services such as Amazon EMR or Athena

  - **Use cases for HDFS**:

    - **Big data processing**: Ideal for running Hadoop-based analytics, including MapReduce, Spark jobs, and machine learning workloads, providing scalable and high-throughput storage optimized for processing large datasets in a distributed environment
    - **Batch processing**: High-throughput data processing for ETL jobs and large-scale data transformations
    - **Data warehousing**: Used as a storage layer in Hadoop-based data warehouses

- **Make your decision**:

  - Choose Amazon S3 if the following is the case:

    - You need highly scalable, durable, and globally accessible storage
    - Your data access patterns are varied, and you require flexibility in storage classes
    - You want to minimize infrastructure management and focus on cloud-native solutions

  - Choose HDFS if the following is the case:

    - You are operating within a Hadoop ecosystem and need high-throughput access to large datasets
    - You prefer or require an on-premises solution with control over the underlying hardware
    - Your use case involves large-scale batch processing with specific performance requirements

Getting ready

The technical requirements for this recipe are as follows:

- **AWS account**: An active AWS account to provide and manage S3 buckets and EMR clusters
- **Network connectivity**: Secure network connectivity between your on-premises environment and AWS, if you need to transfer data

- **Data transfer tools**: Tools such as AWS DataSync or S3 Transfer Acceleration for efficient data movement

How to do it...

Here's a step-by-step guide to help you make an informed choice. Let's start with the steps to implement Amazon S3 storage:

1. **Set up Amazon S3 bucket**:

    I.   In the AWS Management Console, navigate to **S3** and create a bucket.

    II. Define the bucket region, configure permissions, and set up policies as per your security requirements.

2. **Transfer data to S3**:

    I.   Use the AWS **Command Line Interface (CLI)** or S3 console to upload data.

    - **Command example**: `aws s3 cp /path/to/data s3://your-bucket-name/ --recursive`

    II. For large-scale data transfer, use AWS DataSync or AWS Snowball to migrate data efficiently from on-premises systems.

3. **Enable life cycle policies**:

    I.   Define storage policies to automatically move older data to cheaper storage classes, such as S3 Glacier, or delete unneeded objects.

    II. Configure versioning and replication as needed for backup and disaster recovery.

4. **Integrate with AWS services**:

    I.   Set up AWS Glue, Athena, or EMR to interact with the data stored in S3 for analysis and processing.

    II. Use Amazon S3 Select to retrieve specific data directly from S3 objects, reducing data transfer and processing overhead.

5. **Monitor and manage storage**:

    I.   Use Amazon CloudWatch and S3 Storage Lens to monitor usage, performance, and storage costs.

    II. Set up alerts for storage thresholds or unexpected data transfers.

The steps to implement HDFS storage on Amazon EMR are as follows:

1. **Set up an EMR cluster with HDFS:**

   I.   In the AWS Management Console, navigate to **EMR** and create a cluster (`https://console.aws.amazon.com/emr`).

   II. Choose **Hadoop** as the framework and configure the cluster with HDFS.

   III. Select your instance types (for example, `r5.2xlarge` for master and core nodes) and configure the number of nodes based on your processing requirements.

2. **Configure HDFS on EMR:**

   I.   HDFS is automatically configured as the default storage for the EMR cluster.

   II. Data stored on HDFS will be local to the cluster and will not persist after cluster termination unless backed up.

3. **Upload data to HDFS:**

   I.   After the cluster is running, **Secure Shell** (**SSH**) into the Master Node.

   II. Use the following command to upload data to HDFS:

   ```bash
   hadoop fs -put /local/data/path /hdfs/path
   ```

   III. For large datasets, use **AWS DataSync** (`https://docs.aws.amazon.com/datasync/`) or **DistCp** (`https://docs.aws.amazon.com/prescriptive-guidance/latest/patterns/migrate-data-from-an-on-premises-hadoop-environment-to-amazon-s3-using-distcp-with-aws-privatelink-for-amazon-s3.html`) to efficiently migrate data to HDFS.

4. **Managing and monitoring HDFS:**

   I.   Use the **Hadoop ResourceManager** and **HDFS NameNode** dashboards in EMR to monitor disk space, replication status, and node health.

   II. Adjust the replication factor for fault tolerance (default is 3):

   ```bash
   hadoop fs -setrep -R 3 /hdfs/path
   ```

5. **Job execution on HDFS:**

   I.   Submit your data processing jobs (for example, MapReduce and Spark) on the EMR cluster using the EMR console or via the command line.

   II. The jobs will read and write data directly from and to HDFS.

6. **Backup and data persistence**: As HDFS data is tied to the EMR cluster, back up your data to Amazon S3 before shutting down the cluster:

```bash
hadoop distcp hdfs:///path s3://your-s3-bucket/
```

By following these action steps, you can successfully implement and manage either HDFS or Amazon S3 based on your storage needs.

See also

- *HDFS configuration*: https://docs.aws.amazon.com/emr/latest/ReleaseGuide/emr-hdfs-config.html

Migrating on-premises HDFS data using AWS DataSync

Migrating large datasets from an on-premises HDFS environment to Amazon S3 can be complex, but AWS DataSync simplifies and accelerates this process. In this recipe, you'll learn how to use AWS DataSync to seamlessly transfer data from Hadoop HDFS to Amazon S3, ensuring a secure and cost-effective migration.

AWS DataSync automates the tasks involved in data transfers, such as managing encryption, handling scripts, optimizing networks, and ensuring data integrity. It supports one-time migrations, ongoing workflows, and automatic replication for disaster recovery, offering transfer speeds up to 10 times faster than open source tools.

AWS DataSync supports the following for HDFS:

- Copying files and folders between Hadoop clusters and AWS storage
- DataSync agents running external to the cluster
- Transferring over internet, Direct Connect, or VPN
- End-to-end data validation
- Incremental transfers, filtering, and scheduling
- Kerberos authentication
- Hadoop in-flight encryption
- Hadoop at-rest encryption **transparent data encryption** (TDE) when using simple authentication

Getting ready

To successfully execute DataSync, the following prerequisites must be met:

- Ensure that you have an AWS account
- Your Hadoop cluster should be operational, and you should have administrative access to it
- Network connectivity between your Hadoop environment and AWS
- **DataSync agent**: Installation and configuration of the DataSync agent on an on-premises server with access to the HDFS cluster
- **Network connectivity**: Ensure network connectivity between the DataSync agent and your AWS VPC
- **S3 bucket**: An S3 bucket to store the migrated data
- **IAM role**: IAM role for DataSync with permissions to access the S3 bucket and perform data transfer operations
- DataSync IAM sample policy

How to do it...

To move data from Hadoop HDFS to Amazon S3 using AWS DataSync, you'll go through a series of steps that involve setting up DataSync, configuring your Hadoop environment, and executing the data transfer. Here are the comprehensive steps to walk you through the process:

1. **Create a DataSync IAM policy**: Create an IAM role with a policy that grants DataSync the necessary permissions to access your S3 bucket. The policy should look something like this:

```
{
  "Version": "2012-10-17",
  "Statement": [
    {
      "Effect": "Allow",
      "Action": [
        "s3:PutObject",
        "s3:GetObject",
        "s3:ListBucket",
        "s3:DeleteObject"
      ],
      "Resource": [
        "arn:aws:s3:::your-s3-bucket-name",
        "arn:aws:s3:::your-s3-bucket-name/*"
      ]
    },
```

```
{
  "Effect": "Allow",
  "Action": [
    "datasync:StartTaskExecution",
    "datasync:ListTasks",
    "datasync:DescribeTaskExecution"
  ],
  "Resource": "*"
}
]
```

2. **Deploy and configure the DataSync agent**: Deploy the DataSync agent as a **virtual machine (VM)** in your environment. The agent is available in various formats, such as **Open Virtual Appliance (OVA)** for VMware, Hyper-V, and AWS EC2.

    I. Open the AWS DataSync console `https://console.aws.amazon.com/datasync/`.

    II. In the left navigation pane, choose **Agents**, and then choose **Create agent**. The following page will open:

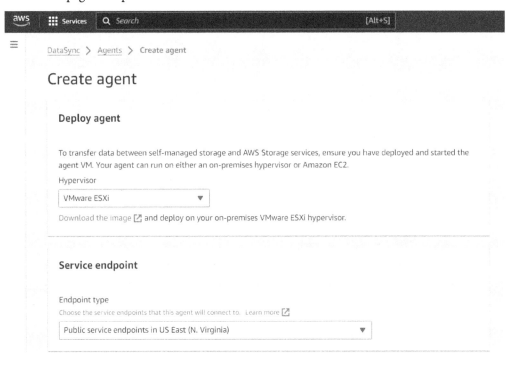

Figure 13.5 – Creating a DataSync agent

3. **Activate the agent**: After installing the agent, an activating agent is needed. To activate the agent and secure the connection between an agent and the DataSync service, several network ports should be opened by the firewall. For example, if your Hadoop network was inside the on-premises firewall, open the outbound traffic (**Transmission Control Protocol (TCP)** ports `1024-1064`) from the DataSync agent to the **Virtual Private Cloud (VPC)** endpoint. You also had to open the TCP port `443` for the entire subnet where the VPC endpoint is located because dynamic IP is assigned to **Elastic Network Interface (ENI)** for data transmission of DataSync. For more detailed network requirements, you can refer to the network requirement document (`https://docs.aws.amazon.com/datasync/latest/userguide/agent-requirements.html`):

    Once deployed, activate the agent using the AWS Management Console by entering the agent's IP address. This step registers the agent with your AWS account.

4. **Configure network settings**: Ensure the agent can connect to both your HDFS cluster and the AWS DataSync service by configuring the necessary network settings, such as **Domain Name System (DNS)** and routing.

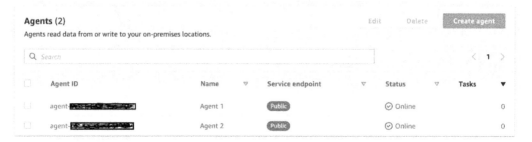

Figure 13.6 – DataSync agent running

5. **Create a DataSync task**:

    I.   **Define source location (HDFS)**:

    i.   In the AWS Management Console, navigate to the DataSync service and create a new task.

    ii. Select HDFS as the source location and enter the required details, such as the HDFS NameNode address and the path to the data you want to migrate.

    II. **Define destination location (Amazon S3)**:

    i.   Select Amazon S3 as the destination location.

    ii. Choose the S3 bucket you created earlier and specify the target directory where the data will be stored.

III.   **Configure data transfer settings**:

    i.   Set transfer options, including bandwidth throttling, task scheduling, and data integrity checks.

    ii.   Enable encryption and configure other security settings to protect your data during transfer.

6. **Start and monitor the data transfer**:

    I.   **Start the DataSync task**:

        i.   Start the task manually or schedule it to run at a specific time.

        ii.   Monitor the progress of the data transfer through the AWS Management Console, where you can view logs and statistics.

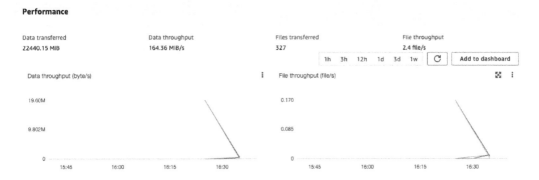

Figure 13.7 – DataSync data transfer monitor

    II.   **Verify data integrity**:

        i.   Once the transfer is complete, verify that all data has been accurately copied to Amazon S3.

        ii.   Check for any errors or discrepancies in the transfer logs and resolve them as needed.

7. **Verify data transfer**:

    I.   After the transfer is complete, verify the data in your S3 bucket to ensure everything is transferred correctly and completely.

    II.   Perform any necessary data validation or integrity checks.

8. **Post-transfer cleanup and optimization**:

    I.   If it was a one-time transfer, you might want to delete the DataSync task or agent.

    II.   For recurring transfers, review the performance and make any necessary adjustments.

9. **Security and compliance**:

   I.   Ensure that your data transfer complies with your organization's data security and compliance policies.

   II. Consider encrypting data both in transit and at rest.

Migrating data from HDFS to Amazon S3 using AWS DataSync is a straightforward process that significantly reduces the complexity and time required for large-scale data transfers. By following the steps outlined in this recipe, you can efficiently and securely move your data to the cloud, taking advantage of Amazon S3's scalability, durability, and integration with other AWS services.

See also

- *AWS DataSync FAQs*: `https://aws.amazon.com/datasync/faqs/`

- *Prerequisites to start running the labs*: `https://cloudone-datamigr.awsworkshop.io/20_prerequisites.html`

Migrating the Hive Metastore to AWS

The **Hive Metastore** is a crucial component within Hadoop, acting as a centralized storehouse for metadata related to Hive tables, schemas, and partitions. When transitioning your Hadoop cluster to the AWS cloud, you can opt to either establish a dedicated Hive Metastore on AWS or utilize the managed AWS Glue Data Catalog.

Migrating to the AWS Glue Data Catalog provides benefits such as schema versioning and efficient integration with Amazon EMR, especially for transient clusters. This migration ensures high availability, fault tolerance, and better data governance.

When it comes to data discovery and management, two data catalogs are among the most popular choices:

- **Hive Metastore:** This repository houses essential information about Hive tables and their underlying data structures, including partition names and data types. Hive, an open source data warehousing and analytics tool built on Hadoop, can be deployed on platforms such as EMR.

- **AWS Glue Data Catalog**: A fully managed service by AWS, the Glue Data Catalog offers a blend of flexibility and reliability. It's particularly well suited for those starting with metastore creation or management, as it reduces the need for dedicated resources and hands-on configuration. It's designed for high availability and fault tolerance, ensuring data durability through replication and automatic scaling based on usage. It also provides granular control over features such as encryption and access management.

Choose migration methods as follows:

- **Direct migration**:

  - Use AWS Glue ETL jobs to extract metadata from the Hive Metastore and load it into the Glue Data Catalog

  - Straightforward for smaller datasets

- **Federation**:

  - Connect AWS Glue to your existing Hive Metastore for seamless access without full migration

  - Ideal for maintaining on-premises metastore or gradual migration

- **Third-party tools**:

  - Explore tools such as AWS **Schema Conversion Tool (SCT)** for specific migration scenarios

Getting ready

Before you begin the migration process of your Hive Metastore to the AWS Glue Data Catalog, ensure that you have met the following prerequisites:

- **AWS account**: You need an active AWS account with the necessary permissions to create and manage AWS Glue, Amazon S3, and related resources.

- **Existing Hive Metastore**: Ensure you have access to your existing Hive Metastore, which contains the metadata for your Hive tables and schemas.

- **Amazon S3 bucket**: Create an Amazon S3 bucket in your AWS account to store any data that needs to be persisted during the migration.

- **AWS Glue Data Catalog**: Optionally, set up the AWS Glue Data Catalog, which will serve as your centralized metadata repository in the cloud. This managed service provides a scalable and reliable alternative to a self-managed Hive Metastore.

- **Network connectivity**: Ensure that your on-premises Hadoop cluster, where the Hive Metastore resides, has network connectivity to AWS services, including AWS Glue and Amazon S3.

- **Access and permissions**: Verify that you have the appropriate IAM roles and policies in place, allowing the migration tool to access the Hive Metastore, the AWS Glue Data Catalog, and Amazon S3.

By fulfilling these prerequisites, you'll be well prepared to migrate your Hive Metastore to the AWS Glue Data Catalog, leveraging the benefits of a fully managed, cloud-native metadata repository.

How to do it...

Here is a step-by-step guide to help you implement metadata federation using AWS Glue and Hive Metastore:

1. **Create an AWS Glue connection to the Hive Metastore**:

    I. **Set up a JDBC connection in AWS Glue to the Hive Metastore database (MySQL, PostgreSQL, and so on)**: Create a Glue connection that allows Glue to access your Hive Metastore using JDBC:

```bash
aws glue create-connection --name hive-metastore-connection \
--connection-input '{
    "Name": "hive-metastore-connection",
    "Description": "JDBC connection to Hive Metastore",
    "ConnectionType": "JDBC",
    "ConnectionProperties": {
        "JDBC_CONNECTION_URL": "jdbc:mysql://<hive-metastore-host>:3306/hive_metastore",
        "USERNAME": "hiveuser",
        "PASSWORD": "hivepassword"
    },
    "PhysicalConnectionRequirements": {
        "AvailabilityZone": "us-east-1a",
        "SubnetId": "subnet-0bb1c79de3EXAMPLE"
    }
}'
```

> **Note to replace**
> - `<hive-metastore-host>`: The host or IP where your Hive Metastore database is located
> - `hiveuser`: Your Hive Metastore database username
> - `hivepassword`: Your Hive Metastore database password

    II. Verify the connection to ensure it's set up correctly:

```bash
aws glue get-connections --name hive-metastore-connection
```

2. **Configure AWS Glue to use the connection**:

   I. Create an AWS Glue job that will access the Hive Metastore using the connection you just set up:

```bash
aws glue create-job --name GlueFederationJob \
--role arn:aws:iam::<your-account-id>:role/GlueServiceRole \
--command '{"Name": "glueetl", "ScriptLocation": "s3://your-
bucket/scripts/hive-metadata-federation.py"}' \
--connections '{"Connections": ["hive-metastore-
connection"]}' \
--default-arguments '{
    "--TempDir": "s3://your-bucket/temp/",
    "--enable-glue-datacatalog": "true"
}'
```

   Let's break this down:

   - `ScriptLocation`: The S3 location of your ETL script, which we will define next

   - `--connections`: This specifies the JDBC connection to your Hive Metastore

   - `--enable-glue-datacatalog`: This enables the use of the AWS Glue Data Catalog as the metadata repository

   II. Create the ETL script that reads from the Hive Metastore and performs metadata federation. Save this script in the S3 bucket specified in `ScriptLocation`.

   You can find the full ETL script (`hive-metadata-federation.py`) here: `https://github.com/PacktPublishing/Data-Engineering-with-AWS-Cookbook/tree/main/Chapter13`.

```
# Reading from the Hive Metastore
datasource = glueContext.create_dynamic_frame.from_catalog(
    database="hive_database_name",
    table_name="hive_table_name"
)
# Writing data to S3 (you can modify the output path or
format)
glueContext.write_dynamic_frame.from_options(
    frame=datasource,
    connection_type="s3",
    connection_options={"path": "s3://your-bucket/output-
path"},
    format="parquet"
)
job.commit()
```

Let's break this down:

- database and table_name: Replace with your Hive Metastore's database and table names

- output-path: Specify the S3 bucket where you want to store the output

3. **Set up AWS Glue crawlers (optional)**: If you want AWS Glue to automatically crawl your Hive Metastore tables and update the Glue Data Catalog, then perform the following steps:

   I.   Create a Glue crawler that crawls the data in your Hive Metastore:

   ```bash
   aws glue create-crawler --name hive-crawler \
   --role arn:aws:iam::<your-account-id>:role/GlueServiceRole \
   --database-name hive_database \
   --targets '{
       "JdbcTargets": [{
           "ConnectionName": "hive-metastore-connection",
           "Path": "hive_metastore"
       }]
   }'
   ```

   II.   Run the crawler to populate the Glue Data Catalog with metadata from your Hive Metastore:

   ```bash
   aws glue start-crawler --name hive-crawler
   ```

4. **Run the AWS Glue job**: Once the Glue job and the script are ready, run the job to begin metadata federation:

   ```bash
   aws glue start-job-run --job-name GlueFederationJob
   ```

5. **Monitor job execution**: You can monitor the job's progress either in the AWS Glue console or by using the following command:

   ```bash
   aws glue get-job-runs --job-name GlueFederationJob
   ```

   This will give you the status of your Glue job, including whether it succeeded or failed.

6. **Verify the federated metadata**: Once the Glue job finishes, do the following:

   I.   **Check the AWS Glue Data Catalog**: You will see the federated metadata from your Hive Metastore in the Glue Data Catalog.

   II.   **Validate the results**: Make sure that all your Hive tables, schemas, and partitions are correctly federated and accessible via the Glue Data Catalog.

By following these steps, you have successfully federated your Hive Metastore metadata with the AWS Glue Data Catalog. This allows you to leverage AWS Glue's scalable, fully managed capabilities without needing to fully migrate your Hive Metastore. The Glue Data Catalog now acts as a centralized metadata store, facilitating efficient analytics workflows and seamless access across AWS services, such as Amazon Athena, EMR, and Redshift.

Migrating and running Apache Oozie workflows on Amazon EMR

Apache Oozie is a popular workflow scheduler for Hadoop ecosystems, orchestrating complex data processing tasks and dependencies. When migrating from your on-premises Hadoop cluster to Amazon EMR, you can seamlessly continue using Oozie or explore alternative AWS services for workflow orchestration.

If your migration strategy is "lift and shift" and your ETL scripts are set up to interact with HDFS for both input and output, then your existing scripts – including those for Hive, EMR, and Spark – should operate effectively in EMR without significant modifications. However, if you've chosen to re-architect your system during the move to AWS and switch to using Amazon S3 as your persistent storage layer instead of HDFS, you'll need to update your scripts. They must be adapted to work with Amazon S3 (using the s3:// protocol) via **Elastic MapReduce File System** (**EMRFS**).

In addition to migrating your Hive and Spark scripts, if Apache Oozie is your chosen tool for orchestrating ETL job workflows, it's essential to also strategize for its migration. Let's explore the available options for this process and the methodologies for migrating to Oozie.

Getting ready

Before you begin migrating Apache Oozie to Amazon EMR, ensure that you meet the following prerequisites:

- **Active AWS account**: You'll need an active AWS account with the necessary permissions to create and manage EMR clusters and related resources

- **Oozie workflows**: Access to your existing Oozie workflows, including workflow definitions (XML files), coordinator definitions (if applicable), and any associated scripts or configurations

- **EMR cluster**: An EMR cluster configured with the desired Hadoop version and components, including Oozie

- **S3 bucket**: An S3 bucket to store your Oozie workflow definitions and any required libraries or dependencies

- **Network connectivity**: Ensure network connectivity between your EMR cluster and any external systems or data sources required by your Oozie workflows

- **IAM roles**: Configure IAM roles with appropriate permissions for EMR to access S3 and other AWS services used by your workflows

- Existing Oozie workflows ready for migration

How to do it...

The following crucial phase in your migration process involves thorough testing and validation. This ensures that your cluster setup is functioning correctly and that the data migration was accurate and complete:

1. **Launch an Elastic MapReduce (EMR) cluster**:

   I.   Use the AWS Management Console, **Command Line Interface (CLI)**, or **Software Development Kit (SDK)** to create an EMR cluster.

   II.   Choose a compatible EMR release that supports Oozie (for example, `emr-5.30.0` or later).

   III. Optionally, customize software settings and instance types based on workload requirements.

   You can launch an EMR cluster using the AWS Management Console, AWS CLI, or AWS SDK. Here is a CLI example:

   ```
   aws emr create-cluster --name "Oozie Cluster" \
       --release-label emr-5.30.0 \
       --applications Name=Hadoop Name=Hive Name=Pig \
       --use-default-roles \
       --instance-type m5.xlarge \
       --instance-count 3
   ```

   - **Release version**: Choose an EMR release that supports Oozie (for example, `emr-5.30.0` or later)

   - **Instance types**: Customize instance types and the number of instances based on your workload requirements

2. **Install Oozie**:

   I.   Oozie isn't installed by default on EMR. Use bootstrap actions during cluster creation to install Oozie manually. Follow the AWS documentation (`https://docs.aws.amazon.com/emr/latest/ReleaseGuide/emr-oozie.html`) for specific installation steps and configuration options.

II. Create a bootstrap action script to install Oozie and its dependencies. Here's an example script:

```
!/bin/bash
sudo yum install -y oozie oozie-client
sudo service oozie start
```

III. Add this script during the EMR cluster creation:

```
aws emr create-cluster --name "Oozie Cluster with Bootstrap" \
   --release-label emr-5.30.0 \
   --applications Name=Hadoop Name=Hive Name=Pig \
   --bootstrap-actions Path=s3://your-bucket/path-to-bootstrap.sh \
   --use-default-roles \
   --instance-type m5.xlarge \
   --instance-count 3
```

IV. Follow the AWS documentation (https://docs.aws.amazon.com/emr/latest/ReleaseGuide/emr-oozie.html) to configure Oozie properly, including setting up oozie-site.xml and ensuring that the Oozie server is running on the master node of the cluster.

3. **Prepare workflows**:

I. Review workflows for compatibility with EMR's Hadoop distribution and configuration.

II. Address any version-specific differences or dependencies.

III. Test workflows in a staging environment before migrating to production.

4. **Transfer data**:

I. Move the data required for workflows to S3 or other accessible storage within EMR.

II. Ensure proper permissions and access from the cluster.

i. **Move data to S3**: Transfer the datasets required by your workflows to Amazon S3, making them accessible from the EMR cluster:

```
aws s3 cp /local-data-path s3://your-bucket/data-path
--recursive
```

ii. **Permissions**: Ensure that the EMR cluster has the correct IAM role permissions to access the S3 bucket where your data is stored.

5. **Modify the default EMR role**:

    I.   Go to the IAM console: `https://console.aws.amazon.com/iam/`.

    II.   In the left sidebar, select **Roles**.

    III. Search for the role named `EMR_EC2_DefaultRole`.

    IV. Under the role's **Permissions** tab, choose **Add inline policy** to define custom permissions for S3 access.

You can define the permissions using an IAM policy that grants the EMR cluster access to specific S3 buckets. Here is a sample policy that grants read and write access to S3:

```json
{
    "Version": "2012-10-17",
    "Statement": [
        {
            "Effect": "Allow",
            "Action": [
                "s3:GetObject",
                "s3:ListBucket"
            ],
            "Resource": [
                "arn:aws:s3:::your-bucket-name",
                "arn:aws:s3:::your-bucket-name/*"
            ]
        },
        {
            "Effect": "Allow",
            "Action": [
                "s3:PutObject",
                "s3:DeleteObject"
            ],
            "Resource": "arn:aws:s3:::your-bucket-name/*"
        }
    ]
}
```

> **Note**
> Please update `your-bucket-name` with the bucket you are planning to use.

6. **Submit workflows**:

    I.   Use Oozie's command-line tools (`oozie job -submit`) or web **user interface** (**UI**) to submit workflows to the EMR cluster.

    II.   Monitor workflow execution through Oozie's web UI or logs:

        i.   **Submit workflows**: Use Oozie's command-line tools or web UI to submit workflows to your EMR cluster. Here's an example of submitting a workflow using the command line:

```
oozie job -oozie http://<EMR-MasterNode-DNS>:11000/oozie
-config /path/to/job.properties -run
```

        ii.   **Monitor execution**: Monitor the execution of your workflows through Oozie's web UI or by reviewing logs generated in the cluster.

Alternative options to move out of Oozie

Consider **managed workflow services** or **AWS Step Functions** for simplified orchestration if appropriate for your workloads.

Migrating an Oozie database to the Amazon RDS MySQL

Apache Oozie, a widely used workflow scheduler in the Hadoop ecosystem, orchestrates a variety of Hadoop jobs, including Hive, Pig, Sqoop, Spark, DistCp, Linux shell actions, and more. It stands out in the Hadoop community for its scalability and reliability.

Oozie operates with two key components: **workflow jobs**, which allow you to map out workflow steps in the form of **Directed Acyclic Graphs** (**DAGs**), and the **Oozie Coordinator**, designed for scheduling these workflow jobs based on events or timed triggers.

Using XML definitions, Oozie enables the creation of workflows and has been available on Amazon EMR since the 5.0.0 release.

Like Hive, Oozie also relies on a Metastore database, a crucial aspect to consider during migration. When moving Oozie workflows to EMR, it's essential to transfer both the workflow definition files and the Metastore database.

This recipe provides a step-by-step walkthrough of migrating your Oozie database to Amazon **Relational Database Service** (**RDS**) for MySQL. You'll learn how this migration enhances scalability, reliability, and maintenance by automating tasks such as backups, patching, and recovery. By moving to RDS, you can seamlessly integrate your Oozie workflows with AWS services, reduce management overhead, and create more efficient and secure data pipelines. Follow along to learn how to make this transition and reap the benefits of a managed database service for your Oozie workflows.

Getting ready

Before migrating your Apache Oozie database to Amazon RDS MySQL, ensure you have the following prerequisites in place:

- **AWS account**: An active AWS account with permissions to create and manage RDS instances and associated resources.

- **Existing Oozie database**: Access to your current Oozie database, which could be running on MySQL or another supported database system.

- **MySQL client**: Install a MySQL client on your local machine to interact with both your current Oozie database and the RDS MySQL instance.

- **Database dump**: Ensure you can export a dump of your existing Oozie database. This is typically done using tools such as `mysqldump` for MySQL databases.

- **VPC and security groups**: Basic understanding of AWS VPCs and security groups, as they will be needed to configure your RDS instance's network access.

- **Backup plan**: A backup strategy in place before migrating, ensuring that you have a fallback option in case of any issues during the migration process.

- **Amazon RDS MySQL instance**: You will need to create an Amazon RDS MySQL instance where the Oozie database will be migrated.

- **IAM roles and permissions**: Ensure your AWS account has IAM roles and permissions to create, manage, and access RDS instances.

- **Network configuration**: Ensure your RDS instance is configured to allow connections from your current Oozie environment and any management machines.

- **Database migration tools**: Install database migration tools such as `mysqldump` for exporting your database and a `mysql` client for importing it into RDS.

How to do it...

1. **Setting up the Amazon RDS MySQL instance**: Create a RDS instance with the following steps:

   I.   Go to the AWS Management Console and navigate to the RDS service.

   II.   Click on **Create database** and choose **MySQL** as the database engine.

   III. Select **Standard create** and configure your instance details, such as database instance class, storage, and VPC settings.

   IV.   Set the database name, username, and password. This will be used later for migration.

Here's an example using AWS CLI:

```
aws rds create-db-instance \
    --db-instance-identifier oozie-db-instance \
    --db-instance-class db.m5.large \
    --engine mysql \
    --allocated-storage 20 \
    --master-username admin \
    --master-user-password yourpassword \
    --vpc-security-group-ids sg-xxxxxxx \
    --availability-zone us-west-2a \
    --db-name oozie
```

2. **Configure security groups**: Ensure that your RDS instance is accessible from your current Oozie environment. Modify the security group associated with the RDS instance to allow incoming MySQL traffic (port 3306) from your IP range.

3. **Migrating an Oozie database to the Amazon RDS**: For the migration of an Oozie database to the Amazon RDS MySQL engine, follow these steps, which will lead you through the export and import procedure:

   I.   Access the Oozie server node in your on-premises cluster, go to the directory containing oozie-setup.sh, and run the specified Oozie command to export the Metastore database:

   ```
   ./oozie-setup.sh export /<path>/<oozie-exported-db>.zip
   ```

   II.   In EMR, the oozie-setup.sh file is in the /usr/lib/oozie/bin/ directory.

   III. Next, upload the exported database ZIP file to Amazon S3, from which you can import the database using the following command:

   ```
   aws s3 cp <oozie-exported-db>.zip s3://<bucket-name-
   path>/<oozie-exported-db>.zip
   ```

   IV.   Then, SSH into the EMR master node using Putty to download the file:

   ```
   aws s3 cp s3://<bucket-name-path>/<oozie-exported-db>.zip
   <oozie-exported-db>.zip
   ```

   V.   Following that, it's necessary to establish the Oozie database in Amazon RDS and assign the necessary permissions. Access your database with root privileges and run the following commands in the MySQL prompt:

   ```
   mysql> create database oozie default character set utf8;
   mysql> grant all privileges on oozie.* to
   'oozie'@'localhost' identified by 'oozie';
   mysql> grant all privileges on oozie.* to 'oozie'@'%'
   identified by 'oozie';
   ```

VI.   Once you've successfully set up the database with all the essential permissions, the next step involves importing the database file. This can be done using the same `oozie-setup.sh` utility, as demonstrated in the following command:

```
./oozie-setup.sh import <oozie-exported-db>.zip
```

4.   **Update Oozie configuration**: Once the database is prepared with all the imported metadata, it's necessary to update the Oozie configuration in EMR to redirect it to the new Amazon RDS database. To do this, apply the following modifications to the `oozie-site.xml` configuration file:

```
<property>
    <name>oozie.service.JPAService.jdbc.driver</name>
    <value>com.mysql.jdbc.Driver</value>
</property>
<property>
    <name>oozie.service.JPAService.jdbc.url</name>
    <value>jdbc:mysql://<amazon-rds-host>:3306/oozie</value>
</property>
<property>
    <name>oozie.service.JPAService.jdbc.username</name>
    <value><mysql-db-username></value>
</property>
<property>
    <name>oozie.service.JPAService.jdbc.password</name>
        <value><mysql-db-password></value>
</property>
```

> **Note**
>
> Ensure that you substitute the placeholders with original values in the previous code with your specific database connection details:
>
> - `<amazon-rds-host>`
> - `<mysql-db-username>`
> - `<mysql-db-password>`

To apply the updates to Oozie, the final step involves restarting the Oozie service. This can be done using the following command:

```
sudo restart oozie
```

The outlined steps assist in transferring the Oozie Metastore database to a remote database on EMR RDS. Following this, your next task is to relocate all workflow definition files to EMR.

5. **Migrating Oozie workflow definitions**:

   For each workflow definition, you will have a set of essential files, including the following:

   - `job.properties`
   - `workflow.xml`
   - `coordinator.xml`
   - Any other related dependent files

   To back up these files, you can archive them into a single ZIP file. Upload this archive to Amazon S3, and then transfer it to EMR using the `aws s3 cp` command to integrate it into your system. Remember to update the workflow configuration file to align with the EMR connection.

Subsequently, you can submit your jobs to EMR in the same manner as you did in your on-premises setup. The Hue interface is available for monitoring your Oozie workflows on EMR.

See also

- *Converting Oozie workflows to AWS Step Functions with AWS Schema Conversion Tool*: `https://docs.aws.amazon.com/SchemaConversionTool/latest/userguide/big-data-oozie.html`

Setting up networking – establishing a secure connection to your EMR cluster

Amazon EMR provides a managed Hadoop framework in the cloud, enabling powerful big data processing and analytics. However, ensuring the security of your EMR cluster is crucial to protect sensitive data and prevent unauthorized access. This recipe will guide you through the essential considerations and steps to establish a secure connection to your EMR cluster, safeguarding your valuable information.

In this recipe, we introduce a more secure method for engineers to access Amazon EMR cluster instances located in a private subnet, along with the traditional approach of using a bastion host or jump server with open SSH inbound ports.

Getting ready

Before setting up a secure network connection, ensure you have the following prerequisites in place:

- **Active AWS account**: You'll need an active AWS account with the necessary permissions to create and manage EMR clusters and related resources
- **EMR cluster**: An existing or newly created EMR cluster in your AWS account

- **SSH client**: An SSH client installed on your local machine (for example, PuTTY or OpenSSH)
- **Key pair**: An SSH key pair generated within your AWS account for secure authentication

To ensure secure communication with your EMR cluster, carefully configure the following network settings:

- **Public subnet**: If you need to access the EMR cluster from the internet, ensure it's launched in a public subnet with appropriate security groups to allow inbound SSH traffic (typically on port 22)
- **Private subnet:** If your cluster is in a private subnet, consider using a bastion host or AWS Systems Manager Session Manager for secure access
- **Security groups**: Configure security groups to restrict inbound and outbound traffic to your EMR cluster based on your specific needs
- **IAM roles**: Assign appropriate IAM roles to your EMR instances and users to control access to AWS services and resources

How to do it...

You have three options for establishing a secure connection to your EMR cluster, each offering different levels of security and flexibility. We will discuss each method in detail.

Method 1 – SSH connection (for clusters in public subnets)

1. **Retrieve master node public DNS**:

    I.   Open the Amazon EMR console.

    II.   Navigate to your cluster details and locate the **Master public DNS** field.

2. **Connect using SSH**:

    I.   Open your SSH client.

    II.   Enter the master node's public DNS as the hostname.

    III. Specify the appropriate username (for example, `hadoop` or `ec2-user`, depending on your EMR configuration).

    IV.   Select the private key file associated with the key pair you created for the cluster.

    V.   Click on **Connect** or initiate the SSH connection.

Method 2 – bastion host (for clusters in private subnets)

1. **Launch bastion host**:

     I.   Create an EC2 instance (bastion host) in a public subnet within the same VPC as your EMR cluster.

     II.   Configure security groups to allow inbound SSH access from your IP address and outbound SSH access to the EMR master node's private IP.

2. **Connect to bastion host**: Use SSH to connect to the bastion host's public IP address using its associated key pair.

3. **Connect to EMR master node**: From the bastion host, use SSH to connect to the EMR master node's private IP address using the EMR cluster's key pair.

Method 3 – AWS Systems Manager Session Manager (recommended for private subnets)

AWS Systems Manager offers a consolidated UI, enabling you to monitor and control your Amazon EC2 instances effectively. Within this, Session Manager enhances security and auditability, for instance, management.

When integrated with IAM, Systems Manager facilitates centralized access control to your EMR cluster. However, by default, Systems Manager lacks the necessary permissions to execute actions on cluster instances. To enable this functionality, you must assign an IAM role to the instance that carries the required access permissions. Before beginning this process, it's important to establish an IAM service role for your cluster's EC2 instances, ensuring it adheres to the principle of least privilege in its access policy. The steps are as follows:

1. **Enable Session Manager**:

     I.   Ensure Session Manager is enabled for your EMR cluster (it's often enabled by default for newer EMR versions).

     II.   Configure an IAM role with necessary permissions for Session Manager access.

     III.   Connect using Session Manager:

          i.   Open the AWS Systems Manager console.

          ii.   Navigate to **Session Manager** and click on **Start session**.

          iii. Select your **EMR master node** instance from the list.

          iv.   A browser-based shell session will open, providing secure access to your EMR cluster without needing SSH keys or open ports.

IV. Create an IAM service role specifically for EMR cluster EC2 instances (known as the Amazon EMR role for EC2) and associate it with the AWS-managed Systems Manager core instance policy (`AmazonSSMManagedInstanceCore`):

```
{
  "Version": "2012-10-17",
  "Statement": [
    {
      "Effect": "Allow",
      "Action": [
        "ssm:DescribeInstanceProperties",
        "ssm:DescribeSessions",
        "ec2:describeInstances",
        "ssm:GetConnectionStatus"
      ],
      "Resource": "*"
    },
    {
      "Effect": "Allow",
      "Action": [
        "ssm:StartSession"
      ],
      "Resource": [
        "arn:aws:ec2:${Region}:${Account-Id}:instance/*"
      ],
      "Condition": {
        "StringEquals": {
          "ssm:resourceTag/ClusterType": [
            "QACluster"
          ]
        }
      }
    }
  ]
}
```

Now, attach the least privilege policy to the IAM principal (role or user).

You can set up the AWS **Systems Manager Agent (SSM Agent)** on your Amazon EMR cluster nodes using bootstrap actions. The SSM Agent allows Session Manager to control and configure these nodes.

Using Session Manager doesn't incur extra costs for managing Amazon EC2 instances, though other features might have additional charges (check the Systems Manager pricing page for details). The agent receives and executes instructions from the Session Manager service in the AWS cloud according to user requests. By installing the Systems Manager plugin on your local machine, you can use dynamic port forwarding. Also, IAM policies help in centrally controlling access to the EMR cluster.

2. **Configuring SSM Agent on an EMR cluster**: To set up the SSM Agent on your cluster, follow these steps:

I. When you are launching the EMR cluster, go to the **Bootstrap Actions** section and select **Add bootstrap action**. Choose **Custom action**.

II. Include a bootstrap action that executes a script from Amazon S3. This script will install and set up the SSM Agent on your Amazon EMR cluster instances.

III. The SSM Agent requires a localhost entry in the host's file. This is necessary for redirecting traffic from your local computer to the EMR cluster instance when you're using dynamic port forwarding.

IV. The bootstrap script is as follows:

```bash
#!/bin/bash
## Name: SSM Agent Installer Script
## Description: Installs SSM Agent on EMR cluster EC2
instances and update hosts file
##
sudo yum install -y https://s3.amazonaws.com/ec2-downloads-
windows/SSMAgent/latest/linux_amd64/amazon-ssm-agent.rpm
sudo status amazon-ssm-agent >>/tmp/ssm-status.log


## Update hosts file
echo "\n ########## localhost mapping check ########## \n"
> /tmp/localhost.log
lhost=`sudo cat /etc/hosts | grep localhost | grep
'127.0.0.1' | grep -v '^#'`
v_ipaddr=`hostname --ip-address`
lhostmapping=`sudo cat /etc/hosts | grep $v_ipaddr | grep -v
'^#'`
if [ -z "${lhostmapping}" ];
then
echo "\n ########## IP address to localhost mapping NOT
defined in hosts files. add now ########## \n " >> /tmp/
localhost.log
sudo echo "${v_ipaddr} localhost" >>/etc/hosts
else
echo "\n IP address to localhost mapping already defined in
hosts file \n" >> /tmp/localhost.log
fi
echo "\n ########## IP Address to localhost mapping check
complete and below is the content ########## " >> /tmp/
localhost.log
```

```
sudo cat /etc/hosts >> /tmp/localhost.log
echo "\n ########## Exit script ########## " >> /tmp/
localhost.log
```

V.   In the **Security options** section, navigate to **Permissions** and select **Custom**. Then, for the **EMR role**, choose the IAM role that you previously created.

VI.   Once the cluster has successfully launched, go to the **Session Manager** console and select **Managed instances**. Then, choose your cluster instance. From the **Actions** menu, click on **Start session**.

The following figure shows AWS Systems Manager:

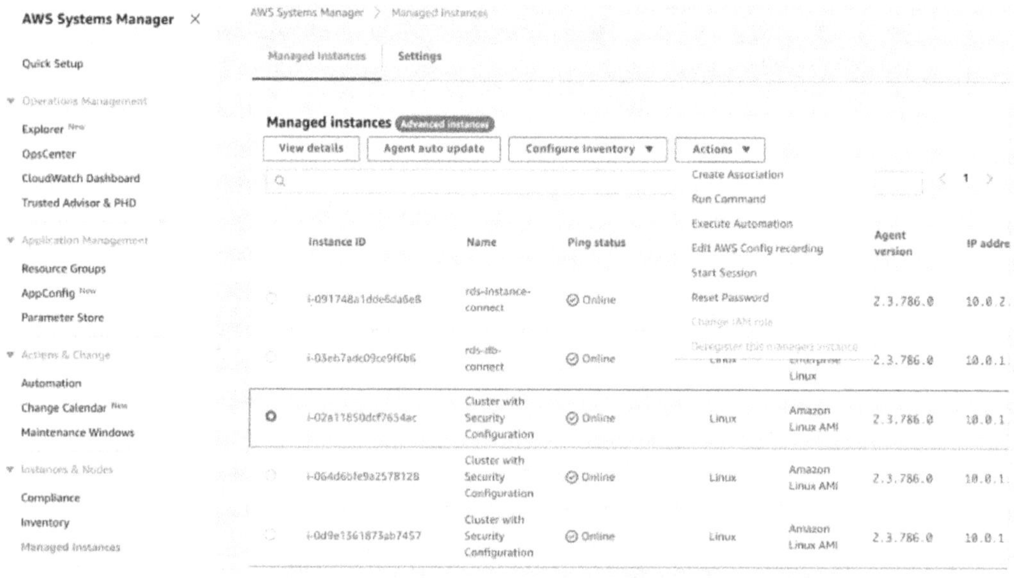

Figure 13.8 – AWS Systems Manager

VII. To access web UIs of Hadoop applications, such as YARN Resource Manager and the Spark Job Server, on the Amazon EMR primary node, you can establish a **secure tunnel** between your computer and the primary node using Session Manager.

3. **Install AWS CLI**: Please use the instructions at https://docs.aws.amazon.com/cli/v1/userguide/cli-chap-install.html to set up AWS CLI. Once done, run the following command:

```
aws ssm start-session --target "Your Instance ID"
--document-name AWS-StartPortForwardingSession --parameters
"portNumber"=["8080"],"localPortNumber"=["8158"]
```

In an environment with a multi-tenant Amazon EMR cluster, you can limit access to the cluster instances using specific Amazon EC2 tags.

For instance, in the example code provided, an IAM principal (either an IAM user or role) is permitted to initiate a session on any instance (as indicated by the resource **Amazon Resource Name (ARN)**: `arn: aws:ec2:::instance/*`) under the condition that the instance is tagged as `TestCluster` (denoted by `ssm:resourceTag/ClusterType: TestCluster`).

If the IAM principal attempts to start a session on an instance that either lacks a tag or has a tag different from `ClusterType: TestCluster`, the outcome will display a message indicating that they are not authorized to perform the `ssm:StartSession` action:

```
{
    "Version": "2012-10-17",
    "Statement": [
        {
            "Effect": "Allow",
            "Action": [
                "ssm:DescribeInstanceProperties",
                "ssm:DescribeSessions",
                "ec2:describeInstances",
                "ssm:GetConnectionStatus"
            ],
            "Resource": "*"
        },
        {
            "Effect": "Allow",
            "Action": [
                "ssm:StartSession"
            ],
            "Resource": [
                "arn:aws:ec2:${Region}:${Account-Id}:instance/*"
            ],
            "Condition": {
                "StringEquals": {
                    "aws:username": "${aws:username}"
                },
                "StringLike": {
                    "ssm:resourceTag/ClusterType": [
                        "TestCluster"
                    ]
                }
            }
        }
    ]
}
```

```
        ]
    }
```

You can modify the default login settings to limit root access in user sessions. Normally, sessions start with `ssm-user`, but you can opt to use an operating system account's credentials by tagging an IAM user or role with `SSMSessionRunAs` or specifying a username. This is enabled through updates to Session Manager preferences. An example configuration allows the `appdev2` IAM user to start sessions as `ec2-user` rather than the standard `ssm-user`.

See also

- *Best Practices for Securing Amazon EMR*: `https://aws.amazon.com/blogs/big-data/best-practices-for-securing-amazon-emr/`

- *AWS Systems Manager*: `https://aws.amazon.com/systems-manager/`

Performing a seamless HBase migration to AWS

HBase provides a Hadoop-based solution for managing large-scale, sparse datasets. This non-relational, distributed database excels at random read/write access to data with high volume and variety, making it a valuable tool for big data applications.

Migrating an on-premises Apache HBase database to AWS allows organizations to leverage the scalability, flexibility, and cost-effectiveness of the cloud. AWS provides several options for migrating to HBase, including moving to HBase on Amazon S3 (running on Amazon EMR), or migrating to a fully managed service such as Amazon DynamoDB. This recipe outlines the detailed steps to perform this migration.

Getting ready

Before beginning the migration process, ensure the following prerequisites are met:

- **AWS account**: Ensure you have an AWS account with permissions to create and manage resources such as Amazon EMR, S3, and DynamoDB

- **Existing HBase cluster**: Access to your on-premises HBase cluster, including the HBase tables, configuration files, and data

- **Data backup**: Back up your HBase data to ensure that you have a fallback option in case of any issues during migration

- **AWS CLI and SDK**: Install the AWS CLI and AWS SDK if you plan to automate or script parts of the migration process

- **Network connectivity**: Ensure there is secure and reliable network connectivity between your on-premises environment and AWS

The technical requirements are as follows:

- **Amazon S3 bucket**: Create an Amazon S3 bucket where you will store HBase snapshots or data files during the migration process

- **Amazon EMR cluster**: If migrating to HBase on Amazon EMR, ensure you can provision an EMR cluster with HBase and Hadoop installed

- **HBase compatibility**: Ensure that the version of HBase you are migrating from is compatible with the version of HBase available on Amazon EMR or that necessary adjustments are planned

- **IAM roles and permissions**: Set up IAM roles with permissions to access S3, DynamoDB (if applicable), and other AWS services required during the migration process

How to do it...

The migration process involves exporting data from your on-premises HBase cluster, transferring it to AWS, and then importing it into the target AWS service. The detailed steps, including sample commands and code snippets, are as follows:

1. **Migrate HBase on EMR**: Export HBase data:

    I.   **Take a snapshot**:

    i.   Create a snapshot of your HBase tables to capture the current state of your data.

    ii.   Use the HBase shell to create snapshots:

    ```
    echo "snapshot 'tableName', 'snapshotName'" | hbase shell
    ```

    iii. Replace tableName with your HBase table's name and snapshotName with your desired snapshot name.

    II.   **Export the snapshot**:

    i.   Use the ExportSnapshot utility to export the snapshot to your on-premises filesystem or directly to an S3 bucket:

    ```
    hbase org.apache.hadoop.hbase.snapshot.ExportSnapshot \
    -snapshot snapshotName \
    -copy-to hdfs://namenode:8020/path-to-backup
    ```

    ii.   Utilize the ExportSnapshot tool provided by HBase to extract data from the snapshot. This tool converts the snapshot data into sequence files, which can be processed further.

2. **Transfer data to Amazon S3**:

    I.   **Copy data to S3**: If you have exported the snapshot to a local filesystem, use the AWS CLI to copy the snapshot files to Amazon S3:

```
aws s3 cp /local-snapshot-path/ s3://your-s3-bucket/
snapshot-path/ --recursive
```

> **Note**
>
> Replace /local-snapshot-path/ with the local directory containing the snapshot files and -s3-bucket/snapshot-path/ with your S3 bucket and path.

3. **Set up Amazon EMR cluster**:

    I.   **Launch the EMR cluster**: Launch an EMR cluster configured with HBase and Hadoop using the AWS Management Console:

```
aws emr create-cluster --name "HBase Cluster" \
--release-label emr-6.3.0 \
--applications Name=HBase Name=Hadoop \
--use-default-roles \
--instance-type m5.xlarge \
--instance-count 3
```

    II.   **Configure HBase on EMR**: Once the cluster is up, connect to the master node and configure HBase to restore from the snapshot stored in S3.

4. **Import HBase data to EMR**:

    I.   **Restore snapshot to HBase on EMR**: Use the CloneSnapshot utility to restore the snapshot to HBase on EMR:

```
hbase org.apache.hadoop.hbase.snapshot.CloneSnapshot \
-snapshot snapshotName \
-restore-to tableName
```

5. **Validate and test the migration**:

    I.   **Verify data integrity**: Perform checks to ensure that all data has been migrated successfully and that the data integrity is maintained.

    II.   **Run tests**: Run tests on the migrated data to confirm that your applications can interact with it as expected in the new environment.

Migrating HBase to DynamoDB on AWS

Migrating from HBase to DynamoDB can provide significant advantages in terms of scalability, managed infrastructure, and cost-effectiveness. HBase is a powerful, distributed, NoSQL database often used in on-premises or Hadoop-based environments. However, managing an HBase cluster can be complex and resource-intensive. Amazon DynamoDB offers a fully managed, serverless alternative with built-in replication, scalability, and integration with other AWS services, making it a strong candidate for workloads that need high availability and elasticity.

Suppose a media streaming company that stores user session logs and interaction data in an HBase cluster on an on-premises Hadoop environment is looking to migrate to a cloud-based solution. The company needs the ability to handle sudden spikes in traffic (such as during new content releases), while also reducing the overhead of managing their own HBase clusters. Migrating to Amazon DynamoDB provides the elasticity to handle these spikes, along with automatic scaling and reduced operational complexity.

Getting ready

Before beginning the migration process, ensure the following prerequisites are met:

- **AWS account**: Ensure you have an AWS account with appropriate permissions to create and manage DynamoDB, IAM, and data migration services

- **HBase data analysis**: Review your existing HBase data model, including tables, columns, and key structures, as these will need to be restructured in DynamoDB

- **AWS CLI installed**: Install and configure the AWS CLI to interact with AWS services

- **DynamoDB table design**: Understand DynamoDB's table structure, primary keys, and global/local secondary indexes, as they differ from HBase's schema

How to do it...

1. **ExportHBase snapshot**: First, you need to create and export a snapshot of your HBase tables. To create a snapshot in HBase, please follow the steps as explained in the previous recipe, *Seamless HBase migration to AWS*, under *Migrate HBase on EMR* (*step 1*).

2. **Transfer data to Amazon S3**: Next, you need to transfer the exported snapshot data to an Amazon S3 bucket. Please follow the same steps as explained previously.

3. **Set up AWS Glue for data transformation**: Amazon DynamoDB and HBase have different data models, so you may need to transform the data before importing it into DynamoDB. AWS Glue is a suitable service for this task:

    I.   **Create a Glue crawler**: Set up a Glue crawler to catalog the data you uploaded to S3. This will automatically infer the schema and make the data accessible for ETL processes. Use the AWS Management Console or CLI to create the crawler:

    ```
    aws glue create-crawler \
    --name your-crawler-name \
    --role your-glue-role \
    --database-name your-database-name \
    --targets S3Targets=[{Path=s3://your-s3-bucket/snapshot-
    path/}]
    ```

    II. **Run the Glue ETL job**: After the crawler has created the metadata catalog, use a Glue ETL job to transform and load the data into DynamoDB. You can use Python scripts in AWS Glue to perform this transformation.

    You can find the full code on the following path: `https://github.com/PacktPublishing/Data-Engineering-with-AWS-Cookbook/blob/main/Chapter13/dynamodb-import-glue-job.py`.

    ```
    dynamic_frame = glueContext.create_dynamic_frame.from_
    catalog(database = "your-database-name", table_name = "your-
    table-name")


    # Transformation logic here
    glueContext.write_dynamic_frame.from_options(frame =
    dynamic_frame, connection_type = "dynamodb", connection_
    options = {"dynamodb.output.tableName": "your-dynamodb-
    table-name"})
    ```

4. **Import data into Amazon DynamoDB**: Finally, load the transformed data into your DynamoDB table:

    I.   **Create a DynamoDB table**: Ensure you have a DynamoDB table created that matches the schema requirements of your transformed data. An example CLI command to create a DynamoDB table is as follows:

    ```
    bash


    aws dynamodb create-table \
    --table-name your-dynamodb-table-name \
    --attribute-definitions
    AttributeName=PrimaryKey,AttributeType=S \
    ```

```
--key-schema AttributeName=PrimaryKey,KeyType=HASH \
--provisioned-throughput
ReadCapacityUnits=5,WriteCapacityUnits=5
```

II. **Load data into DynamoDB**: Use AWS Glue, as shown in the preceding script, to load the transformed data directly into DynamoDB.

III. **Verify the data import**: After the ETL job completes, verify the data in DynamoDB using the AWS Management Console, CLI, or by running queries against your DynamoDB table.

5. **Validate and optimize**:

   I. **Validation**:

   i. Run queries and validate that all the data has been successfully imported into DynamoDB.

   ii. Check for data integrity and consistency across your new DynamoDB tables.

   II. **Optimization**:

   i.   Adjust DynamoDB read/write capacity based on your application's needs.

   ii. Set up indexes and optimize the table design for efficient querying.

   iii. Use Amazon CloudWatch and other AWS tools to monitor the performance and health.

Migrating from HBase to Amazon DynamoDB offers operational simplicity, scalability, and cost benefits. By following this recipe, you can effectively move your data from an on-premises HBase cluster to a fully managed DynamoDB environment. This transition not only reduces infrastructure management but also improves your application's availability and elasticity on AWS.

See also

- *Comparing the Use of Amazon DynamoDB and Apache HBase for NoSQL*: https://d1.awsstatic.com/whitepapers/AWS_Comparing_the_Use_of_DynamoDB_and_HBase_for_NoSQL.pdf

- *Migrating and restoring Apache HBase tables on Apache HBase on Amazon S3*: https://docs.aws.amazon.com/whitepapers/latest/migrate-apache-hbase-s3/migrating-and-restoring-apache-hbase-tables-on-apache-hbase-on-amazon-s3.html

- *Introduction to Apache HBase Snapshots*: https://blog.cloudera.com/introduction-to-apache-hbase-snapshots/

- *Amazon DynamoDB*: https://aws.amazon.com/dynamodb/

Index

packtpub.com

Subscribe to our online digital library for full access to over 7,000 books and videos, as well as industry leading tools to help you plan your personal development and advance your career. For more information, please visit our website.

Why subscribe?

- Spend less time learning and more time coding with practical eBooks and Videos from over 4,000 industry professionals

- Improve your learning with Skill Plans built especially for you

- Get a free eBook or video every month

- Fully searchable for easy access to vital information

- Copy and paste, print, and bookmark content

Did you know that Packt offers eBook versions of every book published, with PDF and ePub files available? You can upgrade to the eBook version at packtpub.com and as a print book customer, you are entitled to a discount on the eBook copy. Get in touch with us at customercare@packtpub.com for more details.

At www.packtpub.com, you can also read a collection of free technical articles, sign up for a range of free newsletters, and receive exclusive discounts and offers on Packt books and eBooks.

Other Books You May Enjoy

If you enjoyed this book, you may be interested in these other books by Packt:

Data Engineering with Databricks Cookbook

Pulkit Chadha

ISBN: 978-1-83763-335-7

- Perform data loading, ingestion, and processing with Apache Spark

- Discover data transformation techniques and custom user-defined functions (UDFs) in Apache Spark

- Manage and optimize Delta tables with Apache Spark and Delta Lake APIs

- Use Spark Structured Streaming for real-time data processing

- Optimize Apache Spark application and Delta table query performance

- Implement DataOps and DevOps practices on Databricks

- Orchestrate data pipelines with Delta Live Tables and Databricks Workflows

- Implement data governance policies with Unity Catalog

Data Engineering with Google Cloud Platform

Adi Wijaya

ISBN: 978-1-83508-011-5

- Load data into BigQuery and materialize its output
- Focus on data pipeline orchestration using Cloud Composer
- Formulate Airflow jobs to orchestrate and automate a data warehouse
- Establish a Hadoop data lake, generate ephemeral clusters, and execute jobs on the Dataproc cluster
- Harness Pub/Sub for messaging and ingestion for event-driven systems
- Apply Dataflow to conduct ETL on streaming data
- Implement data governance services on Google Cloud

Packt is searching for authors like you

If you're interested in becoming an author for Packt, please visit `authors.packtpub.com` and apply today. We have worked with thousands of developers and tech professionals, just like you, to help them share their insight with the global tech community. You can make a general application, apply for a specific hot topic that we are recruiting an author for, or submit your own idea.

Share Your Thoughts

Now you've finished *Data Engineering with AWS Cookbook*, we'd love to hear your thoughts! Scan the QR code below to go straight to the Amazon review page for this book and share your feedback or leave a review on the site that you purchased it from.

`https://packt.link/r/1-805-12728-4`

Your review is important to us and the tech community and will help us make sure we're delivering excellent quality content.

Download a free PDF copy of this book

Thanks for purchasing this book!

Do you like to read on the go but are unable to carry your print books everywhere?

Is your eBook purchase not compatible with the device of your choice?

Don't worry, now with every Packt book you get a DRM-free PDF version of that book at no cost.

Read anywhere, any place, on any device. Search, copy, and paste code from your favorite technical books directly into your application.

The perks don't stop there, you can get exclusive access to discounts, newsletters, and great free content in your inbox daily

Follow these simple steps to get the benefits:

1. Scan the QR code or visit the link below

https://packt.link/free-ebook/9781805127284

2. Submit your proof of purchase
3. That's it! We'll send your free PDF and other benefits to your email directly